# Intermediate Statistical Methods

# Intermediate
# Statistical Methods

G. BARRIE WETHERILL

*Professor of Statistics,*
*University of Kent at Canterbury*

LONDON  NEW YORK

CHAPMAN AND HALL

First published 1981 by
Chapman and Hall Ltd
11 New Fetter Lane, London EC4P 4EE

Published in the USA by
Chapman and Hall
in association with Methuen, Inc.
733 Third Avenue, New York NY 10017

© 1981 G Barrie Wetherill

ISBN 0 412 16440 X (cased edition)
ISBN 0 412 16450 7 (paperback edition)

**British Library Cataloguing in Publication Data**

Wetherill, George Barrie
  Intermediate statistical methods.

  1. Mathematical statistics
  I. Title
    519.5    QA 276    80–40895

  ISBN 0–412–16440–X
  ISBN 0–412–16450–7 Pbk

# Contents

*Sections and Exercises marked by an asterisk are more difficult.

# *Preface*

This book began many years ago as course notes for students at the University of Bath, and later at the University of Kent. Students used draft versions of the chapters, which were consequently revised. Second and third year students, as well as those taking MSc courses have used selections of the chapters. In particular, Chapters 1 to 7 (only) have been the basis of a very successful second-year course, the more difficult sections being omitted. The aims of this particular course were:-

(a) to cover some interesting and useful applications of statistics with an emphasis on applications, but with really adequate theory;
(b) to lay the foundations for interesting third-year courses;
(c) to tie up with certain areas of pure mathematics and numerical analysis.

Students will find Chapter 1 a useful means of revising the $t$, $\chi^2$ and $F$ procedures, which is material assumed in this text, see Section 1.1. Later sections of Chapter 1 cover robustness and can be omitted by second-year students or at a first reading. Chapter 2 introduces some simple statistical models, so that the discussion of later chapters is more meaningful.

The basic ideas of inference given in Chapter 3 are not stressed much at the second-year level. However, the method of maximum likelihood, dealt with in Chapters 3 and 4 is, in my view, very important. Students should approach this topic gradually, and allow the ideas time to digest before attempting complicated exercises. Numerical application of the method of maximum likelihood is essential.

In Chapter 5, I come out of maximum likelihood into least squares. This seems to me to be a better basis than simply to state least-squares theory. The matrix algebra here is a useful teaching point, and I find that in spite of courses on algebra, students' hair still stands on end when you actually *use* a matrix! This chapter can be used, by comparison with Chapter 2, to show just how useful matrices are. In teaching the course, I cover only the early and later parts of the chapter with students initially, and then go to Chapter 6 and numerical exercises. I then come back to Chapter 5, which makes more sense after students have used regression, or had some discussion of applications of regression current in the department. I have usually asked all students to analyse and submit for assessment at least one multiple regression exercise, with the data randomly produced according to a common 'story'. At this stage also we have used small projects, such as the use of regression analysing the prices of cars of different ages, etc.

Chapter 7 fits in nicely at this point, and illustrates the theory of Chapter 5 once more. Chapters 8 and 9 can be omitted, if necessary, on a first reading, although Chapter 8 contains some very useful material.

In Chapter 10, the theory of Chapter 5 is applied to one- and two-way analysis of variance, using generalized inverses. This leads directly on to Chapters 11 to 16, which cover cross classifications, nested designs and various practical points in the use of analysis of variance. At this point some collateral material on the design of experiments is required.

My treatment of cross classifications in Chapter 16 is not standard, but it seems preferable to any others I have seen, as a general treatment. There is no danger of students labelling meaningless main effects as significant. There is no difficulty in seeing what to do in the unbalanced case, and there is no 'mixed model muddle'.

For the last chapter, I revert to maximum likelihood and introduce the idea of the generalized linear model. This could lead on directly to an exploration of the vast territory now opened up by this model, through the use of GLIM.

It is easy to think of topics which could have been included. The analysis of covariance gets only a passing mention. There is no mention of missing observations. The book leads directly on to a consideration of block designs, factorial experiments and response surfaces. However, what I have tried to do is to lay a sure foundation for a treatment of these topics. The book is very definitely open-ended.

Numerical application of the techniques is essential to a thorough understanding of them. I have included a minimum in the chapters, but in Appendix B there is a list of published data sets which can be used as a basis for further study. Many of these contain considerable detail of the experimental background, and would repay study.

I am indebted to a large number of people for help in the vast amount of work involved in the production of this book. Chief among these is Professor D. R. Cox, FRS, who set me on the right road many years ago when I went to Birkbeck College to lecture in his department. He has also given me specific advice on a number of issues.

The separate chapters have been issued as notes in courses, and many students have provided helpful comments and suggestions. I have discussed some technical issues with a number of people, including Dr E. E. Bassett, Dr J. Kollerström, Professor S. C. Pearce, Dr D. A. Preece and Dr B. J. Vowden. Several people have commented in detail on the manuscript or helped with calculations, checking, etc., including Mr C. Daffin, Mr T. M. M. Farley, Mr N. Garnham, Mr R. O'Neill and Mr P. Gilbert. Mr K. B. Wetherill has helped with calculations, checking and the preparation of diagrams, and Mr G. Wetherill has also done some calculations.

My secretaries, Mrs M. A. Swain and Mrs M. Wells, have worked uncomplainingly through vast amounts of typing with amendments and corrections, and I am grateful for their help and co-operation.

Last, but not least, I wish to thank my wife for putting up with me for the large stretches of time for which it was clear that my only interest was 'Chapter $X$'. Indeed, the whole family suffered somewhat, so that the book was given an alternative title, 'Interminable Statistical Methods'! However, here it is!

*Canterbury,*                                                                                    G. B. W.
*May,* 1980.

# *Notation conventions*

**General**

A standard notation is that random variables are denoted by capital letters, $Y_1, Y_2, \ldots$, and a particular realization of these by lower case letters, $y_1, y_2, \ldots$. While this convention has certain advantages there are some difficulties in applying it consistently in a book of this nature. It is essential to see, for example, that certain sums of squares which can be calculated are valid summaries of the data. On the other hand, it is also important to see what properties these sums of squares have when the arguments are regarded as random variables. Fortunately, all relationships established for data, $y_1, y_2, \ldots$, are also true for the random variables $Y_1, Y_2, \ldots$, while the reverse is not necessarily true.

In this book, therefore, the standard convention will be used with the following exception. If we write the sum of squares due to regression

$$S_{\mathrm{par}} = \sum(y_i - \bar{y})^2 - \{\sum(y_i - \bar{y})(x_i - \bar{x})\}^2 / \sum(x_i - \bar{x})^2$$

(with the notation of Chapter 2), then this is a property of the data which we can calculate. If now we write

$$E(S_{\mathrm{par}}) = \sigma^2 + \beta^2 \sum(x_i - \bar{x})^2$$

then we take it to be implied that the arguments $y_i$ of $S_{\mathrm{par}}$ have been redefined as random variables $Y_i$. We shall also take this to apply to quantities such as $\hat{\beta}$ in Chapter 2.

A further point on this convention is as follows. In some cases we ought to use the words:

We define random variables $Y_1, Y_2, \ldots, Y_n$ which are independently and identically distributed, with a distribution $N(\mu, \sigma^2)$. A particular realization of this is $y_1, y_2, \ldots, y_n, \ldots$.

Instead we shall use the shorthand:

Observations $y_1, y_2, \ldots, y_n$ are drawn independently from a distribution $N(\mu, \sigma^2)$.

## Dot notation

Given $y_{ij}$, $i = 1, 2, \ldots, r$,  $j = 1, 2, \ldots, c$,

$$y_{i.} = \sum_{j=1}^{c} y_{ij}, \qquad y_{.j} = \sum_{i=1}^{r} y_{ij}, \qquad y_{..} = \sum_{i=1}^{r} \sum_{j=1}^{c} y_{ij}$$

$$\bar{y}_{i.} = \frac{1}{c} \sum_{j=1}^{c} y_{ij} \qquad \bar{y}_{.j} = \frac{1}{r} \sum_{i=1}^{r} y_{ij}, \qquad \bar{y}_{..} = \frac{1}{rc} \sum_{i=1}^{r} \sum_{j=1}^{c} y_{ij}$$

## Sums of squares

$$CS(x, y) = \sum (x_i - \bar{x})(y_i - \bar{y})$$
$$CS(x, x) = \sum (x_i - \bar{x})^2$$
$$S_{\min} = \text{sum of squared deviations from a model}$$
$$S_{par} = \text{sum of squares due to the model}$$

## Likelihood

$$l = \text{likelihood} \qquad L = \log(\text{likelihood})$$

## The normal distribution

1. $\phi(x) = \displaystyle\int_{-\infty}^{x} \frac{1}{\sqrt{(2\pi)}} \exp\left(-\frac{x^2}{2}\right) dx$

2. If $X$ normal, $E(X) = \mu$,  $V(X) = \sigma^2$, use $N(\mu, \sigma^2)$.

3. $u_p$ defined by $p = \Pr\{X > u_p \mid X \sim N(0, 1)\}$
   $$= 1 - \phi(u_p)$$

4. The $i$th expected order statistic in a sample of $n$ is denoted $u_{i,n}$.

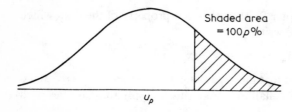

## The $t$-distribution

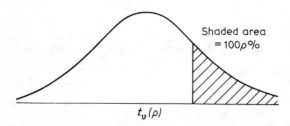

## The $\chi^2$ distribution

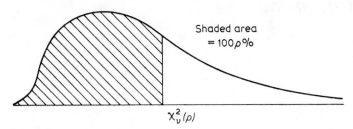

Shaded area
$= 100p\%$

$\chi^2_\nu(p)$

## The $F$-distribution

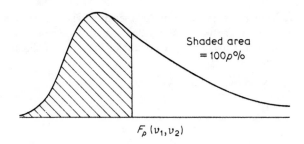

Shaded area
$= 100p\%$

$F_p(\nu_1, \nu_2)$

# Acknowledgements

I am indebted to the following for permission to use previously published material in this book: Biometrika Trustees; Cambridge University Press; John Wiley and Sons; S C Pearce; O L Davies and ICI Ltd; V Chew; E J Williams; C A Bennett and N L Franklin.

# CHAPTER ONE

# Some properties of basic statistical procedures

## 1.1 Problems of statistics

This text assumes a knowledge of elementary statistics up to confidence intervals and significance tests using the $t$, $\chi^2$ and $F$ procedures; this material is covered in many textbooks, such as the first six chapters of Wetherill (1972). Some theoretical results which are required are set out in Appendix A and a list of notation is given at the beginning of the book. The text covers maximum likelihood, regression and analysis of variance. Sections or problems marked with an asterisk are more difficult and should be omitted at a first reading.

In our study, we shall find that the $t$, $\chi^2$ and $F$ procedures come up repeatedly in new and different settings, and so in this chapter we shall study the assumptions made when using these procedures, and we shall examine how sensitive they are to these assumptions. We shall also introduce some terminology and notation. All of the statistical procedures we shall discuss are based on statistical models (see Chapter 2 for an illustration), and it is important that at the start we realize that models are unlikely to hold true exactly. Consider the following example.

*Example* 1.1   For an investigation of the effects of dust on tuberculosis, two laboratories each used sixteen rats. The set of rats at each laboratory was divided at random into two groups, A and B, the animals in group A being kept in an atmosphere containing a known percentage of dust while those in group B were kept in a dust-free atmosphere. After three months, the animals were killed, and their lung weights measured, the results being given in Table 1.1.                                                                 □ □ □

In Example 1.1 the experimental conditions assigned to the two groups of rats in each laboratory are examples of what we shall call *treatments*. The rats are examples of what we shall call *experimental units*. An essential point about a treatment is that it can be allocated to different experimental units by the experimenter. The chief variable of interest in Example 1.1 is the lung

Table 1.1 *Lung weights (g)*

| Laboratory 1 | | Laboratory 2 | |
| --- | --- | --- | --- |
| A | B | A | B |
| 5.44 | 5.12 | 5.79 | 4.20 |
| 5.36 | 3.80 | 5.57 | 4.06 |
| 5.60 | 4.96 | 6.52 | 5.81 |
| 6.46 | 6.43 | 4.78 | 3.63 |
| 6.75 | 5.03 | 5.91 | 2.80 |
| 6.03 | 5.08 | 7.02 | 5.10 |
| 4.15 | 3.22 | 6.06 | 3.64 |
| 4.44 | 4.42 | 6.38 | 4.53 |

weight, and this is an example of what we shall call a *response variable*.

One aim of the experiment described in Example 1.1 is to examine evidence of a difference in the pattern of the response variable due to the treatments. We call the difference

$$\text{mean response (A)} - \text{mean response (B)}$$

a *treatment comparison*; we are interested in estimating this quantity, and we are also interested in whether the treatment comparison differs between the two laboratories. In carrying out the analysis we must bear in mind the possibility of differences in mean and variance, and also the possibility of straggling observations arising for some reason.

As a start at the analysis, we plot the data out as in Fig. 1.1, and calculate the sample means and standard deviations, denoted $\bar{x}$ and $s$, where

$$\bar{x} = \frac{1}{n}\sum x_i$$

and

$$s^2 = \sum (x_i - \bar{x})^2/(n-1)$$

and $x_i$, $i = 1, 2, \ldots, 8$, denote the observations.

Figure 1.1 reveals a substantial amount of information about the data. In both laboratories, the mean of treatment A data set is higher than the mean of

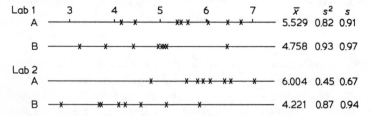

Fig. 1.1 The data of Example 1.1 plotted out.

treatment B data set. However, the difference between the A and B means is much greater for laboratory 2 than for laboratory 1. The differences in dispersion do not look great, bearing in mind the random scatter present.

A closer look at Fig. 1.1 reveals some suspiciously large gaps in the data for laboratory 1, 4.44 to 5.36 for A, and 5.12 to 6.43 for B. If these gaps indicate straggling observations, the treatment difference for laboratory 1 should be rather larger than it appears, and more nearly equal to the treatment difference for laboratory 2. It is known that experiments on biological populations such as Example 1.1 can yield results including stragglers, due to various causes. For example, we are not told about the initial weights of the rats.

These rather subjective impressions from Fig. 1.1 are obviously very useful in indicating what further steps are necessary in analysing the data, but they need to be made much more precise. We shall not proceed with a full analysis of the data, but use this example to illustrate some of the techniques we shall be using subsequently. In order to proceed we set up a model in which each data set is thought of as having been drawn from a normal population, and then the following questions arise.

*Problems of point estimation*    What are the best estimates of the mean and variance for each data set?

*Problems of interval estimation*    Between what limits are the true treatment comparisons likely to lie, considering the data from each laboratory separately?

*Problems of hypothesis testing*    Are the data sets consistent with there being no true treatment difference or no laboratory difference? Are the data sets consistent with having a homogeneous variance?

A number of similar questions can be listed for Example 1.1, and we would find that an analysis of the data would be likely to use the $t$, $\chi^2$ and $F$ procedures for confidence intervals and significance tests based on these. In the next section we set out the assumptions made by these procedures. At this point we must define our notation. Capital letters, $X_1, X_2, \ldots$, will denote random variables, and lower case will denote data. However, any such convention gives rise to problems in a book of this type and readers are referred to the section on 'notation conventions' in the preface.

## 1.2 The $t$, $\chi^2$ and $F$ procedures

All of the confidence-interval and significance-test procedures set out in this section are based on samples drawn from normal populations. One defence for this is the central limit theorem (Result 4 of Appendix A), but we need to know how important this and other assumptions are when making inferences. In the review of the procedures given below all assumptions made are in italics. The $t, \chi^2$ and $F$ distributions are defined in Appendix A.

*(a) Single-sample problems*
A single set of data $x_1, x_2, \ldots, x_n$ is presented, and *we assume that it is a random sample of independent observations from an $N(\mu, \sigma^2)$ distribution.* Two different problems arise.

(i) *Procedures for a single mean*  We conclude from Result 8 of Appendix A that if $\mu$ is the (true) expectation of the $x$'s then the statistic

$$(\bar{x} - \mu)\sqrt{n}/s \tag{1.1}$$

can be regarded as an observation on a random variable with a $t$-distribution on $(n - 1)$ degrees of freedom.

Hence a $100(1 - \alpha)$ per cent confidence interval for $\mu$ is given by

$$\bar{x} \pm t_{(n-1)}\left(\frac{\alpha}{2}\right)s/\sqrt{n} \tag{1.2}$$

where $t_{(n-1)}(\alpha/2)$ is the one-sided $100\alpha/2$ tail percentage point for the $t$-distribution on $(n - 1)$ degrees of freedom given in Appendix C; see also the preface for the notation used. A significance test of whether there is evidence that the expectation $\mu$ differs from some hypothesized value $\mu_0$ proceeds by calculating the expression (1.1) with $\mu = \mu_0$, and referring to $t$-tables on $(n - 1)$ degrees of freedom.

*Example* 1.2  A 99 per cent confidence interval for the expectation $\mu_{1,A}$ for laboratory 1 treatment A is given by using expression (1.2), which leads to

$$5.529 \pm (3.50 \times 0.91/\sqrt{8}) = (4.40, 6.66)$$

where 3.50 is the one-sided 0.5 per cent point of $t$ on 7 degrees of freedom, and the other figures are taken from Fig. 1.1.  □ □ □

*Example* 1.3  A large number of trials of treatment A in laboratory 1, in Example 1.1, have yielded an average of 5.70. Are the results given in Example 1.1 consistent with this value of the mean?

By substituting values from Fig. 1.1 we obtain

$$\frac{(\bar{x} - \mu)\sqrt{n}}{s} = \frac{(5.529 - 5.700)\sqrt{8}}{0.9054} = -0.534$$

and we refer this to tables of $t$. The result is not significant at any interesting level, and we conclude that the data are consistent with the previous value of the mean.  □ □ □

(ii) *Procedures for a single variance*  We conclude from Result 7 of Appendix A that if $\sigma^2$ is the (true) population variance of the $x$'s then the statistic

$$(n - 1)s^2/\sigma^2 \tag{1.3}$$

can be regarded as an observation on a $\chi^2_{(n-1)}$ random variable, where $(n - 1)$ denotes the number of degrees of freedom: again see Appendix C for percentage points and the preface for notation. Hence a $100(1 - \alpha)$ per cent confidence interval for $\sigma^2$ is given by

$$\left\{ \frac{(n - 1)s^2}{\chi^2_{(n-1)}(1 - \alpha/2)}, \quad \frac{(n - 1)s^2}{\chi^2_{(n-1)}(\alpha/2)} \right\} \tag{1.4}$$

where $\chi^2_{(n-1)}(p)$ denotes the $100p$ percentage point of the cumulative distribution of $\chi^2$, given in Appendix C. A significance test of whether there is evidence that the population variance differs from a hypothesized value $\sigma_0^2$ proceeds by calculating the expression (1.3) with $\sigma^2 = \sigma_0^2$, and referring to $\chi^2_{(n-1)}$ tables.

*Example* 1.4    A 99 per cent confidence interval for the variance $\sigma^2_{1,A}$ for laboratory 1 treatment A is given by

$$\left[ \frac{7 \times 0.82}{20.28}, \frac{7 \times 0.82}{0.989} \right] = (0.283, 5.80)$$

where 0.989 and 20.28 are the 0.5 and 99.5 per cent points of $\chi^2_7$.    □ □ □

*Example* 1.5    Previous experience in laboratory 1 for treatment A (Example 1.1), has established a value 0.42 for the variance. Are the results given in Example 1.1 consistent with this value of $\sigma^2$ ?

By substituting values from Fig. 1.1 we obtain

$$\frac{(n - 1)s^2}{\sigma_0^2} = \frac{7(0.82)}{0.42} = 13.67$$

From tables we find that $\chi^2_7(5\%) = 14.07$, so that even on a one-sided test this result is not quite significant at the 5 per cent level; a one-sided test could be justified if the previous results were carried out by more experienced personnel than the result given in Example 1.1. However, the value of $\chi^2$ is large enough to raise suspicions of there being a difference in variance.    □ □ □

(b)  *Two-sample problems*
Two sets of data are presented, $x_1, \ldots, x_n$ and $y_1, \ldots, y_m$, and *we assume that all observations are independently and normally distributed, with distributions* $N(\mu_x, \sigma_x^2)$, $N(\mu_y, \sigma_y^2)$ *respectively*. Again two different problems arise.

(i) *The comparison of two variances*   We   conclude   from   Result   10   of
Appendix A that

$$\frac{s_x^2 \sigma_y^2}{s_y^2 \sigma_x^2} \tag{1.5}$$

can   be   regarded   as   an   observation   on   a   random   variable   with   an
$F$-distribution on $((n-1), (m-1))$ degrees of freedom. Hence a $100(1-\alpha)$
per cent confidence interval for the ratio $\sigma_y^2/\sigma_x^2$ is given by

$$F_{\alpha/2}((n-1), (m-1))s_y^2/s_x^2 \qquad F_{(1-\alpha/2)}((n-1), (m-1))s_y^2/s_x^2 \tag{1.6}$$

where $F_p(v_1, v_2)$ is the $100p$ percentage point of the $F$-distribution for
$(v_1, v_2)$ degrees of freedom (d.f.), given in Appendix C.

A significance test of the hypothesis that the two samples come from
populations with equal variances is obtained by calculating the ratio

$$\text{(larger variance)/(smaller variance)}$$

and referring to $F((n-1), (m-1))$ tables.

*Example* 1.6   A 99 per cent confidence interval for the ratio $\sigma_{1,B}^2/\sigma_{1,A}^2$ for
laboratory 1 in Example 1.1 is as follows:

$$\left[ \frac{0.93}{8.89 \times 0.82}, \frac{8.89 \times 0.93}{0.82} \right] = (0.128, 10.083)$$

where 8.89 is the 99.5 percentage point for $F(7, 7)$, and the 0.5 percentage
point is obtained using Result 9 of Appendix A,

$$F_{\alpha/2}(v_1, v_2) = 1/\{F_{(1-\alpha/2)}(v_2, v_1)\} \qquad \qquad \square \; \square \; \square$$

*Example* 1.7   In order to test whether the data for laboratory 1 are consistent
with the variance of treatment A results being equal to the variance of
treatment B results, we calculate

$$0.93/0.82 = 1.13$$

and refer to $F$-tables on $(7, 7)$ d.f. The result is not significant at the 5 per cent
level, so that the data are consistent with the population variances being
equal.                                                                      $\square \; \square \; \square$

It   should   be   emphasized   that   the   $F$-test   is   designed   to   compare   *two*
variances. If we use the test for the largest variance ratio of, say, four variances, it is clear that significance values will be greater than those given in
$F$-tables. In Example 1.1 the largest variance ratio is

$$0.93/0.45 = 2.07$$

and we find from tables that $F_{95\%}(7, 7) = 3.79$, so that the result is not significant. If the variance ratio had been greater than 3.57, we could not have drawn any conclusion, since the appropriate value for comparing four variances is also greater than 3.57. A different test should really be applied for comparing more than two variances (see Chapter 13).

(ii) *Comparison of two means*    The question here is whether two sample means $\bar{x}$ and $\bar{y}$ are consistent with a common value of the population mean. A test proceeds by assuming $\sigma_x^2 = \sigma_y^2 = \sigma^2$, and we obtain a combined estimate of $\sigma^2$ by using

$$s^2 = \{(n-1)s_x^2 + (m-1)s_y^2\}/(m+n-2)$$

It can be shown that $(m+n-2)s^2/\sigma^2$ is distributed as $\chi^2$ on $(m+n-2)$ d.f. The test proceeds by recognizing that

$$V(\bar{X} - \bar{Y}) = \sigma^2\left(\frac{1}{m} + \frac{1}{n}\right)$$

so that

$$(\bar{X} - \bar{Y})\sqrt{(mn)}/\sqrt{(m+n)}$$

is distributed $N(0, \sigma^2)$ if $\mu_x = \mu_y$. Therefore by using Definition 6 of Appendix A we refer

$$\frac{(\bar{x} - \bar{y})\sqrt{(mn)}}{s\sqrt{(m+n)}} \qquad (1.7)$$

to $t$-tables on $(m+n-2)$ d.f.

*Example* 1.8    As an illustration, we test the hypothesis that there is no difference between the means of treatments A and B in laboratory 1. The calculations are as follows:

Laboratory 1

|  | Treatment A | Treatment B |  |
|---|---|---|---|
| Total | 44.23 | 38.06 | |
| Mean | 5.5288 | 4.7575 | |
| $\sum x_i^2$ | 250.2743 | 187.6130 | |
| $(\sum x_i)^2/n$ | 244.5388 | 181.0704 | |
| CSS | 5.7355 | 6.5426 | (d.f. = 7 each) |

| | | |
|---|---|---|
| Pooled CSS | 12.2781 | (d.f. = 14) |
| Estimate $\sigma^2$ | 0.8770 | (12.2781/14) |
| Variance (diff. means) | 0.2193 | (0.8770/4) |
| Standard error | 0.4682 | |
| $\bar{x}_A - \bar{x}_B$ | 0.7713 | |

$$t = 0.7713/0.4682 = 1.64 \text{ on } 14 \text{ d.f.}$$

A one-sided significance test is probably appropriate here as treatment A could not have a mean less than that of treatment B. Even so, the result is not significant, so that there is no clear evidence of a treatment difference in laboratory 1 results.                                                    □ □ □

---

### Exercises 1.2

1.   Obtain a 95% confidence interval for $(\mu_A - \mu_B)$ for laboratory 1.

2.*  How would you carry out a significance test of the hypothesis that $(\mu_{1,A} - \mu_{2,A} - \mu_{1,B} + \mu_{2,B})$ is zero? What interpretation does this test have? What assumptions do you have to make?

---

### 1.3  Standard assumptions and their plausibility

In the remainder of this chapter, which may be omitted at a first reading, we discuss the main assumptions underlying the statistical techniques discussed in the book, and the consequences for the statistical procedures when the assumptions fail. In the present section we concentrate on the assumptions themselves. The material in this and later sections of this Chapter is largely based on Scheffé (1959, Chapter 10) and Pearson and Please (1975), but the presentation below should be sufficient for most readers.

The main assumptions were in italics in the previous section, and they are summarized as follows:

(a) The observations are normally distributed.
(b) The observations are independently distributed and also identically distributed within samples.
(c) For the procedure for the comparison of two means given above, it is necessary to assume that the two population variances are equal.

*Normality*

We have already referred to the Central Limit Theorem (Result 4 of Appendix A) as being one defence for the assumption of normality. Further, there is a great deal of empirical evidence for normality. However, deviations from normality do commonly occur.

Deviations from normality are best thought of in terms of the values of coefficients of skewness and kurtosis, defined as $\gamma_1$ and $\gamma_2$ respectively, where

$$\gamma_1 = E(X - \mu)^3/\sigma^3 \qquad \gamma_2 = E(X - \mu)^4/\sigma^4 - 3$$

Both of these are independent of scale and origin, and are defined so that $\gamma_1$ and $\gamma_2$ are both zero for normal distributions. The coefficients are negative and positive for the shapes shown in Fig. 1.2; for further discussion see

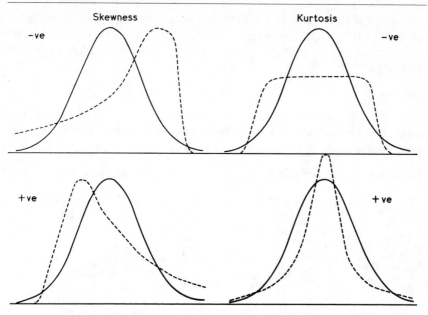

Fig. 1.2 Skewness and kurtosis.

Table 1.2 *Values of skewness and kurtosis coefficients*

| Distribution | $\gamma_1$ | $\gamma_2$ |
|---|---|---|
| Normal | 0 | 0 |
| Uniform | 0 | −1.2 |
| Exponential | 2 | 6 |
| $\chi^2_8$ | 1 | 1.5 |
| $\chi^2_{18}$ | 0.67 | 0.67 |
| Log normal $(\sigma = \frac{1}{2})$ | 1.75 | 5.89 |
| $(\sigma = 1)$ | 6.18 | 1.11 |

Table 1.3 *Values of skewness and kurtosis coefficients found in data*

| Source | Type | $\gamma_1$ | $\gamma_2$ |
|---|---|---|---|
| Bell Telephone Labs. | Engineering data | −0.7 to +0.9 | −0.4 to +1.8 |
| Australia | Brides' ages | 2.0 | 6.3 |
|  | Grooms' ages | 2.0 | 5.3 |
|  | Mothers' ages | 0.3 | −0.6 |
|  | Fathers' ages | 0.7 | 0.6 |
| Greenwich | Barometric heights | 0.5 | 0.4 |
| Student (1927) | Routine chemical analysis |  | 'usually +ve' |

*Source*: Scheffé (1959)

elementary texts such as Wetherill (1972). Values of the coefficients for some common distributions are given in Table 1.2. Values found in some real data sets are shown in Table 1.3 and Fig. 1.3; for some other values see Stigler (1977).

*Independence*
Lack of independence can be checked by plotting the data out in suitable ways, or by calculating correlation coefficients. In discussing deviations from independence later in this chapter, we shall consider a sequence of random

**1. Resistance in granular carbon**

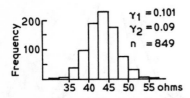

**2. Minimum sound intensity**

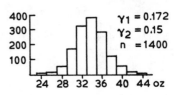

**3. Length of telephone conversations**

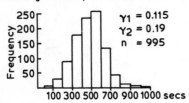

**4. Telephone jacks: tension in ounces**

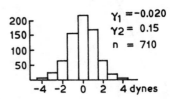

**5. Vacuum tubes: gain in db**

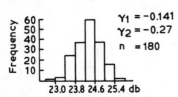

**6. Telephone poles: depth of sapwood**

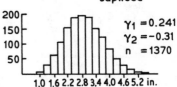

**7. Malleable iron: yield strength**

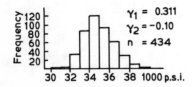

**8. Steel torsion bar springs: life in 1000 cycles**

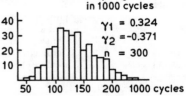

**9. Steel torsion bar springs**

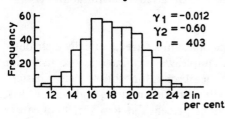

$\gamma_1 = 0.424$
$\gamma_2 = 0.15$
$n = 71$

**10. Duck cloth: warp strength**

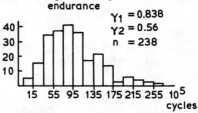

$\gamma_1 = -0.421$
$\gamma_2 = -0.25$
$n = 1000$

**11. Malleable iron: elongation**

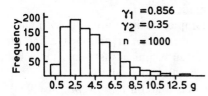

$\gamma_1 = -0.012$
$\gamma_2 = -0.60$
$n = 403$

**12. Rotating bend fatigue tests: endurance**

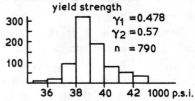

$\gamma_1 = 0.838$
$\gamma_2 = 0.56$
$n = 238$

**13. Indian cotton: fibre strength**

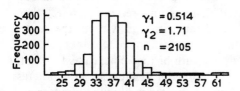

$\gamma_1 = 0.856$
$\gamma_2 = 0.35$
$n = 1000$

**14. Ferritic malleable iron: yield strength**

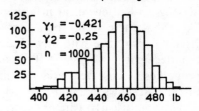

$\gamma_1 = 0.478$
$\gamma_2 = 0.57$
$n = 790$

**15. Sitka spruce: specific gravity**

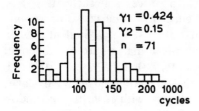

$\gamma_1 = 0.514$
$\gamma_2 = 1.71$
$n = 2105$

Fig. 1.3 Skewness and kurtosis of some real data sets. (Quoted by permission from *Biometrika*, 1975, **62**, 225).

variables $X_i$, $i = 1, 2, \ldots, n$, in which successive observations are correlated but all other covariances are zero,

$$C(X_i, X_j) = \begin{cases} \rho\,\sigma^2 & |i - j| = 1 \\ 0 & |i - j| \geq 2 \end{cases} \tag{1.8}$$

This is called *serial correlation*, and it is a very simple model but convenient for our purposes. It might arise, for example, in automatic instruments for making on-line chemical measurements in a process; in such cases it is likely that there would be a carry-over effect from the last observation. It could also arise with any observations taken serially. 'Student' (1927) reported that in a series of 100 routine chemical analyses for five different properties he observed serial correlations of 0.27, 0.31, 0.19, 0.09, 0.09, and this at least shows that non-zero serial correlation is common in his particular application. Clearly, careful experimental design, with strict use of randomization, could eliminate or reduce serial correlation in an experiment, since the order in which observations were made could be randomized.

*Outliers*
The assumption that all observations of a sample are identically distributed is sometimes found to be false. For example, we may find on close inspection of a given body of data that most of the observations follow a clear pattern, but a few observations differ markedly from this pattern, and we call these observations 'outliers'. Outliers could be present due to a variety of causes, and some possibilities will be listed in the discussion below. The way in which we deal with outliers differs according to the type of study, and depends on the way in which the observations were made.

In many areas of engineering, for example, outliers are very common indeed, and arise from factors such as transcription errors, misreading of instruments, blunders in experimental procedure, and 'adjustments' made by operators, and cover-up values inserted because no value was taken, etc. Actual evidence of the frequency of outliers is difficult to find; Stigler (1977) merely comments on the basis of 24 real data sets, that outliers are present in small quantities. In such cases the best course of action is to try and trace the conditions and experimental procedure of the aberrant observations through laboratory notebooks, etc., to see if there is any reason to reject them. Failing this, an analysis can be carried out with and without the suspect observations, to see if there is any vital difference in the conclusions made.

In some other fields, such as in agriculture or medicine, cases frequently arise where the 'outliers' are the observations of greatest scientific interest. As an example, much harm was done in the early study of plant growth hormones by rejecting as outliers all plants that grew abnormally; it is now known that this is an occasional effect of hormones. In such cases it is the outliers which show up our lack of understanding of the underlying scientific theory.

We have spoken above of 'engineering data', etc., but it is difficult to generalize; it depends on the conditions of the experiment. As another interesting illustration, consider a problem regarding outliers which arose in tracking an artificial satellite. A station recorded two angles, azimuth and elevation, continuously, and occasional ranges; but there were outliers in the measurements of the two angles. On deeper investigation it turned out that the angle measurements 'jumped' whenever a range was taken. Clearly, it would not be satisfactory just to reject the outliers!

Some automatic rules have been devised for the rejection of outliers, but it is not always wise to use them. Another approach has been to devise test statistics which are not so sensitive to outliers, such as trimmed or Winsorized means. There has been a good deal of discussion in statistical circles on how to deal with this problem.

*Other points*

Another assumption for the two-sample *t*-test was equality of the two population variances. Clearly, this could be checked by using the *F*-test in this problem, and methods will be given in Chapter 13 for testing homogeneity of variance more generally. We should note here that for two significance tests in sequence certain problems arise, and the final significance level is affected. The consequences of lack of homogeneity of variance are explored in Section 1.9.

Finally, published data sets often show evidence of too heavy rounding, transformation of the raw data before submission for analysis, etc., and any data set should be examined carefully for these features. Certain data sets in this text show these factors.

In the following notes we shall proceed as follows. First we shall examine very briefly some methods of detecting non-normality. Next we shall examine the effect of skewness and kurtosis on the tests mentioned in Section 1.2. After this we shall examine the effect of serial correlation.

*First time readers are advised to go directly to chapter 2.*

## 1.4* Tests of normality
Four commonly used tests of normality are described below.

*(a) Probability plotting*
A very good visual (but subjective) test is to plot the cumulative distribution of the observations on normal probability paper [for an explanation of normal probability paper see elementary texts, such as Wetherill (1972), or use the alternative method given below]. Any marked curvature indicates non-normality, and any point not in line with others indicates a possible outlier.

An equivalent way of carrying out the test is to plot the ordered observations against expected normal order statistics, which are defined as follows. Let $Z_1, Z_2, \ldots, Z_n$ be a random sample of $N(0, 1)$ deviates, then we write the ordered $Z_i$'s as $Z_{(i)}$,

$$Z_{(1)} \leq Z_{(2)} \leq \ldots \leq Z_{(n)}$$

so that $Z_{(1)}$ is the smallest value and $Z_{(n)}$ the largest. The expected values of

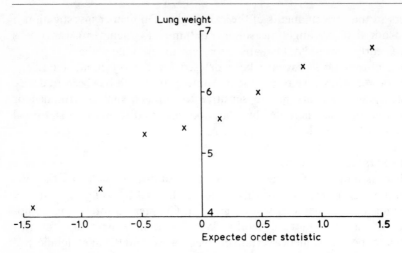

Fig. 1.4 Normal plot of laboratory 1, treatment A data (Example 1.1).

these are the expected order statistics, and we write

$$E\{Z_{(i)}\} = u_{i,n}$$

Tables of $u_{i,n}$ are given in Appendix C Table 9 and in *Biometrika Tables for Statisticians*, Vol. 1, for $n \leq 50$. For $n > 50$ we use the approximation

$$u_{i,n} = \Phi^{-1}\left(\frac{i}{(n+1)}\right)$$

A plot for treatment A, laboratory 1, of Example 1.1 is shown in Fig. 1.4.

From this graph we see again the point noted earlier, that the lowest two observations in this set tend to straggle, but nothing much can be concluded from a single sample size of 8. However, the plot is easy to draw and it can reveal quite a lot about the data, and it is widely used.

*(b) Calculation of $\gamma_1$ and $\gamma_2$*

Another procedure is to calculate the coefficients of skewness and kurtosis, $\gamma_1$ and $\gamma_2$, and then carry out significance tests on these, to test the hypotheses $\gamma_1 = 0$ and $\gamma_2 = 0$. If $n > 150$, an approximate method of carrying out a test on skewness is to treat estimates of $\gamma_1$ as normally distributed with expectation zero and variance $6/n$. The equivalent test for $\gamma_2$ requires $n > 1000$, when we use the approximation

$$V(\hat{\gamma}_2) = 24/n$$

where $\hat{\gamma}_2$ denotes the estimate of $\gamma_2$ based on a randon sample of $n$ observations. These tests may be useful if the sample size is large enough, but for smaller sample sizes, tables of significant values are available in tables such as *Biometrika Tables for Statisticians*, Vol. 1.

### (c) The Shapiro–Wilk test

Shapiro and Wilk (1965) produced a test for normality which has been shown to be one of the most powerful available over a reasonable class of alternative hypotheses. It has also been shown to be effective in small sample sizes ($n < 20$). The test arises out of considerations of the probability plot, and readers interested in a detailed derivation and explanation should see the 'Further reading' section at the end of the chapter for references. The test is carried out as follows.

Let the sample be $x_1, \ldots, x_n$, then we order these, to obtain

$$x_{(1)} \leq x_{(2)} \leq \cdots \leq x_{(n)}$$

Then for even $n$, say, $n = 2k$, we calculate

$$b = \sum_{i=1}^{k} a_{n-i+1}(x_{(n-i+1)} - x_{(i)}) \tag{1.9}$$

where the constants $a_i$ are tabulated in Appendix C. For odd sample sizes, say $n = 2k + 1$, we still calculate the expression (1.9), and the value $y_{(k+1)}$ does not enter the calculations. The test statistic is obtained by calculating

$$W = b^2 \left/ \sum_{i=1}^{n} (x_i - \bar{x})^2 \right. \tag{1.10}$$

Small values indicate non-normality, and a table of significant values is given in Appendix C.

*Example* 1.9    For laboratory 1, treatment A the ordered values are

|       |       |       |       |       |       |       |       |
|-------|-------|-------|-------|-------|-------|-------|-------|
| 4.15, | 4.44, | 5.36, | 5.44, | 5.60, | 6.03, | 6.46, | 6.75  |

and the values of $a_i$ are obtained from Appendix C,

$$0.6052, \qquad 0.3164, \qquad 0.1743, \qquad 0.0561$$

for $i = 1, \ldots, 4$ respectively. Therefore Equation (1.9) is

$$0.6052(6.75 - 4.15) + 0.3164(6.46 - 4.44)$$
$$+ 0.1743(6.03 - 5.36) + 0.0561(5.60 - 4.44) = 2.3384$$

and Equation (1.10) is

$$W = (2.3384)^2/5.7343 = 0.954$$

From Appendix C we find that this value is not significant. We therefore conclude that there is no evidence that the data for laboratory 1, treatment A are non-normal.                                                                $\square\ \square\ \square$

## (d) $\chi^2$ goodness-of-fit test

The $\chi^2$ goodness-of-fit test is described in elementary texts such as Wetherill (1972). It can be applied in a straightforward way to testing for normality by breaking up the range of the variable into intervals, and then comparing observed frequencies in cells with those of the best fitting normal distribution. However, differences from normality tend to show up more clearly in the tails of the distribution, and it is in the tails that the goodness-of-fit test comes off badly, due to the need to combine cells to avoid small frequencies. The $\chi^2$ goodness-of-fit test is not a very good test of normality.

## (e) Comparison of tests

For a comparative study of these and other tests see Shapiro *et al.* (1968). A good procedure would be to use the probability plot generally, and calculate $W$ if the plot looks suspicious. However, tests of skewness and kurtosis are useful on occasions. Notwithstanding the success of the $W$-test, the general conclusion is that departures from normality have to be large to be detected in sample sizes less than about 25.

---

### Exercises 1.4

(These exercises take a considerable amount of time, and could be done as a class exercise where the book is being used in a course).

1. Look up the percentage points of the $\chi^2$-distribution in *Biometrika Tables*, or any similar table, and plot these on probability paper for 10, 30 and 100 degrees of freedom. The plots should be made on the same piece of graph paper, and the axis rescaled for each one so that the 1 per cent points nearly coincide.

2. Similarly, plot the percentage points of the $t$-distribution on probability paper for 6, 30 and 100 degrees of freedom.

3. Obtain a table of 'random normal deviates' and make a few plots using the expected order statistics method for random samples of size 15 and 25.

---

### 1.5* Moments of $\bar{x}$ and $s^2$

Suppose now we have a sample known to be independently drawn from a (non-normal) population with $E(X) = \mu$, $V(X) = \sigma^2$, and coefficients of skewness and kurtosis $\gamma_1$ and $\gamma_2$. Consider the random variable

$$Y = \sum_{1}^{n}(X_i - \mu)/n$$

then we have

$$E(Y) = 0 \qquad V(Y) = \sigma^2/n$$

and

$$E(Y^3) = \frac{1}{n^3} E\{\sum (X_i - \mu)\}^3 = \gamma_1 \sigma^3/n^2$$

so that

$$\gamma_1(Y) = \frac{\gamma_1 \sigma^3}{n^2} \frac{n^{3/2}}{\sigma^3} = \gamma_1/\sqrt{n} \tag{1.11}$$

Similarly, we find that

$$\gamma_2(Y) = \gamma_2/n \tag{1.12}$$

These results merely confirm the powerful effect of the central limit theorem (see Appendix A, Result 4) and $Y$ becomes more nearly normal as $n$ increases. For example, even if $\gamma_1$ and $\gamma_2$ were of the order of 2 or 3, then the coefficients of the skewness and kurtosis of the mean of a sample size 30 or more would be quite small.

If we now examine $s^2$, we have

$$s^2 = \frac{1}{n-1} \sum_1^n (X_i - \bar{X})^2 = \frac{1}{(n-1)} \sum \left\{ (X_i - \mu) - \frac{1}{n} \sum_1^n (X_j - \mu) \right\}^2$$

$$= \frac{1}{(n-1)} \left\{ \left(1 - \frac{1}{n}\right) \sum (X_i - \mu)^2 - \frac{2}{n} \sum \sum_{i<j} (X_i - \mu)(X_j - \mu) \right\}$$

Hence we find that

$$E(s^2) = \sigma^2 \tag{1.13}$$

and

$$V(s^2) = \frac{\gamma_2(n-1) + 2n}{n(n-1)} \sigma^4$$

$$= \frac{2}{(n-1)} \left\{ 1 + \frac{\gamma_2(n-1)}{2n} \right\} \sigma^4 \tag{1.14}$$

For other moments of $s^2$, see Church (1925). This result contrasts sharply with the results obtained above for the sample mean, which showed no effect of $\gamma_2$ in the variance.

If we now assume that $\gamma_1$ and $\gamma_2$ are both zero, but that successive observations have a correlation $\rho$, we obtain

$$E(Y) = 0$$

$$V(Y) = \frac{\sigma^2}{n} \left\{ 1 + 2\rho \left( 1 - \frac{1}{n} \right) \right\} \tag{1.15}$$

$$E(s^2) = \sigma^2 \left( 1 - \frac{2\rho}{n} \right) \qquad (1.16)$$

We shall consider the implications of these results below.

---

**Exercises 1.5**

1. Check Equation (1.14) for independent random variables $X_i$ defined as in Section 1.5, by proceeding as follows. Show that

$$s^4 = \frac{1}{(n-1)^2} \left\{ \left( 1 - \frac{1}{n} \right)^2 \sum_i (X_i - \mu)^4 \right.$$

$$+ 2 \left( 1 - \frac{1}{n} \right)^2 \sum_{i<j} \sum (X_i - \mu)^2 (X_j - \mu)^2$$

$$+ \frac{4}{n^2} \sum_{i<j} \sum (X_i - \mu)^2 (X_j - \mu)^2$$

$$\left. + \text{ terms involving } (X_i - \mu) \right\}$$

Hence show that

$$E(s^4) = \frac{(3 + \gamma_2)\sigma^4}{n} + \frac{(n^2 - 2n + 3)\sigma^4}{n(n-1)}$$

Thus show that

$$V(s^2) = E(s^4) - \sigma^4 = \frac{\{\gamma_2(n-1) + 2n\}}{n(n-1)} \sigma^4$$

---

**1.6\* The effect of skewness and kurtosis on the _t_-test**

From the results given in Appendix A and in Section 1.5, we conclude that even for extreme non-normality, the distribution of the sample mean $\bar{x}$ tends rapidly to $N(\mu, \sigma^2/n)$, where $E(X) = \mu$, $V(X) = \sigma^2$, for almost any population distribution of $X$. Further, the presence of skewness and kurtosis do not affect $E(s^2)$, although $V(s^2)$ is affected. From all this we conclude that the distribution of

$$t = (\bar{X} - \mu)\sqrt{n}/s \qquad (1.17)$$

is very little affected by non-zero $\gamma_1$ and $\gamma_2$; the distribution of $t$ will be asymptotically $N(0, 1)$ for a very wide range of population distributions of $X$. Some confirmation of these findings can be gained from the results of Geary

(1947). He obtained expansions for the moments of $t$ and we have the following results:

$$E(t) = -\tfrac{1}{2}\gamma_1/\sqrt{n} - O(n^{-3/2})  \tag{1.18}$$

$$V(t) = 1 + 2/n + 7\gamma_1^2/(4n) + O(n^{-2})  \tag{1.19}$$

$$\gamma_1(t) = -2\gamma_1/\sqrt{n} - O(n^{-3/2})  \tag{1.20}$$

$$\gamma_2(t) = 2(3 - \gamma_2 + 6\gamma_1^2)/n + O(n^{-2})  \tag{1.21}$$

These results suggest that $\gamma_1$ is the most important parameter, and if this is positive, $E(t)$ and $\gamma_1(t)$ will be negative, so that the probability of falling below a given lower normal-theory percentage point will be above that intended, and of falling above the upper percentage point will be below that intended. In two-sided tests, the effects will tend to cancel, and in any case the effect rapidly vanishes as $n$ increases.

Pearson and Please (1975) carried out extensive computer simulation trials of the effect of deviations from assumption on common significance tests. Their charts confirm the results given above, and they show that for $n = 10$, $\gamma_1 = 0.6$ produces some rather serious discrepancies with a one-sided test, while $\gamma_2$ has little effect.

For the two-sample $t$-test let us proceed on the basis that the two means being compared come from a similar shaped non-normal population. The effect of skewness and kurtosis then tends to cancel, and the expansions equivalent to Equations (1.18) to (1.21) do not contain terms in $\gamma_1$ and $\gamma_2$ for equal sample sizes. Again this result is confirmed by Pearson and Please (1975) and their graphs show no effect even for extreme values of $\gamma_1$ and $\gamma_2$.

## 1.7*  The effect of skewness and kurtosis on inferences about variances
We expect from Equations (1.13) and (1.14) that a $\chi^2$-test is almost independent of skewness, but strongly dependent on kurtosis. The graphs of Pearson and Please (1975) confirm this, and show that for either one- or two-sided tests, $n = 10$, a $\gamma_2 = \pm 0.3$ has serious effects. As expected from the results (1.13) and (1.14), the effects do not disappear as sample size increases.

In order to demonstrate this, we see from Equation (1.14) that asymptotically

$$V(s^2) = \frac{2}{(n-1)}(1 + \tfrac{1}{2}\gamma_2)\sigma^4  \tag{1.22}$$

and the distribution of $s^2$ tends to the normal for large $n$. Therefore the probability of a one-sided test for $\sigma$ being significant at what would be the $100\alpha$ percentage level under normality is

$$\Pr(\text{significant}) = 1 - \Phi(u_\alpha/K)$$

Table 1.4 *Probability of $s^2$ being significant at the nominal 5% level of the one-sided $\chi^2$-test in large samples*

| $\gamma_2$ | $-1.5$ | $-1$ | $-0.5$ | 0 | 0.5 | 1.0 | 2.0 |
|---|---|---|---|---|---|---|---|
| Prob. | 0.0005 | 0.010 | 0.029 | 0.050 | 0.071 | 0.09 | 0.12 |

where

$$\Phi(x) = \int_{-\infty}^{x} \frac{1}{\sqrt{(2\pi)}} \exp(-\tfrac{1}{2}z^2)\,dz \tag{1.23}$$

and

$$\alpha = \int_{u_\alpha}^{\infty} \frac{1}{\sqrt{(2\pi)}} \exp(-\tfrac{1}{2}z^2)\,dz \qquad K = (1 + \tfrac{1}{2}\gamma_2)^{1/2}$$

This leads to the values shown in Table 1.4.

This table shows that quite small values of the coefficient of kurtosis, well within the range of values commonly experienced (see Table 1.3), can have serious effects on inferences from the $\chi^2$-test.

In order to study the effect of skewness and kurtosis on the $F$-test, it is easier to deal with a transformation of $F$ which is more nearly normal. Specifically, we consider equal sample sizes $n$ for the two variances and let

$$Y = \tfrac{1}{2}\log(s_1^2/s_2^2) \tag{1.24}$$

then we have

$$E(Y) = 0 \tag{1.25}$$

$$V(Y) = (\gamma_2 + 2)/2n + O(n^{-2}) \tag{1.26}$$

This leads us to expect similar effects to those shown in the $\chi^2$-test, and Pearson and Please's numerical results show this.

## 1.8* The effect of serial correlation

In order to see the effect of serial correlation on the $t$-test we resort to asymptotic arguments. From Equations (1.15) and (1.16) it follows that the asymptotic distribution of expression (1.17) is $N(0, 1 + 2\rho)$ instead of $N(0, 1)$. This means, for example, that if $\rho > 0$, the actual probability of exceeding the $100\alpha$ percentage point is increased. Table 1.5 gives some values.

It is easy to see that the effect of serial correlation is just as severe with the two-sample $t$-test. Further, there are reasons to believe that the effect may be even worse in small samples.

Table 1.5 *Probability of exceeding nominal two-sided 5% limits on the one-sample $t$-test (large samples). Serial correlation present*

| $\rho$ | $-0.3$ | $-0.2$ | $-0.1$ | 0.0 | 0.1 | 0.2 | 0.3 |
|---|---|---|---|---|---|---|---|
| Prob. | 0.002 | 0.011 | 0.028 | 0.050 | 0.074 | 0.098 | 0.12 |

Unfortunately, it is rather difficult to examine the effect of serial correlation on inferences about variances. However, it is apparent from Equation (1.16) that we may expect very serious effects on any tests that we do.

## 1.9* The effect of unequal variances on the two-sample $t$-test

The two-sample $t$-test is described in Section 1.2 $b$(ii). Now we have

$$V(\bar{X} - \bar{Y}) = \frac{\sigma_X^2}{n} + \frac{\sigma_Y^2}{m}$$

$$= \sigma_X^2 \left( \frac{1}{n} + \frac{\lambda}{m} \right)$$

where

$$\lambda = \sigma_Y^2 / \sigma_X^2$$

However, the denominator of the test involves calculating

$$s^2 = \{(n-1)s_X^2 + (m-1)s_Y^2\}/(m+n-2)$$

so that

$$E(s^2) = \sigma_X^2 \{n - 1 + \lambda(m-1)\}/(m+n-2).$$

This shows that unless $\lambda = 1$, the denominator of the two-sample $t$-test is not based on an unbiased estimate of the variance of the numerator.

Again we use an asymptotic argument. Let $m/n = R$, and let $\mu_X = \mu_Y$, and let $m$ and $n$ be large. Then we find that

$$V(t) = \frac{\sigma_X^2 \{(1/n) + (\lambda/m)\}(m+n-2)mn}{\sigma_X^2 \{n - 1 + \lambda(m-1)\}(m+n)}$$

$$\simeq \frac{R + \lambda}{1 + \lambda R}$$

Some indication of how this affects nominal significance levels is shown in Table 1.6.

Clearly, if the sample sizes are very different, lack of equality of variances can be serious, and this shows that from this point of view there would be an advantage in keeping sample sizes at least approximately equal.

Table 1.6 *Probability of exceeding nominal 5% limits (large samples)*

| $R$ | $\lambda$ | | | | | | |
|---|---|---|---|---|---|---|---|
| | 0* | $\frac{1}{5}$ | $\frac{1}{2}$ | 1 | 2 | 5 | $\infty$* |
| 1 | 0.050 | 0.050 | 0.050 | 0.050 | 0.050 | 0.050 | 0.050 |
| 2 | 0.17 | 0.12 | 0.080 | 0.050 | 0.029 | 0.014 | 0.006 |
| 5 | 0.38 | 0.22 | 0.12 | 0.050 | 0.014 | 0.002 | $1 \times 10^{-5}$ |
| $\infty$* | 1.00 | 0.38 | 0.17 | 0.050 | 0.006 | $1 \times 10^{-5}$ | 0 |

*Unattainable limiting cases to show bounds.

**1.10 Discussion**

The assumption of normality can be avoided by using different forms of test
from those described in Section 1.2. For example, Wetherill (1972, Chapter 7)
contains a description of very efficient tests of means which do not assume
normality. However, it has been shown that these tests are more sensitive to
their assumptions than the $t$-test is to its assumptions; see Wetherill (1960).
In fact inferences about means appear to be rather insensitive to the distributional form of the observations. In the case of the two-sample $t$-test sensitivity
to the assumption of equality of variance is not too serious, as we can test
this assumption before carrying out the $t$-test. The main burden of the
sections above is that inferences about variances are very sensitive to the
assumptions of normality and independence, so that, if possible, we should
check up carefully on these points when using the $\chi^2$ or $F$-tests.

We shall meet the $t$, $\chi^2$ and $F$ tests continually in the following chapters.
A great deal of statistics is concerned with devising and testing 'statistical
models' which are equations containing unknown parameters and random
components, constructed to explain given sets of data. In one very broad
class called *linear models*, the $t$, $\chi^2$ and $F$ tests are used to test hypotheses
about the parameters of the model, and the equivalent confidence interval
procedures are also used frequently. The reader would be advised to master
the basic tests given in Section 1.2 before proceeding. In the next chapter,
some very simple statistical models will be discussed, and the points made
above about use of the basic tests will be illustrated.

---

**Exercises 1.10**

1. Carry out an analysis of the Example 1.1 data, and write a report on your
conclusions. You should include:

    (a) confidence intervals for the variance;

    (b) a test of whether there is evidence of a difference between the two
        treatments in laboratory 2;

    (c) confidence intervals for the treatment difference for the two laborato-
        ries;

    (d) a test of whether there is evidence of a difference between laboratories;

    (e) tests for non-normality.

**Further reading**

For an explanation and illustration of the idea of a sampling distribution, and of the $t$, $\chi^2$ and $F$ procedures see Wetherill (1972, Chapters 1–6). For further reading on the sensitivity of the procedures to departures from assumption, see Scheffé (1959, Chapter 10), Pearson and Please (1975), and Stigler (1977). For further material on the Shapiro–Wilk test and other tests for non-normality, see Shapiro and Francia (1972), Shapiro, Wilk and Chen (1968), and Dyer (1974). For further reading on outliers, see Anscombe (1960), Tukey (1962), Grubbs (1969); and Section 13.5.

# CHAPTER TWO

# Regression and the linear model

## 2.1  Linear models

A key feature in most statistical analyses is a statistical model and it will be helpful to look at examples of some simple models, and then discuss some terminology.

*Example* 2.1    Part of the monitoring of progress of patients with kidney failure involves injecting a substance into the bloodstream and measuring how the resulting level of fibrin in the blood varies with time. For a particular patient the figures were as shown in Table 2.1.

Table 2.1

| Time after injection (in 5-min units) | log (fibrin level) |
|---|---|
| $t_i$ | $y_i$ |
| 8 | 1.4037 |
| 12 | 1.4102 |
| 19 | 1.3813 |
| 20 | 1.3618 |
| 25 | 1.3367 |
| 32 | 1.3205 |
| 36 | 1.3123 |
| 42 | 1.3261 |
| 48 | 1.2887 |
| 52 | 1.2729 |
| 100 | 1.1948 |

□ □ □

In Example 2.1 the response variable is log (fibrin level). The time measurement for each response variable is an example of what we shall call an *explanatory variable*. We shall use the term explanatory variable to cover variables describing the conditions under which the response variable was made, or variables taken to explain variation in the response variable.

In this example, the statistician's aim is to see how fibrin level is related to

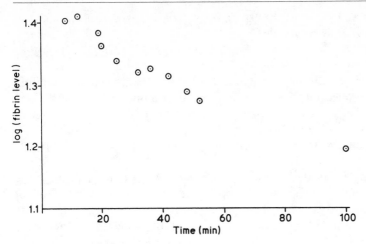

Fig. 2.1 Graph of Example 2.1 data.

time, and a useful initial step is to plot the observations on a graph, as in
Fig. 2.1. Clearly there is a roughly linear relationship between $y$ and $t$, but
this is obscured by some fluctuations, which we can think of as being the
haphazard results of a variety of biological factors and errors of measurement.

It is then resonable to consider these fluctuations as observations on
random variables, and we obtain a model for the process of the form

$$Y_i = \alpha + \beta t_i + e_i \qquad (2.1)$$

where $e_i$ is a random 'error' term and $\alpha$ and $\beta$ are unknown constants whose
values we wish to determine. This model means that the value of the response
variable is made up of two additive components; one due to the time at which
the response variable was measured, and the other component due to
random fluctuations. Logically, since $e_i$ is an observation on a random
variable, so is $y_i$; however, the explanatory variable $t_i$, the time at which the
$i$th observation is made, is a variable under the experimenter's control and
is non-random.

There is an alternative way of expressing the information in Equation (2.1),
in terms of the expectation of the random variable $Y_i$, of which $y_i$ is the
observed value. The 'error' term resulting in the observed value $e_i$ is conven-
tionally taken to have expectation zero, any constant non-zero expectation
could in any case be incorporated into $\alpha$. Taking expectations throughout
Equation (2.1) now gives

$$E(Y_i) = \alpha + \beta t_i \qquad (2.2)$$

This last equation illustrates what we call a *regression model*, which is
defined as a relationship between the expected values of a response variable,

and the observed values of the explanatory variables. (Another term for $t$ in this case is a *regressor variable*.)

*Example 2.2* (*A weighing example.*) We have two objects, A and B, whose weights $\theta_A$ and $\theta_B$ are unknown, and (in the best mathematical tradition) we have three measurements of weight, one of A alone ($y_1$), one of B alone ($y_2$), and one of the two together ($y_3$). Unfortunately, the unreliability of the weighing machine makes the three measurements incompatible. How should one combine them to determine $\theta_A$ and $\theta_B$? ☐ ☐ ☐

In this example the idea of random variation is introduced through the unreliability of the weighings; if we repeated the weighing of A we might not obtain exactly $y_1$ again. We therefore think of $y_1$ as an observation made on a random variable $Y_1$, and similarly $Y_2$, $Y_3$. Our model then becomes:

$$Y_1 = \theta_A + \varepsilon_1$$
$$Y_2 = \theta_B + \varepsilon_2$$
$$Y_3 = \theta_A + \theta_B + \varepsilon_3$$

where the $\varepsilon$'s represent the random fluctuations. To complete the specification of the model we need only the (joint) distribution of the $\varepsilon$'s. From the description given in Example 2.2 it appears that there is good reason to suppose that the variances of the random fluctuations are equal. A natural and simple assumption is therefore that they are uncorrelated and have equal variance ($\sigma^2$ say) and zero mean, implying that the extent of the inaccuracy does not vary with the weight involved, and that there is no tendency towards over- or underestimating the weight. A model with these features can alternatively be written as

$$E(Y_1) = \theta_A$$
$$E(Y_2) = \theta_B$$
$$E(Y_3) = \theta_A + \theta_B$$
$$V(Y_i) = \sigma^2, \qquad i = 1, 2, 3; \qquad Y\text{'s uncorrelated.}$$

Now in Example 2.1 the model was *linear in the parameters*, and also *linear in the explanatory variables*. In order to make this terminology clearer we develop Example 2.1 a little. The reader may have noticed that the scatter of points in Fig. 2.1 is arguably more appropriate to a smooth curve than to a straight line. In an example of this kind a curve would in general be expected rather than a line, since the model of Equation (2.2) implies that the log-fibrin level changes monotonically and linearly as time increases. In fact we would expect there to be some minimum level, which is gradually attained as time increases, and from which the only departures are those caused by the underlying fluctuations. One way of reflecting this feature would be to use instead of the original model

$$E(Y_i) = \alpha + \beta e^{\gamma t_i}$$

where one would expect $\gamma$ to be negative. This model is not linear in the (explanatory) variable. However, if the rate $\gamma$ is known, it is linear in the unknown parameters $\alpha$ and $\beta$. If $\gamma$ is not known, the model is not linear in the parameters.

In statistical literature there is a great deal of discussion of what is called the *linear model*, and by this is meant a model which is linear in the parameters, whether or not it is also linear in the variables. This class of models – linear in the parameters – is a very important class of models, and much of this book is devoted to it.

We can now see immediately that the model in Example 2.2 is linear in the parameters. The parameters are $\theta_A$ and $\theta_B$, and the expectations of $Y_1$, $Y_2$ and $Y_3$ are all linear functions of them.

In general, suppose that observations are taken on random variables $Y_1, Y_2, \ldots, Y_n$, whose expectations are linear functions of $p$ parameters $\theta_1, \theta_2, \ldots, \theta_p$; to be specific, let

$$E(Y_i) = a_{i1}\theta_1 + \ldots + a_{ip}\theta_p$$

Then writing

$$\mathbf{Y} = \begin{bmatrix} Y_1 \\ \vdots \\ Y_n \end{bmatrix} \quad \text{and} \quad \boldsymbol{\theta} = \begin{bmatrix} \theta_1 \\ \vdots \\ \theta_p \end{bmatrix}$$

thus obtaining vectors of observed random variables and parameters, where the $a_{ij}$ may be known constants, or linear or non-linear functions of explanatory variables, we can write

$$E(\mathbf{Y}) = \mathbf{A}\boldsymbol{\theta}, \tag{2.3}$$

where $\mathbf{A} = \{a_{ij}\}$ is a matrix of known coefficients. This is the general form of the model linear in the parameters.

There is no great difficulty in providing *ad hoc* solutions to the problems posed in Examples 2.1 and 2.2. In Example 2.1 one could estimate $\alpha$ and $\beta$ by drawing a straight line passing close to all the points, and in Example 2.2 a suggestion is made in Exercise 2.1.2. However, since the two examples are seen to be special cases of a more general model, it is clearly more satisfying to develop theory which covers both these cases and more.

## Exercises 2.1
1. Write down simple examples of regression models in the four categories, linear and non-linear in both the parameters and the explanatory variables.

2. In Example 2.2, show that $\{\frac{1}{2}Y_1 + \frac{1}{2}(Y_3 - Y_2)\}$ is an unbiased estimator of $\theta_A$. What is its variance? Show further that any unbiased estimator of $\theta_A$ which is a linear function of the $Y$'s can be written in the form

$$cY_1 + (1 - c)(Y_3 - Y_2)$$

Find the value of $c$ for which the variance of this estimator is a minimum. (Such an estimator is called the minimum variance linear unbiased estimator (MVLUE) of $\theta_A$).

3.* In Example 2.2, what condition is necessary on the constants $a_i$ for $\{a_1 Y_1 + a_2 Y_2 + a_3 Y_3\}$ to be an unbiased estimator of $\alpha\theta_A + \beta\theta_B$? Find the MVLUE of $\alpha\theta_A + \beta\theta_B$ and show that it can be written as $\alpha g_A + \beta g_B$, where $g_A$ and $g_B$ are the MVLUEs of $\theta_A$ and $\theta_B$ respectively.

## 2.2 The method of least squares

Suppose that we are faced with a particular practical situation and that we have checked in an *ad hoc* manner, such as by graphing, that a model linear in the parameters of the form of Equation (2.3) seems to be appropriate; then two problems arise. The first is to obtain good estimates of the unknown parameters $\theta$, and the second is to check whether the model is compatible with the data, or whether there are features which suggest that some modifications to the model are desirable. We shall deal with the second problem more thoroughly later, but, as a start, we shall use the *method of least squares* to obtain estimates of parameters in a simple case. This method turns out to have extremely good properties, but at the moment we present it simply as intuitively reasonable.

Now the model for Example 2.1 is in the form

$$Y = \alpha + \beta x + \varepsilon \qquad (2.4)$$

where the $x$'s are thought of as being exact measurements, and the $y$'s are subject to random fluctuations. It is an essential feature of this model that the explanatory variable $x$ is subject to, at most, a negligible amount of random variation. The terms $\varepsilon$ are taken as being error terms, which are independent of $x$; they are frequently taken to be normally distributed random variables with zero expectation and constant variance $\sigma^2$. It follows that a reasonable method of estimating $\alpha$ and $\beta$ in Equation (2.4) is to fit a line $y = \alpha + \beta x$ which minimizes the sum of squared distances of the observed points $y_i$, from the values fitted by the line $(\alpha + \beta x_i)$. In Fig. 2.2 this means fitting the line so as to minimize the sum of squares of the lengths of the dotted lines.

One alternative here would be to minimize the sum of the absolute distances, but this would not be very convenient mathematically. Another possibility would be to minimize the sum of squares of the perpendicular distances of the observed points $(y_i, x_i)$ from the fitted line. However, as the random fluctua-

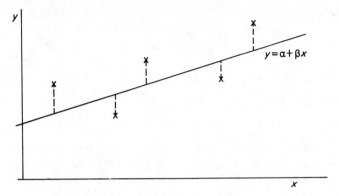

Fig. 2.2 The method of least squares.

tions are operating in the $y$-direction, it is most reasonable to minimize the sum of squared distances in this direction.

Let us denote the estimates of $\alpha$ and $\beta$ obtained from the data as $\hat{\alpha}$ and $\hat{\beta}$ respectively; then if the observed points are $(y_i, x_i)$, $i = 1, 2, \ldots, n$, the distances to the fitted line are

$$y_i - \hat{\alpha} - \hat{\beta} x_i$$

so that we wish to choose $\hat{\alpha}$ and $\hat{\beta}$ to minimize

$$S = \sum_1^n (y_i - \alpha - \beta x_i)^2 \tag{2.5}$$

By differentiating we obtain

$$\frac{\partial S}{\partial \alpha} = -2 \sum (y_i - \alpha - \beta x_i)$$

$$\frac{\partial S}{\partial \beta} = -2 \sum x_i (y_i - \alpha - \beta x_i)$$

If we put these equal to zero we obtain, for the least squares estimates $\hat{\alpha}$ and $\hat{\beta}$,

$$n\hat{\alpha} + \hat{\beta} n\bar{x} = n\bar{y} \qquad \text{where } \bar{y} = \sum y_i / n$$

and

$$\hat{\alpha} n\bar{x} + \hat{\beta} \sum x_i^2 = \sum y_i x_i$$

Hence

$$\hat{\beta} = \frac{\sum y_i x_i - n\bar{y}\bar{x}}{\sum x_i^2 - n\bar{x}^2} = \frac{\sum (y_i - \bar{y})(x_i - \bar{x})}{\sum (x_i - \bar{x})^2} \tag{2.6}$$

and

$$\hat{\alpha} = \bar{y} - \hat{\beta}\bar{x} \tag{2.7}$$

It is easy to check that these estimates give a minimum for $S$.

If we use the notation

$$CS(x, x) = \sum(x_i - \bar{x})^2$$
$$= \sum x_i^2 - (\sum x_i)^2/n \qquad (2.8)$$

then Equation (2.8) can be expressed in words;

$$\begin{bmatrix} \text{Corrected sum} \\ \text{of squares} \end{bmatrix} = \begin{bmatrix} \text{Uncorrected sum} \\ \text{of squares} \end{bmatrix} - \begin{bmatrix} \text{Correction} \end{bmatrix}$$

where $CS(x, x)$ is the corrected sum of squares of $x$, etc. The sum of squares $CS(x, x)$ is corrected in the sense that the observations are taken from their mean as origin. Similarly, we have

$$CS(x, y) = \sum(x_i - \bar{x})(y_i - \bar{y})$$
$$= \sum x_i y_i - (\sum x_i)(\sum y_i)/n \qquad (2.9)$$

or in words,

$$\begin{bmatrix} \text{Corrected sum} \\ \text{of products} \end{bmatrix} = \begin{bmatrix} \text{Uncorrected sum} \\ \text{of products} \end{bmatrix} - \begin{bmatrix} \text{Correction} \end{bmatrix}$$

The calculation of these terms is illustrated below. With this notation the least squares estimate of $\beta$ is

$$\hat{\beta} = CS(x, y)/CS(x, x) \qquad (2.10)$$

Having obtained estimates of $\alpha$ and $\beta$, the only other parameter remaining to be estimated is $\sigma^2$, the variance of the error terms. Or putting it in a different way, it is all very well to estimate the straight line $y = \alpha + \beta x$, but we need some measure of the scatter about it. Clearly, the minimized sum of squared distances, Equation (2.5), is one measure of scatter about the line, and we shall see later how this can be used to obtain an estimate of $\sigma^2$. We shall denote the minimized sum of squared distances by $S_{\min}$.

Now if we put the estimates $\hat{\alpha}$ and $\hat{\beta}$ back into Equation (2.5) we have

$$\begin{aligned} S_{\min} &= \sum\{y_i - (\bar{y} - \hat{\beta}\bar{x}) - \hat{\beta}x_i\}^2 \\ &= \sum\{y_i - \bar{y} - \hat{\beta}(x_i - \bar{x})\}^2 \\ &= \sum(y_i - \bar{y})^2 - 2\hat{\beta}\sum(y_i - \bar{y})(x_i - \bar{x}) + \hat{\beta}^2\sum(x_i - \bar{x})^2 \\ &= \sum(y_i - \bar{y})^2 - \{\sum(y_i - \bar{y})(x_i - \bar{x})\}^2/\sum(x_i - \bar{x})^2 \end{aligned}$$

by using Equation (2.6). Thus we have

$$S_{\min} = CS(y, y) - \{CS(x, y)\}^2/CS(x, x) \qquad (2.11)$$

For the data of Example 2.1 the calculations for the estimates and for the

Table 2.2 *Calculations for Example* 2.1

|  | $y$ | $yx$ | $x$ |
|---|---|---|---|
| Total | 14.609 | — | 394 |
| Mean | 1.328 1 | — | 35.818 2 |
| USS or USP | 19.441 255 | 507.773 6 | 20 686 |
| Correction | 19.402 080 | 523.267 8 | 14 112.36 |
| CSS or CSP | 0.039 175 | − 15.494 2 | 6 573.636 4 |

minimized sum of squares are shown in Table 2.2.

$$\hat{\beta} = -15.494\ 2/6573.64 = -0.002\ 357$$
$$\hat{\alpha} = 1.328\ 09 - 35.818\ 2 \times (-0.0023\ 57) = 1.412\ 5$$

The regression line is therefore

$$y = 1.412\ 5 - 0.002\ 357x$$

Also we have

$$\{CS(x, y)\}^2/CS(x, x) = 0.036\ 520$$

so that

$$S_{min} = 0.002\ 655$$

We shall use this quantity later.

---

### Exercises 2.2

1. Given Table 2.2 show that for lines of slope $\beta$ through the mean $(\bar{y}, \bar{x})$, the sum of squares is

$$S = CS(y, y) - 2\beta CS(x, y) + \beta^2 CS(x, x)$$

Hence draw a graph of $S$ versus $\beta$, for values of $\beta$ near the least-squares estimate.

2. As a check on the strength of some large castings, a small test-piece was produced at the same time as each casting. To see whether the test-piece gave a reliable indication of the strength of the whole casting, 11 castings were chosen at random and they and their associated test-pieces broken. The following were the breaking stresses (tons/sq. in.):

| Casting | $(y)$ | 45 | 67 | 61 | 77 | 71 | 51 | 45 | 58 | 48 | 62 | 36 |
|---|---|---|---|---|---|---|---|---|---|---|---|---|
| Test-piece | $(x)$ | 39 | 86 | 97 | 102 | 74 | 53 | 62 | 69 | 80 | 53 | 48 |

Plot the results and estimate graphically the regression line of $y$ on $x$. Also obtain the regression line fitted by least squares.

## 2.3 Properties of the estimators and sums of squares

*The estimators of the parameters*
Before we can discuss the properties of the estimates obtained in Section 2.2, it will be necessary to state precisely our assumptions about the model. Specifically, we assume that the observations $y$ are independently and normally distributed with expectation

$$E(Y) = \alpha + \beta x \qquad (2.12)$$

and variance

$$V(Y) = \sigma^2 \qquad (2.13)$$

which is constant, independent of $x$.

Now it turns out easier to restate the model by writing instead of Equation (2.12)

$$E(Y) = \gamma + \beta(x - \bar{x}) \qquad (2.14)$$

This is the same model as (2.12) since it is linear in $x$, and if we do this, the least-squares estimate for $\beta$ is unaltered and equal to that in Equation (2.10), but instead of Equation (2.7) we have

$$\hat{\gamma} = \bar{y} \qquad (2.15)$$

For this model, we see that since the $x_i$ are constants, and only the $Y_i$ are random variables, both $\hat{\gamma}$ and $\hat{\beta}$ are linear combinations of normally distributed quantities, and so are normally distributed by Result 3 of Appendix A. The expectation and variance of $\hat{\gamma}$ and $\hat{\beta}$ can be obtained directly (see Exercise 2.3.1), and the results are as follows:

$$E(\hat{\gamma}) = \gamma \qquad E(\hat{\beta}) = \beta \qquad (2.16)$$
$$V(\hat{\gamma}) = \sigma^2/n$$

$$V(\hat{\beta}) = \sigma^2/CS(x, x) \qquad (2.17)$$

Furthermore, by Exercise 2.3.2. we find that

$$C(\hat{\gamma}, \hat{\beta}) = 0$$

so that $\hat{\gamma}$ and $\hat{\beta}$ are independently and normally distributed. (Strictly speaking, zero covariance does not imply independence in general, but it does with normally distributed random variables: see Result 6 of Appendix A.) From these results, confidence intervals and significance tests for $\gamma$ and $\beta$ can be carried out as for any normally distributed quantity using the $t$-distribution (see Definition 6 of Appendix A) provided a suitable estimate of $\sigma^2$ is available. An example will be given later in the chapter, but first we must consider estimation of $\sigma^2$.

In Chapter 5 we shall show that $S_{min}/\sigma^2$ has a $\chi^2$-distribution on $(n-2)$ degrees of freedom. Therefore since the expectation of $\chi^2_{(n-2)}$ variate is $(n-2)$, an unbiased estimate of $\sigma^2$ is provided by

$$\hat{\sigma}^2 = S_{min}/(n-2) \qquad (2.18)$$

and confidence intervals and significance tests for $\sigma^2$ can be carried out using the $\chi^2_{(n-2)}$ distribution. Therefore in Example 2.1, an unbiased estimate of $\sigma^2$ is given by $0.002\,655/9 = 0.000\,295$, since $n-2=9$.

*The sums of squares*

Another interesting property is obtained as follows. Write the fitted values for the data as

$$\hat{y}_i = \bar{y} + \hat{\beta}(x_i - \bar{x}) \qquad (2.19)$$

then by summing we obtain

$$\sum \hat{y}_i = \sum \bar{y}$$

and the average of the fitted values is $\bar{y}$. Therefore the corrected sum of squares due to the fitted values (due to the fitted regression) is from Equation (2.19)

$$\sum(\hat{y}_i - \bar{y})^2 = \hat{\beta}^2 \, CS(x, x) \qquad (2.20)$$

and by substituting for $\hat{\beta}$ we see that

$$\hat{\beta}^2 CS(x, x) = [CS(y, x)]^2/CS(x, x)$$

which is one of the terms of Equation (2.11). Therefore if we rewrite Equation (2.11) in the form

$$CS(y, y) = \{CS(x, y)\}^2/CS(x, x) + S_{min}$$

we can interpret this equation

$$\begin{bmatrix} \text{Total variation} \\ \text{in } y \end{bmatrix} = \begin{bmatrix} \text{Corrected sum of squares} \\ \text{due to regression} \end{bmatrix} + \begin{bmatrix} \text{Sum of squared} \\ \text{deviations from} \\ \text{the regression} \end{bmatrix} \quad (2.21)$$

That is, we have separated the original variation in $y$ (measured in terms of the corrected sum of squares) into a part attributable to the regression line, and a part due to deviations from the line (the dotted parts of Fig. 2.2).

It must be emphasized that Equation (2.21) is true as a summary of the data. We now study the probability properties of these sums of squares, and so regard the arguments of the terms in Equation (2.21) as random variables. The distribution of the second term on the right-hand side of Equation (2.21) will be covered by a general theorem given in Chapter 5, as mentioned above. We therefore consider the corrected sum of squares due to regression.

First, from Equation (2.17), we have

$$E(\hat{\beta}^2) = V(\hat{\beta}) + E^2(\hat{\beta}) = \{\sigma^2/CS(x, x)\} + \beta^2$$

so that

$$E\left[\begin{array}{c} \text{corrected sum of squares} \\ \text{due to regression} \end{array}\right] = E\{\sum(\hat{Y}_i - \bar{Y})^2\}$$

$$= CS(x, x)E(\hat{\beta}^2) = \sigma^2 + \beta^2 CS(x, x) \quad (2.22)$$

The distribution of this sum of squares, due to the regression, is easy to obtain assuming that $\beta = 0$. The random variable $\hat{\beta}$ is normally distributed, with a variance $\sigma^2/CS(x, x)$ and expectation zero provided that $\beta = 0$, so that

$$\hat{\beta}^2 CS(x, x)/\sigma^2$$

is the square of a standard normal variable, and so has a $\chi_1^2$-distribution. (If $\beta \neq 0$, the distribution takes a more complicated form called the non-central $\chi^2$-distribution.)

*The analysis-of-variance table*

Therefore we can summarize the properties found for the terms of Equation (2.11) in the following 'analysis-of-variance' table. This table presents a partition of the amount of variation in the data, as represented by the corrected sum of squares of $y$, into components to which a meaning can be attributed. In the main part of the table the variables $y$ and $x$ denote actual data. In writing the $E$(MS) column the foregoing response variables in the table are assumed to be random variables; see the section on 'notation conventions' in the preface.

The corrected sums of squares in Table 2.3 add up because of Equation (2.11). The mean squares are defined by the relation

$$\text{Mean square} = \text{CSS}/(\text{degrees of freedom})$$

It follows from the results given above that the mean squares have expectations given in the final column. The mean square for deviations gives an estimate of $\sigma^2$ by Equation (2.18), irrespective of the value of $\beta$, and the expectation of the mean square due to regression is given by Equation (2.22). If $\beta = 0$ the expectation of the mean square due to regression is equal to the

Table 2.3 *Analysis of variance for regression*

| Source | CSS | d.f. | Mean square | E(MS) |
|--------|-----|------|-------------|-------|
| Due to regression | $\{CS(x, y)\}^2/CS(x, x)$ | 1 | $CS(x, y)^2/CS(x, x)$ | $\sigma^2 + \beta^2 CS(x, x)$ |
| Deviations | $S_{\min}$ | $(n-2)$ | $S_{\min}/(n-2)$ | $\sigma^2$ |
| Total sum of squares | $CS(y, y)$ | $(n-1)$ | | |

expectation of the mean square for deviations from regression, but if $\beta \neq 0$ the expectation of the mean square due to regression is greater than $\sigma^2$, the mean square for deviations.

The use of the degrees of freedom can be justified by the argument from expectations, but they are actually related to the relevant $\chi^2$-distributions. If $\beta = 0$, the total sum of squares has a distribution proportional to $\chi^2_{(n-1)}$; see Result 7 of Appendix A. The residual sum of squares has a distribution proportional to $\chi^2_{(n-2)}$, by a general theorem stated in Chapter 5. The distribution of the sum of squares due to regression has been discussed above, and is $\chi^2_1$ if $\beta = 0$.

Another more intuitive explanation of the degrees of freedom is as follows. The total sum of squares has one degree of freedom subtracted from it, for fitting the mean. The top line gives the corrected sum of squares of the fitted values, and has one degree of freedom, for fitting $\beta$. The deviations sum-of-squares line is simply the sum of squared residuals from the fitted model, and it has $(n-2)$ degrees of freedom, since two parameters, $\alpha$ and $\beta$, have been fitted. The deviations line of the analysis of variance table is sometimes labelled *residual*, so that we refer to the residual mean square or the residual sum of squares.

One vital result which follows from the analysis-of-variance table is the estimate of $\sigma^2$, since this enables us to estimate variances of $\hat{\gamma}$ and $\hat{\beta}$, and to calculate confidence intervals for $\gamma$ and $\beta$. We have seen that both $\hat{\gamma}$ and $\hat{\beta}$ are normally distributed, and that their variances are given by Equation (2.17). Therefore 95 per cent confidence limits for $\beta$ are given by

$$\hat{\beta} \pm t_{(n-2)}(2\tfrac{1}{2}\%)\sqrt{(V(\hat{\beta}))} \tag{2.23}$$

and a similar result holds for $\gamma$.

If we wish to predict the mean at a value $x'$, then this is given by

$$\left\{ \begin{array}{c} \text{Estimated mean } y \\ \text{at } x = x' \end{array} \right\} = \bar{y} + \hat{\beta}(x' - \bar{x}) \tag{2.24}$$

This is also a linear combination of normal variables, and so is normally distributed and its variance is

$$\sigma^2 \left[ \frac{1}{n} + \frac{(x' - \bar{x})^2}{CS(x, x)} \right] \tag{2.25}$$

Again, we can estimate this variance using $\hat{\sigma}^2$, and calculate confidence intervals if we wish.

The division of the corrected sums of squares into the two components in the analysis-of-variance table is itself of interest. This gives us a numerical summary of how much of the variation in the data is taken up by the regression line.

Finally, we shall show in Chapter 5 that $\hat{\beta}$ is statistically independent of the deviations sum of squares. Therefore if $\beta = 0$, we have two independent estimates of $\sigma^2$ in Table 2.3. Since two estimates of $\sigma^2$ are compared using an $F$-test, it follows that a significance test of the hypothesis that $\beta = 0$ is provided by referring the ratio

$$\frac{\text{Mean square due to regression}}{\text{Mean square deviations}}$$

to $F$-tables on $(1, n - 2)$ degrees of freedom; see Definition 7 and Result 10 of Appendix A.

An alternative (exactly equivalent) significance test that $\beta = 0$ is given by referring

$$\frac{\hat{\beta}}{\sqrt{[\widehat{V(\hat{\beta})}]}} = \frac{\hat{\beta}\sqrt{[CS(x, x)]}}{\sqrt{[\hat{\sigma}^2]}} \tag{2.26}$$

to the $t$-distribution for $(n - 2)$ degrees of freedom.

*The residuals*
Finally, in order to examine whether or not the model is compatible with the data we need to study the *residuals*, defined as

$$r_i = y_i - \bar{y} - \hat{\beta}(x_i - \bar{x})$$

We shall use the symbols $r_i$ to denote actual values of the residuals. In order to study properties of residuals, it is necessary to define random variables $R_i$ in a similar way:

$$R_i = Y_i - \bar{Y} - \hat{\beta}(x_i - \bar{x}) \qquad i = 1, 2 \ldots, n \tag{2.27}$$

where here $\hat{\beta}$ is thought of as a random variable as well. With these definitions the following results can be established (see Exercise 2.3.3 or Theorem 5.3):

$$E(R_i) = 0 \tag{2.28}$$

$$V(R_i) = \sigma^2 \left\{ 1 - \frac{1}{n} - \frac{(x_i - \bar{x})^2}{CS(x, x)} \right\} \tag{2.29}$$

$$C(R_i, R_j) = -\sigma^2 \left\{ \frac{1}{n} + \frac{(x_i - \bar{x})(x_j - \bar{x})}{CS(x, x)} \right\} \tag{2.30}$$

Since the $R_i$ are linear combinations of normal variables, they are also normally distributed, but do not have constant variance, and are not quite independent.

**Exercises 2.3**
1. Find $E(\hat{y})$, $V(\hat{y})$, $E(\hat{\beta})$ and $V(\hat{\beta})$, for the model defined in Equations (2.13)

and (2.14). [See Wetherill, 1972, pp. 230–1.]

2. Show that

$$C(y_i, \hat{\beta}) = \sigma^2 (x_i - \bar{x})/CS(x, x)$$

and hence that

$$C(\hat{y}, \hat{\beta}) = 0.$$

3. For residuals, defined as in Equation (2.27), show that $E(R_i)$, $V(R_i)$ and $C(R_i, R_j)$ are as given in Equations (2.28) to (2.30).

4. In some applications the intercept of the regression line on the $y$-axis is of interest. Show that this intercept is $(\bar{y} - \hat{\beta}\bar{x})$, and that

$$V(\bar{Y} - \hat{\beta}\bar{x}) = \sigma^2 \left\{ \frac{1}{n} + \frac{\bar{x}^2}{CS(x, x)} \right\}$$

Hence obtain 95 per cent confidence intervals for the intercept in the example in Sections 2.1 and 2.2.

---

## 2.4 Further analysis of Example 2.1

The calculations given in Table 2.2 are sufficient for us to be able to state the analysis-of-variance table, as in Table 2.4.

This table presents a neat summary of the data. Of the total sum of squares of 0.0392, the amount 0.0365 is accounted for by the fitted regression line. This is a reflection of the fact that if the fitted line were drawn in on Fig. 2.1, most of the points would lie very close to it.

The mean square residual is an estimate of $\sigma^2$ so that we have $\hat{\sigma} = \sqrt{(0.000\,295)} = 0.0172$. An estimate of the variance of $\hat{\beta}$ is, from (2.17)

$$\frac{\hat{\sigma}^2}{CS(x, x)} = \frac{0.000\,295}{6573.64} = 0.000\,000\,045$$

and the standard error of $\hat{\beta}$ is

$$\sqrt{0.000\,000\,045} = 0.000\,212$$

Table 2.4 *Analysis of variance for Example 2.1 data*

| Source | CSS | d.f. | Mean square |
|---|---|---|---|
| Due to regression | 0.036 520 | 1 | 0.036 520 |
| Deviations | 0.002 655 | 9 | 0.000 295 |
| Total sum of squares | 0.039 175 | 10 | |

We obtain 95 per cent confidence intervals for $\beta$ from Equation (2.23), and this is

$$-0.002\,357 \pm 2.26 \times 0.000\,212$$
$$= -0.002\,357 \pm 0.000\,479$$
$$= (-0.002\,836, -0.001\,878)$$

A test of significance of the hypothesis that $\beta = 0$ is given by referring the ratio

$$\frac{\text{Mean square due to regression}}{\text{Mean square residual}} = \frac{0.036\,520}{0.000\,295} = 124$$

to $F$-tables for $(1, 9)$ degrees of freedom; since the 0.1 percentage point is only 22.86, this result is very highly significant. There is therefore very strong evidence of a relationship between $y$ and $x$.

An equivalent significance test of the hypothesis that $\beta = 0$ is to refer

$$\frac{\hat{\beta}}{\sqrt{[\text{Estimate } V(\hat{\beta})]}} = -\frac{0.002\,357}{0.000\,212} = -11.1$$

to $t$-tables for 9 degrees of freedom.[†]

Before proceeding too far with significance tests and confidence intervals, it is important to test the assumptions by studying the residuals, which are defined by Equation (2.27).

The fitted equation is

$$y = \bar{y} + \hat{\beta}(x - \bar{x})$$

or

$$y = 1.3281 - 0.002\,357(x - 35.8182)$$

Table 2.5  *Residuals for Example 2.1 data*

| $x_i$ | Fitted values $\hat{y}_i = \hat{\bar{y}} + \hat{\beta}(x_i - \bar{x})$ | $y_i - \hat{y}_i$ |
|---|---|---|
| 8 | 1.3936 | + 0.0101 |
| 12 | 1.3842 | + 0.0260 |
| 19 | 1.3677 | + 0.0136 |
| 20 | 1.3654 | − 0.0036 |
| 25 | 1.3536 | − 0.0169 |
| 32 | 1.3371 | − 0.0166 |
| 36 | 1.3276 | − 0.0153 |
| 42 | 1.3135 | + 0.0126 |
| 48 | 1.2994 | − 0.0107 |
| 52 | 1.2899 | − 0.0170 |
| 100 | 1.1768 | + 0.0180 |

[†] Since a two-sided test is being used the sign is ignored when looking up $t$-tables

which is
$$y = 1.4125 - 0.002\ 357x$$

The residuals are given in Table 2.5.

The principal aims of an analysis of residuals are as follows:

(i)   To check for possible wild or outlying observations.

(ii)  To check whether or not the data is adequately fitted by an equation of the type (2.3)

(iii) To check on the assumption that the variance of $y$ is homogeneous.

(iv)  To check on the assumption that the underlying distribution is normal.

Points (i), (ii) and (iii) can be checked roughly from Fig. 2.1 for the simple regression model we are dealing with in this chapter. With more complicated regression models it is essential to calculate residuals to check any of the points listed. In any case, there are occasions when a more precise check is required than is possible from Fig. 2.1, and we shall therefore discuss each of these points briefly, with Example 2.1 data in mind.

*Outliers*   When checking for outliers we really ought to calculate *standardized residuals*, defined by
$$r_i/\sqrt{[V(r_i)]}$$
where the denominator uses the estimated value of $\sigma^2$. If we treat these standardized values as being approximately normally distributed, we look for any suspiciously large values. In Example 2.1 there are only 11 observations, and so values larger than about 2.0 in modulus would be suspect. In larger data sets this borderline value would have to be increased a little. Many outliers stand out so markedly that the question of a precise borderline is immaterial. Furthermore, we are simply interested in a technique here for picking out observations for closer scrutiny. It is inadvisable to reject data on the grounds of statistical tests for outliers alone (see the discussion in Section 1.10).

In many data sets the variances of the residuals are so nearly homogeneous that the calculation of standardized residuals can be dispensed with, and we can treat the residuals as having a standard deviation approximately equal to $\hat{\sigma}$ (0.0172 here). We shall return to this in Chapter 5.

*Form of model*   There is a definite pattern in the residuals, with the positive and negative ones coming together. This indicates that the underlying model is not linear. In fact, we would expect the log (fibrin level) to tend to an asymptote as time increased. The linear model can only be thought of as a short-term approximation.

*Normality*   We can check for normality as described in Section 1.4. In fact we know from Equation (2.30) that the residuals we are plotting are not

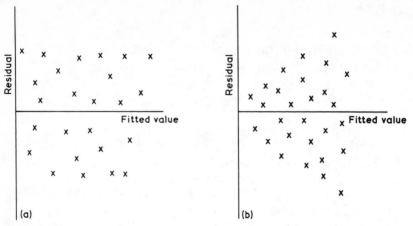

Fig. 2.3 Plot to check homogeneity of variance.

independent, but in most sets of data, it is easy enough to check roughly that the correlations are of order $O(1/n)$ and can be ignored.

*Homogeneity of variance*    The most common form of departure from homogeneity of the variance is dependence of the variance on $E(Y)$. A useful plot is therefore of the residuals against $\hat{y}_i$, and if we obtain a plot such as that given in Fig. 2.3(b), we conclude that the variance is *not* homogeneous. If this check is run on Example 2.1 data, we find that there is no evidence of heterogeneity of the variance.

---

**Exercises 2.4**

1. Carry out a check of the normality assumption for Example 2.1 data.

2. Calculate the standard deviations of the residuals using Equation (2.29) and then calculate standardized residuals.

3. The predicted mean response at $t_i = 20$ is

$$\hat{\gamma} + \hat{\beta}(20 - \bar{x})$$

Calculate the variance of this quantity, and a 95 per cent confidence interval for it. Repeat this at $t = 5, 30, 50, 70$ and plot the results on a graph.

4. Continue the analysis of Exercise 2.2.2 along the lines illustrated in this section.

5. Fibres from the forebody region of a one-year-old Romney lamb were cut into pieces 2 cm long. The diameter of each piece was measured in microns,

Table 2.6

| Group edges | $\log_{10}$ (fibre diameter measured in microns) | | | | | | | | | | | | Total |
|---|---|---|---|---|---|---|---|---|---|---|---|---|---|
| | 1.15 | 1.20 | 1.25 | 1.30 | 1.35 | 1.40 | 1.45 | 1.50 | 1.55 | 1.60 | 1.65 | 1.70 | |
| −0.10–0.00 | 1 | 1 | | | | | | | | | | | 2 |
| 0.00–0.10 | 1 | 6 | 3 | | | | | | | | | | 10 |
| 0.10–0.20 | | 3 | 6 | 1 | | | | | | | | | 10 |
| 0.20–0.30 | | 8 | 13 | 3 | 1 | 1 | | | | | | | 26 |
| 0.40 | | 2 | 5 | 5 | 2 | 2 | 1 | | | | | | 17 |
| 0.50 | | | | 5 | 8 | | | | | | | | 13 |
| 0.60 | | | | 3 | 7 | 2 | | | | | | | 12 |
| 0.70 | | | | | 5 | 13 | 1 | 1 | | | | | 20 |
| 0.80 | | | | | 3 | 6 | 5 | 5 | | | | | 19 |
| 0.90 | | | | | | 3 | 9 | 3 | 2 | | | | 17 |
| 1.00 | | | | | | | 2 | 12 | 1 | 1 | | | 16 |
| 1.10 | | | | | | | 2 | 10 | 4 | 1 | | | 17 |
| 1.20 | | | | | | | | 1 | 3 | 1 | | 1 | 6 |
| Total | 2 | 20 | 27 | 17 | 26 | 27 | 20 | 32 | 10 | 3 | | 1 | 185 |

*Hint*: In order to reduce calculation, scale the axes − 5 to + 5 for fibre diameter, and − 6 to + 6 for breaking load. It will then be found helpful to have a table of the product of these scales for each cell in which there is an observation. (Modified from London BSc, 1958.)

and the breaking load of each piece was then determined in grams. The frequency table for $\log_{10}$ (breaking load) and $\log_{10}$ (fibre diameter) of 185 pieces is given Table 2.6.

(i) Calculate the regression line of $\log_{10}$ (breaking load) on $\log_{10}$ (fibre diameter), and give the analysis-of-variance table.

(ii) Test whether the slope of the regression line differs significantly from 2, the value that would for circular fibres correspond to a proportionality between strength and cross-sectional area.

## 2.5 The regressions of y on x and of x on y

It is essential to notice that in regression analysis we do not treat the variables symmetrically. The random variation is considered to be in the $y$-direction, and the sum of squares is minimized for differences in this direction. This is called the *regression line* of $y$ on $x$; if we calculated the regression line of $x$ on $y$ it would in general be different.

The regression line of $y$ on $x$ is

$$y = \bar{y} + \hat{\beta}_{y|x}(x - \bar{x})$$

where $\hat{\beta}_{y|x} = CS(x, y)/CS(x, x)$. The minimized sum of squares is

$$S_{\min}^{y|x} = CS(y, y) - \{CS(x, y)\}^2/CS(x, x)$$
$$= CS(y, y)[1 - \{CS(x, y)\}^2/\{CS(x, x)\ CS(y, y)\}]$$
$$= CS(y, y)\{1 - \rho_s^2\} \tag{2.31}$$

where $\rho_s$ is the sample correlation coefficient between $y$ and $x$.

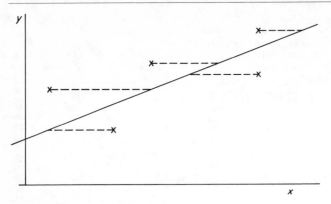

Fig. 2.4 The regression of **x** on **y**.

If we carried out a regression analysis of $x$ on $y$, then we would have to minimize the sum of squares in the $x$-direction; see Fig. 2.4 in comparison with Fig. 2.2. The regression line would be

$$x = \bar{x} + \hat{\beta}_{x|y}(y - \bar{y})$$

where $\hat{\beta}_{x|y} = CS(x, y)/CS(y, y)$. Similarly we find that the minimized sum of squares is as follows:

$$S^{x|y}_{\min} = CS(x, x) - \{CS(x, y)\}^2/CS(y, y)$$

$$= CS(x, x)\{1 - \rho_s^2\}$$

(2.32)

Thus we find that both regression lines go through the point $(\bar{x}, \bar{y})$, and both lines take up the same proportion of the total sum of squares. However, for the lines to be of equal slope we need

$$\hat{\beta}_{y|x} = 1/\hat{\beta}_{x|y}$$

which leads to

$$\frac{CS(x, y)}{CS(x, x)}\frac{CS(x, y)}{CS(y, y)} = 1$$

or

$$\rho_s^2 = 1$$

That is, the two lines only have the same slope if the sample correlation coefficient between $y$ and $x$ is plus or minus unity. This means that all data must fall exactly on the same straight line.

The two regression lines have a different meaning. The regression line of $y$ on $x$ is the locus of expected values of $y$, for given values of $x$. The regression line of $x$ on $y$ is the locus of expected values of $x$, for given values of $y$. Since we are taking expectations in different directions it is obvious that in general the lines will differ.

Clearly, for both regression lines to be valid, rather special conditions need to apply, since expectations must be valid in *both* directions. Often one variable is not a stochastic variable, and this is taken as the explanatory (or regressor) variable, as in Example 2.1. We cannot treat time in Example 2.1 as a random variable, and we cannot, therefore, take expectations in this direction.

If conditions do arise when both lines may be valid the following points should be noted:

(i)   If one variable is to be used as a predictor of the other, then the predictor is used as the explanatory variable.

(ii)  If there is a causal relationship between the two variables and this is thought to be in a particular direction, we take the explanatory variable to be the initial variable.

(iii) If both variables contain random variation then regression may not be appropriate (see Sprent, 1969).

In further work in this text we shall denote the response variable by $y$, and the explanatory variables by $x_1, x_2$, etc.

---

**Exercise 2.5**
1. Calculate and plot both regression lines for Exercise 2.2.2.

---

## 2.6 Two regressor variables
The method of least squares is easily extended to estimate the unknown parameters of more complicated examples of the linear model. Consider the following problem.

*Example* 2.3   The data given below are part of the data collected in an investigation into ways of measuring the weight of plum trees without damaging them. One possibility was to pull up a group of mature trees, weigh each one, and see how far a relationship could be established with the trunk circumferences. A linear relationship was expected with the logarithms of the variables, and the variates given in Table 2.7 are

$w$ : log (circumference at the base of the trunk)
$x$ : log (circumference at the top of the trunk)
$y$ : log (weight of tree above ground)

We are going to try to fit a linear regression model between the dependent variable $y$, and the explanatory variables $w$ and $x$; bearing in mind the simplification made by using Equation (2.14) rather than Equation (2.12),

Table 2.7

| w | x | y |
|---|---|---|
| 1.690 | 1.663 | 2.318 |
| 1.583 | 1.568 | 2.100 |
| 1.693 | 1.643 | 2.225 |
| 1.648 | 1.609 | 2.140 |
| 1.628 | 1.599 | 2.107 |
| 1.600 | 1.603 | 2.049 |
| 1.677 | 1.640 | 2.228 |
| 1.588 | 1.535 | 2.029 |
| 1.645 | 1.606 | 2.140 |

we write this as

$$E(Y) = \alpha + \beta_1 (w - \bar{w}) + \beta_2 (x - \bar{x})$$
$$V(Y) = \sigma^2 \tag{2.33}$$

where $\bar{w} = \sum w_i/n$, $\bar{x} = \sum x_i/n$, $n$ is the number of observations, and the $Y$'s are assumed to be independent.

We now proceed to obtain estimates of $\alpha$, $\beta_1$ and $\beta_2$ as in Section 2.2. The deviations from the model (residuals) are

$$y_i - \hat{\alpha} - \hat{\beta}_1 (w_i - \bar{w}) - \hat{\beta}_2 (x_i - \bar{x}) \tag{2.34}$$

and the method of least squares is to choose $(\hat{\alpha}, \hat{\beta}_1, \hat{\beta}_2)$ to minimize the sum of squares of these deviations,

$$S = \sum_{i=1}^{n} \{y_i - \hat{\alpha} - \hat{\beta}_1 (w_i - \bar{w}) - \hat{\beta}_2 (x_i - \bar{x})\}^2 \tag{2.35}$$

We must therefore choose $(\hat{\alpha}, \hat{\beta}_1, \hat{\beta}_2)$ as the solution of the equations

$$\frac{\partial S}{\partial \alpha} = \frac{\partial S}{\partial \beta_1} = \frac{\partial S}{\partial \beta_2} = 0$$

These equations are called the *normal equations* of least squares, and they are as follows:

$$\sum \{y_i - \hat{\alpha} - \hat{\beta}_1 (w_i - \bar{w}) - \hat{\beta}_2 (x_i - \bar{x})\} = 0 \tag{2.36}$$

$$\sum (w_i - \bar{w}) \{y_i - \hat{\alpha} - \hat{\beta}_1 (w_i - \bar{w}) - \hat{\beta}_2 (x_i - \bar{x})\} = 0 \tag{2.37}$$

$$\sum (x_i - \bar{x}) \{y_i - \hat{\alpha} - \hat{\beta}_1 (w_i - \bar{w}) - \hat{\beta}_2 (x_i - \bar{x})\} = 0 \tag{2.38}$$

Equation (2.36) gives $\hat{\alpha} = \bar{y}$, and by using this and the notation (2.9), Equations (2.37) and (2.38) become

$$\hat{\beta}_1 CS(w, w) + \hat{\beta}_2 CS(w, x) = CS(w, y) \tag{2.39}$$

$$\hat{\beta}_1 CS(x, w) + \hat{\beta}_2 CS(x, x) = CS(x, y) \tag{2.40}$$

leading to

$$\hat{\beta}_1 = \{CS(x, x)CS(w, y) - CS(w, x)CS(x, y)\}/\Delta \tag{2.41}$$

and

$$\hat{\beta}_2 = \{CS(w, w)CS(x, y) - CS(w, x)CS(w, y)\}/\Delta \tag{2.42}$$

where

$$\Delta = CS(w, w)CS(x, x) - \{CS(w, x)\}^2 \tag{2.43}$$

By substituting $\hat{\alpha}, \hat{\beta}_1, \hat{\beta}_2$ back into Equation (2.35), it can be shown that the minimized sum of squared deviations is

$$S_{\min} = CS(y, y) - \hat{\beta}_1 CS(w, y) - \hat{\beta}_2 CS(x, y) \tag{2.44}$$

Again, a general theorem to be discussed later shows that an estimate of $\sigma^2$ is

$$\hat{\sigma}^2 = S_{\min}/(n - 3) \qquad \square \ \square \ \square$$

We shall not proceed with this theory, as the general theory we cover in Chapter 5 derives the results we need in an easier way than by proceeding on the present route. The calculations for Example 2.3 down to $\hat{\alpha}, \hat{\beta}_1, \hat{\beta}_2$ and $\sigma^2$ are summarized below; a full analysis of the data will not be given here (see Chapter 12).

*Example* 2.4  Calculations for Example 2.3.

|       | y        | w         | x         |
|-------|----------|-----------|-----------|
| Total | 19.336   | 14.752    | 14.466    |
| Mean  | 2.148 4  | 1.639 1   | 1.607 3   |
| USS   | 41.611 62| 24.194 62 | 23.263 99 |
| Corr  | 41.542 32| 24.180 17 | 23.251 68 |
| CSS   | 0.069 30 | 0.014 45  | 0.012 31  |

|      | yw        | yx        | xw        |
|------|-----------|-----------|-----------|
| USP  | 31.722 669| 31.105 484| 23.723 547|
| Corr | 31.693 852| 31.079 397| 23.711 381|
| CSP  | 0.028 817 | 0.026 087 | 0.012 166 |

$\hat{\beta}_1 = 1.2473 \qquad \hat{\beta}_2 = 0.8865 \qquad \hat{\alpha} = 2.1484 \qquad \hat{\sigma}^2 = 0.001\ 71$

$\square \ \square \ \square$

### Exercises 2.6
1. Calculate the residuals for Example 2.3 data using Equation (2.34), and carry out analyses of these residuals similar to those given in Section 2.4.

2. Show that

$$V(\hat{\beta}_1) = \sigma^2 CS(x, x)/\Delta \qquad V(\hat{\beta}_2) = \sigma^2 CS(w, w)/\Delta$$
$$C(\hat{\beta}_1, \hat{\beta}_2) = -\sigma^2 CS(x, w)/\Delta$$

where $\Delta$ is given by Equation (2.43).

---

## 2.7 Discussion

In this chapter we have considered some simple examples of statistical models, and we have seen one method of fitting these models. However, there are other methods of estimating unknown parameters, and it is important to know the circumstances under which the method of least squares is satisfactory. We shall give a brief discussion of general theory of estimation, and return later to the method of least squares. For further numerical illustration, see the references given in Appendix B.

# CHAPTER THREE

# Statistical models and statistical inference

## 3.1 Parametric inference

In the previous chapter we have seen how a simple statistical model can be fitted to data by estimating the unknown parameters and then making checks with residuals. After we have done this, various questions can be answered in terms of the fitted model. For instance, in Example 2.1, the question 'Does fibrin level vary with time after injection?' is put in terms of the $F$-test, $\beta = 0$, in Section 2.4. The question 'Between what limits do we expect the mean fibrin level to be at $t = 20$?' is answered by Exercise 2.4.3.

This is the general method of what we call *parametric inference*. A model is built up which involves unknown parameters. We then obtain point estimates of these unknown parameters, as for example, we obtained estimates of $\alpha$ and $\beta$ in Section 2.2. Our questions about the data are then put in terms of the model. This method of inference contrasts with another method called *non-parametric inference*. In this case, significance tests and confidence-interval procedures are carried out without formulating a model which involves distributional parameters. Wetherill (1972, Chapter 7) gives a simple coverage of some non-parametric techniques and we shall not discuss these further in this text.

Generally, the following steps are involved in parametric inference:

(i)  A statistical model for the situation to be considered is proposed. This is usually done after at least a preliminary analysis of some data. Thus in Example 2.1 we show Fig. 2.1 first, after which the model in Equation (2.1) was proposed.

(ii) We obtain point estimates of the unknown parameters, and also obtain estimates of their standard errors. Sometimes confidence limits are calculated and significance tests are carried out at this stage as well. In Example 2.1, point estimates were obtained using the method of least squares, and the calculations are given in Table 2.2. Some further analysis is given in Section 2.4.

(iii) The fit of the proposed model is tested in some way, such as by an examination of residuals. This may lead to a revision of the proposed model, and so to a repeat of step (ii), and so on, until a satisfactory model is obtained. In the analysis of Example 2.1, the data given in Section 2.4, positive and negative residuals tended to come together, showing the possible presence of a non-linear response. This led to a model such as that in Equation (2.2) being proposed.

(iv) Finally, the original problem is posed in terms of the model, and any necessary confidence intervals, significance tests, etc., calculated. Frequently the questions we ask do not involve all of the estimated parameters, and in such a case it is important that if possible, our methods of inference are not conditional on particular values for the unwanted parameters. For example, the question asked about Example 2.1 above, 'Does fibrin level vary with time after injection?', is posed as a question about $\beta$. We want a significance test for $\beta$ here, which is valid irrespective of the values of the other parameters, $\alpha$ and $\sigma^2$. (With respect to this particular question, we say that $\alpha$ and $\sigma^2$ are *nuisance* parameters. If we asked a different question, other parameters might be the nuisance parameters.)

In parametric inference our models must be precise, and specify completely the probability distribution of the observations. Clearly one of the crucial problems is to obtain estimates of the unknown parameters in our models which have – in some sense to be defined – 'good' properties. We have seen one simple method of estimation in a very simple problem. Many other methods of estimation could be devised, even for this very simple problem (see below, for one other), and we now set up some criteria with which to compare alternative estimators. We begin with a particular example.

### 3.2 Point estimates
Suppose we have some data $(y_i, x_i)$, $i = 1, 2, \ldots, n$, to which we wish to fit a simple linear regression model. Let us assume that the random variables $y_i$ are independently and normally distributed with expectation and variance

$$E(Y) = \alpha + \beta(x - \bar{x})$$

$$V(Y) = \sigma^2$$

(3.1)

then the method of least squares leads to the estimator (2.6) for $\beta$, and $\hat{\alpha} = \bar{y}$.

An alternative method of fitting the regression line which can be used is the 'two-group' method. The points $(y_i, x_i)$ are divided into two approximately equal sets by putting into group one all points such that $x_i < x'$, and all other points into group two, for some suitable choice of $x'$. If the group means are $(\bar{y}_{(1)}, \bar{x}_{(1)})$, and $(\bar{y}_{(2)}, \bar{x}_{(2)})$, then the regression line is drawn to join them. Thus the slope of the fitted line is

$$\tilde{\beta} = (\bar{y}_{(1)} - \bar{y}_{(2)})/(\bar{x}_{(1)} - \bar{x}_{(2)}) \qquad (3.2)$$

where we use the symbol $\tilde{\beta}$ in order to distinguish it from the least-squares estimator $\hat{\beta}$.

Which method of fitting the line would we prefer, and why? We shall simplify the problem by comparing only the two estimators of $\beta$.

Each of the estimators has a probability distribution, which we call a *sampling distribution*. These are distributions which we would get by repeatedly (and indefinitely) sampling sets of data $(y_i, x_i)$, $i = 1, 2, \dots, n$ from the same model and using the proposed estimators for each set of data. In Chapter 2 we established that the sampling distribution of $\hat{\beta}$ is normal with expectation and variance

$$E(\hat{\beta}) = \beta$$
$$\qquad (3.3)$$
$$V(\hat{\beta}) = \sigma^2/CS(x, x).$$

The estimator (3.2) is also normally distributed since it is a linear function of normally distributed variables. Furthermore

$$E(\bar{Y}_{(i)}) = \alpha + \beta(\bar{x}_{(i)} - \bar{x})$$

so that

$$E\{\bar{Y}_{(1)} - \bar{Y}_{(2)}\} = \beta(\bar{x}_{(1)} - \bar{x}_{(2)})$$

and

$$E(\tilde{\beta}) = \beta$$

To get the variance, we shall suppose that there are two equal groups of $n/2$ observations each, and then

$$V(\tilde{\beta}) = \frac{1}{\{\bar{x}_{(1)} - \bar{x}_{(2)}\}^2} \{V(\bar{Y}_{(1)}) + V(\bar{Y}_{(2)})\}$$
$$= \frac{4\sigma^2}{n\{\bar{x}_{(1)} - \bar{x}_{(2)}\}^2} \qquad (3.4)$$

A comparison of $\hat{\beta}$ and $\tilde{\beta}$ is therefore straightforward. Both sampling distributions are normal with expectation $\beta$, the true value. For $n = 2$ it is easy to establish that the estimators are equivalent, but for $n > 2$ it can be shown that the variance in Equation (3.3) is less than that in Equation (3.4). The sampling distributions are sketched in Fig. 3.1., and it is obvious that we are likely to be closer to the true value by using $\hat{\beta}$ than $\tilde{\beta}$.

We now turn to a more general description of the point estimation problem and discuss some properties of estimators, which are really properties of the relevant sampling distributions. Suppose that in general we have independent random variables $X_1, X_2, \dots, X_n$, represented by a vector

$$\mathbf{X}' = (X_1, X_2, \dots, X_n)$$

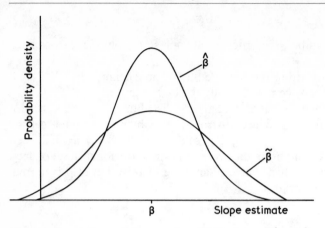

Fig. 3.1 Comparison of the sampling distributions of $\hat{\beta}$ and $\tilde{\beta}$.

from a distribution which depends on an unknown parameter $\theta$, and in order to estimate $\theta$ we calculate some function $t(\mathbf{X})$. We now define unbiasedness, consistency, and efficiency.

*Definition*   A point estimator $t(\mathbf{X})$ of $\theta$ is *unbiased*, if

$$E\{t(\mathbf{X})\} = \theta$$

for all $\theta$ in the range permitted, where the expectation is taken over the sampling distribution of $t(\mathbf{X})$.                    □ □ □

*Example* 3.1   In the simple linear regression example discussed at the beginning of this section, one of the unknown parameters we wish to estimate is $\beta$. Two possible functions suggested for the estimator $t(\mathbf{X})$ are the least-squares estimator $\hat{\beta}$, and the two-group estimator $\tilde{\beta}$. We have seen that

$$E(\hat{\beta}) = E(\tilde{\beta}) = \beta$$

so that both estimators are unbiased.                    □ □ □

*Example* 3.2   If $X_1, X_2, \ldots, X_n$ are independently drawn from $N(\mu, \sigma^2)$, then both the mean and the median are unbiased estimators of $\mu$.   □ □ □

Exact unbiasedness arises largely for reasons of mathematical simplicity, and some common estimators are biased. For example, if under the conditions of Example 3.2 we wish to estimate $\sigma$, we usually calculate

$$s = \sqrt{\left\{ \frac{1}{(n-1)} \sum_{1}^{n} (X_i - \bar{X})^2 \right\}}$$

but Exercise 3.2.1 shows that this is biased. Approximate unbiasedness is a reasonable requirement for an estimator, but this is difficult to define.

*Definition*    The estimator $t(\mathbf{X})$ is said to be a *consistent* estimator of $\theta$ if the probability of making errors of any given size $\varepsilon$, tends to zero as $n$ tends to infinity; that is,

$$\Pr\{|t(\mathbf{X}) - \theta| > \varepsilon\} \to 0$$

as $n \to \infty$, for any fixed positive $\varepsilon$.    □ □ □

Consistency can be regarded as an attempt to define approximate unbiasedness, but it is not a very strong requirement. It does, however, convey the idea that the *location* of the sampling distribution of the estimator $t(\mathbf{X})$ is approximately correct for large $n$.

It states that as $n$ increases indefinitely, the sampling distribution of $t(\mathbf{X})$ becomes more and more concentrated at the true value $\theta$ (see Fig. 3.2).

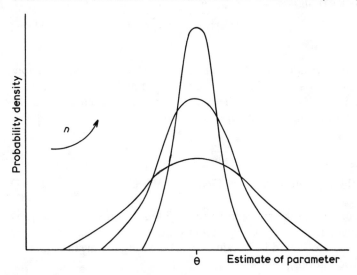

Fig. 3.2 Illustration of consistency.

*Example* 3.3    In sampling from an $N(\mu, \sigma^2)$ population, the sample mean $\bar{X}$ has a distribution $N(\mu, \sigma^2/n)$. Therefore we have for any fixed positive $\varepsilon$,

$$\Pr\{|\bar{X} - \mu| > \varepsilon\} = 2\Pr\{\bar{X} > \mu + \varepsilon\}$$
$$= 2\left\{1 - \Phi\left(\frac{\varepsilon\sqrt{n}}{\sigma}\right)\right\}$$

where $\Phi(x)$ is defined in Equation (1.23). This tends to zero as $n$ tends to infinity, so that $\bar{X}$ is a consistent estimator of $\mu$.    □ □ □

As another example, consider the simple linear regression model discussed above. Both proposed estimators are unbiased, and it is obvious that the variances of the sampling distributions of both $\hat{\beta}$ and $\tilde{\beta}$ shrink indefinitely as $n$ increases, so that both estimators are consistent estimators of $\beta$.

For studying the dispersion of a sampling distribution, the obvious criterion to use is $V\{t(\mathbf{X})\}$. For any two unbiased estimators we define the relative efficiency as follows:

*Definition*   The *relative efficiency* of two unbiased estimators $t_1(\mathbf{X})$ and $t_2(\mathbf{X})$ is

$$V\{t_2(\mathbf{X})\}/V\{t_1(\mathbf{X})\} \qquad \square\,\square\,\square$$

The relative efficiency will usually be a function of $n$, and to avoid this we often use its limiting value.

*Definition*   *The asymptotic relative efficiency*

$$\lim_{n \to \infty} V\{t_2(\mathbf{X})\}/V\{t_1(\mathbf{X})\} \qquad \square\,\square\,\square$$

*Example* 3.4   Independent random variables $X_1, X_2, \ldots, X_n$ are distributed $N(\mu, \sigma^2)$, and we wish to estimate $\mu$. We have seen that $V(\bar{X}) = \sigma^2/n$, and it can be shown that $V(\dot{X}) \simeq \pi\sigma^2/2n$, where $\dot{X}$ is the median. The efficiency of the median relative to the mean is therefore approximately

$$\frac{\sigma^2/n}{\pi\sigma^2/2n} = \frac{2}{\pi}$$

which is also the asymptotic relative efficiency in this case.   $\square\,\square\,\square$

The interpretation of the result in Example 3.4 is that if we estimate $\mu$ by calculating the median of, say, 1000 observations, then we get approximately the same precision as would be obtained by calculating the mean of $(2/\pi) \times 1000 \simeq 637$ observations.

This brief discussion serves to introduce these concepts of unbiasedness, consistency and efficiency, and it will be sufficient for what follows. For a more complete discussion, see the references listed under 'Further reading' at the end of the chapter.

It should also be emphasized that other factors have to be borne in mind when comparing estimators. Thus, notwithstanding Example 3.4, we may wish to use the sample median instead of the sample mean to estimate a normal mean, because the median is less sensitive to the presence of outliers.

In the simple linear-regression case, the two-group estimate of $\beta$ can be shown to be consistent under a wide range of conditions, but less efficient than the least-squares estimate. Nevertheless, one may prefer to use the two-group method on occasions, for instance to avoid untrained staff having to do complicated calculations.

---

**Exercises 3.2**

1. By considering the definition of $V(s)$, or otherwise, show that $E(s) < \sigma$, so that $s$ is not an unbiased estimator of $\sigma$.

2. What criticisms do you have of the use of relative efficiency for comparison of estimators?

---

### 3.3 The likelihood function

In the previous sections we have emphasized that we shall be mainly concerned with parametric inference. The method is to consider a model of a situation, which involves unknown parameters. As far as point estimation is concerned, one way of proceeding would be to conjecture some reasonable estimators and then study their properties, as illustrated in the last section. However, it is clear that we need a general method which produces good estimators. We now proceed to introduce the concept of the likelihood function, which plays a fundamental role in statistical inference. After this we shall introduce a general method of point estimation, based on likelihood, which is used a great deal in practice.

Suppose that in a sampling inspection of a large batch of items, twenty are chosen at random and classified as good or bad. Then if $\theta$ is the proportion of bad items in the batch, the probability distribution of the number $r$ of defective items in the batch is binomial,

$$\Pr(r) = \binom{20}{r} \theta^r (1 - \theta)^{20-r} \qquad r = 0, 1, \dots, 20. \qquad (3.5)$$

If three defectives were observed for a particular batch, the probability of this result is

$$\binom{20}{3} \theta^3 (1 - \theta)^{17}. \qquad (3.6)$$

Now if $\theta$ is unknown, some helpful information is provided by tabulating expression (3.6) for different values of $\theta$; see Table 3.1 and Fig. 3.3 (the probabilities are small outside the range given). We see that values of $\theta$ near to 0.15 are much more likely to have given rise to the observed result than, say, values of $\theta$ above 0.40. We call the function plotted in Fig. 3.3 the *likelihood function*. In Equation (3.5) we have the probability distribution of the

Table 3.1  *Values of* $\binom{20}{3} \theta^3 (1 - \theta)^{17}$

| $\theta$ | Probability | $\theta$ | Probability |
|------|-------------|------|-------------|
| 0.03 | 0.018 | 0.19 | 0.218 |
| 0.05 | 0.060 | 0.21 | 0.192 |
| 0.07 | 0.114 | 0.23 | 0.163 |
| 0.09 | 0.167 | 0.25 | 0.134 |
| 0.11 | 0.209 | 0.27 | 0.106 |
| 0.13 | 0.234 | 0.29 | 0.082 |
| 0.15 | 0.243 | 0.31 | 0.062 |
| 0.17 | 0.236 | 0.33 | 0.041 |
|      |       | 0.35 | 0.032 |

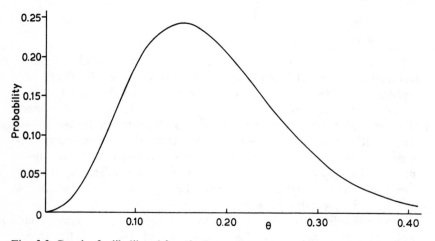

Fig. 3.3 Graph of a likelihood function.

observed sample; $\theta$ is fixed but known, and $r$ runs over the values 0, 1, ... , 20. In expression (3.6) we have the probability of a particular result ($r = 3$, $n = 20$), and if this is plotted for *fixed r* but *varying* $\theta$, we have the likelihood function.

*Definition*  The *likelihood function* is the joint probability of an observed sample, regarded as a function of the unknown parameters. The random variables are taken as fixed at their observed values.  □ □ □

It is easy to see that absolute value of the likelihood function is not relevant when interpreting it. Suppose that in the example given above we inspected 2000 items instead of 20, and found 300 defectives not 3, then the likelihood function would be

$$\binom{2000}{300} \theta^{300} (1 - \theta)^{1700}$$

and all values of the likelihood would be very much smaller than those given in Table 3.1. Indeed, any likelihood can be reduced below any positive quantity, however small, simply by taking more observations.

In fact, in further development of the theories of estimation and of testing of hypotheses, it is *ratios* of likelihoods which occur. Another argument which indicates that it is ratios of likelihoods which are important for interpretation is indicated in Exercise 3.3.5.

Barnard *et al.* (1962, p. 321) state that if $l(x, \theta_1)$ and $l(x, \theta_2)$ are two values of the likelihood for the same result $x$ but different values of the parameter $\theta$, then the likelihood 'gives a measure of relative plausibility of different values of $\theta$, so that we can say that $l(x, \theta_1)/l(x, \theta_2)$ is the odds ratio of $\theta_1$ versus $\theta_2$ on the basis of the result $x$'. The interpretation of the likelihood function therefore is to give some measure of the relative plausibility of different sets of unknown parameters giving rise to the observed set of data. There is a certain amount of asymptotic theory concerning the likelihood function which helps us considerably in interpretation, and we shall discuss this later. For the present we note that merely to plot the likelihood function is a great help.

*Example* 3.5    Three independent observations on a Poisson distribution with an unknown mean $\mu$ are 6, 9 and 11. What is the likelihood function?

The Poisson distribution is

$$\Pr(r) = e^{-\mu} \mu^r / r! \qquad r = 0, 1, 2, \ldots$$

so that the probability of the observed set of results 6, 9, 11 is obtained by multiplying the probabilities of the three separate results, which is

$$e^{-\mu}\frac{\mu^6}{6!} e^{-\mu}\frac{\mu^9}{9!} e^{-\mu}\frac{\mu^{11}}{9!} = e^{-3\mu}\mu^{26}/\{6!\,9!\,11!\}$$

This is the likelihood function, when regarded as a function of $\mu$.     □ □ □

Two points simplify the task of plotting the likelihood. First, likelihoods are nearly always obtained by multiplying the probabilities of independent events, and the likelihood usually simplifies by considering the logarithm. If $l(x, \theta)$ is the likelihood function we write

$$L(x, \theta) = \log l(x, \theta)$$

For example, the log-likelihood for Example 3.5 is

$$L(\mu) = 26 \log \mu - 3\mu - \log(6!\,9!\,11!) \tag{3.7}$$

and a graph of this is shown in Fig. 3.4.

The second simplification is that since likelihoods have relative value only,

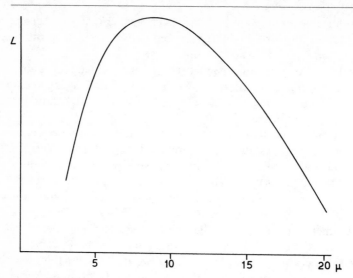

Fig. 3.4 The log-likelihood of Equation (3.7).

for different values of the unknown parameters, terms in the likelihood or log-likelihood not involving the unknown parameters may be dropped. For example, we drop the last term in Equation (3.7).

*Example* 3.6   For another example, suppose we have observations on three independent binomial variates with the same value of $\theta$ but different values of $n$. For $n = 10$ we have $r = 1$; $r = 8$ at $n = 20$; and $r = 9$ at $n = 30$. What is the likelihood function?

The probabilities of the three results are

$$\binom{10}{1}\theta(1-\theta)^9 ; \quad \binom{20}{8}\theta^8(1-\theta)^{12} ; \quad \text{and} \quad \binom{30}{9}\theta^9(1-\theta)^{21}.$$

Therefore the likelihood is

$$l(\theta) = \binom{10}{1}\theta(1-\theta)^9 \binom{20}{8}\theta^8(1-\theta)^{12}\binom{30}{9}\theta^9(1-\theta)^{21}$$

$$= k\theta^{18}(1-\theta)^{42}$$

where $k$ includes all the terms not involving $\theta$.                   ☐ ☐ ☐

For continuous distributions the likelihood is calculated in the same way. Suppose we have a distribution with a probability density function (p.d.f.) $f(x, \theta)$ depending on a single parameter $\theta$, and let the observed results be $x_1, x_2, \ldots, x_n$. Now for any continuous random variable, the probability of

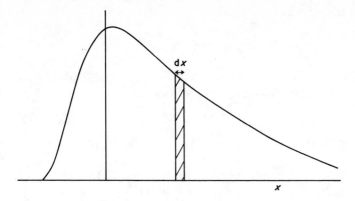

Fig. 3.5 Continuous random variables.

getting exactly $x$ is zero, for any $x$. However, the probability that a result occurs in an interval of length $dx$ centred on $x$ is $f(x, \theta)dx$ (see Fig. 3.5). This corresponds to the practical situation, since continuous variables are only recorded to a specified number of decimal places. If we assume that observations $x_1, x_2, \ldots, x_n$ are independent, we multiply the probabilities, and the likelihood is

$$l(\theta) = \prod_{i=1}^{n} f(x_i, \theta)dx_i$$

$$= k \prod_{i=1}^{n} f(x_i, \theta)$$

since the terms $dx_1\, dx_2 \ldots dx_n$ do not involve $\theta$.

*Example* 3.7    Four observations on a normal distribution with an unknown mean and variance unity are 4.37, 5.56, 3.72, 4.57. What is the likelihood?

The p.d.f. of the observations is

$$\exp\{-\tfrac{1}{2}(x - \theta)^2\}/\sqrt{(2\pi)}$$

so that the likelihood is

$$l(\theta) = k \exp\{-\tfrac{1}{2}(4.37 - \theta)^2 - \tfrac{1}{2}(5.56 - \theta)^2 - \tfrac{1}{2}(3.72 - \theta)^2 - \tfrac{1}{2}(4.57 - \theta)^2\}$$

It is clear that when the squared terms are expanded, there will be terms in $\theta^2, \theta x_i$ and $x_i^2$, and the latter term will drop out since it does not involve $\theta$. If we take the expression

$$x_i - \theta = x_i - \bar{x} + \bar{x} - \theta$$

and square and add, we have

$$\sum_{1}^{n}(x_i - \theta)^2 = \sum(x_i - \bar{x})^2 + n(\bar{x} - \theta)^2 \tag{3.8}$$

When we insert this into $l(\theta)$, the term $\exp\{-\frac{1}{2}\sum(x_i - \bar{x})^2\}$ may be dropped, and since the mean of the observations is 4.55 we have

$$L(\theta) = c - 2(4.55 - \theta)^2 \tag{3.9}$$

where $c$ is a constant.                                    □ □ □

Finally, we discuss a case with two unknown parameters. We shall use $x_1, x_2, \ldots, x_n$ for the data, but it is important to realize that as far as the likelihood is concerned these are fixed quantities.

*Example* 3.8   Independent observations $x_1, \ldots, x_n$ are taken from a normal distribution with unknown mean $\mu$ and unknown variance $\sigma^2$. What is the likelihood function?

The p.d.f. of the observations is

$$\frac{1}{\sqrt{(2\pi)}\sigma}\exp\left\{-\frac{1}{2}\left(\frac{x-\mu}{\sigma}\right)^2\right\}$$

so that the likelihood is

$$l(\mu, \sigma^2) = \frac{k}{\sigma^n}\exp\left\{-\frac{1}{2\sigma^2}\sum(x_i - \mu)^2\right\}$$

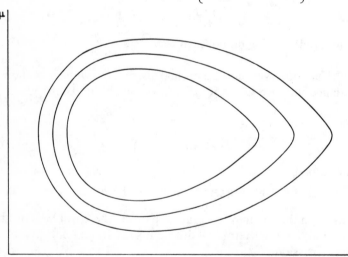

Fig. 3.6 Likelihood surface contours for normal distribution.

and using Equation (3.8) this gives

$$L(\mu, \sigma^2) = c - n \log \sigma - \frac{1}{2\sigma^2}\sum(x_i - \bar{x})^2 - \frac{n}{2\sigma^2}(\bar{x} - \mu)^2 \qquad (3.10)$$

This function has a maximum value at a point $(\hat{\mu}, \hat{\sigma}^2)$ where

$$\hat{\mu} = \bar{x}$$

$$\hat{\sigma}^2 = \sum(x_i - \bar{x})^2/n$$

(see the next section for the derivation of this result), and the contours of the likelihood surface are shown in Fig. 3.6.    □ □ □

---

**Exercises 3.3**
1. Observations 1, 5, 4, 6, 8 are from a geometric distribution $(1 - \theta)^{r-1}\theta$, $r = 1, 2, \ldots$. Plot the log-likelihood function.

2. In a sampling inspection of a large batch of items with an unknown proportion $\theta$ of defective items, no defectives were found in a random sample of 20 items. Plot the log-likelihood function.

3. The time intervals between arrivals at a certain queue are known to be independently exponentially distributed with a p.d.f $\lambda e^{-\lambda t}$. Given observations $t_1, t_2, \ldots, t_n$, obtain the likelihood and log-likelihood functions.

4. Observations $x_1, x_2, \ldots, x_n$ are from a gamma distribution with a p.d.f.

$$\alpha(\alpha x)^4 \quad e^{-\alpha x}/24$$

Obtain the likelihood and log-likelihood functions, and draw their graphs.

5.* Independent random variables $Y_1, \ldots, Y_n$ have a p.d.f. $f(y, \theta)$, and a one-to-one transformation

$$z = z(y)$$

yields $Z_1, \ldots, Z_n$ respectively, with a p.d.f.

$$f(z, \theta) = f(y, \theta)\left|\frac{dy}{dz}\right|$$

Write down the likelihood of $Z_1, \ldots, Z_n$ in terms of the likelihood of $Y_1, \ldots, Y_n$. Hence argue that since inference about $\theta$ cannot reasonably depend on whether $Y$ or $Z$ was observed, it is ratios of likelihoods, rather than say differences, which are relevant (Cox and Hinkley, 1974, p. 12).

## 3.4 The method of maximum likelihood

One of the uses of the likelihood function is in connection with estimation, and it will be useful in discussing this to introduce some terminology. Let us suppose that we have observations which have a probability distribution depending on an unknown parameter $\boldsymbol{\theta} = (\theta_1, \ldots, \theta_k)$, which has $k$ components. The allowable range of $\boldsymbol{\theta}$ is defined by a *parameter space* $\Omega$. Thus in Example 3.8 we have $\boldsymbol{\theta} = (\mu, \sigma^2)$, and our parameter space is $\Omega = \{-\infty < \mu < \infty, \ 0 < \sigma^2 < \infty\}$

It is clear from the definition of likelihood that values of $\boldsymbol{\theta}$ which have a relatively high value of the likelihood, relative to other values of the likelihood function, are more likely to have given rise to the observed data than other values of $\boldsymbol{\theta}$. If we require a single value of $\boldsymbol{\theta}$ to use as our estimate – called a *point estimate* – then it is intuitively clear that the value of $\boldsymbol{\theta}$ which gives the maximum of the likelihood is a 'reasonable' estimate.

*Definition   Method of maximum likelihood*: given the likelihood function $l(\boldsymbol{\theta})$, we choose as a point estimate of $\boldsymbol{\theta}$ that value $\hat{\boldsymbol{\theta}}$ which maximises $l$, such that $\theta \in \Omega$.                                                  □ □ □

It turns out that this method of estimation has many desirable properties, which we shall discuss in the next chapter. The main justifications depend on asymptotic properties, which hold as $n \to \infty$, but there are some very useful small-sample properties as well.

*Example* 3.9   For Example 3.5, the log-likelihood is given by Equation (3.7), or

$$L(\mu) = c + 26 \log \mu - 3\mu$$

so that

$$\frac{dL}{d\mu} = \frac{26}{\mu} - 3$$

and the maximum-likelihood estimate is obtained by solving $dL/d\mu = 0$ which yields $\hat{\mu} = 26/3 = 8.667$.                                      □ □ □

*Example* 3.10   If we observe $r$ successes in $n$ trials when the probability of a success at any trial is $\theta$, the likelihood is

$$l(\theta) = k\theta^r(1 - \theta)^{n-r}$$
$$L(\theta) = c + r \log \theta + (n - r) \log(1 - \theta) \tag{3.11}$$
$$\frac{dL}{d\theta} = \frac{r}{\theta} - \frac{(n-r)}{(1-\theta)}$$

so that the maximum likelihood estimate is

$$\hat{\theta} = r/n$$

☐ ☐ ☐

In both of these examples it can be shown that this estimator is unbiased, and has the minimum possible variance of any unbiased estimator (see Section 3.5).

*Example* 3.11   The maximum likelihood estimates for Example 3.8 can be derived as follows. From (3.10) we have

$$\frac{\partial L}{\partial \mu} = +\frac{1}{\sigma^2}\sum(x_i - \mu)$$

$$\frac{\partial L}{\partial \sigma} = -\frac{n}{\sigma} + \frac{1}{\sigma^3}\sum(x_i - \mu)^2$$

from which $\hat{\mu} = \bar{x}$, and $\hat{\sigma}^2 = \sum(x_i - \bar{x})/n$.

☐ ☐ ☐

We notice here that the maximum-likelihood estimate of $\sigma^2$ is not the estimate we normally use, as the divisor $n$ has replaced the divisor $(n-1)$. This emphasizes the point made above, that the main justifications of the method are asymptotic.

It should be pointed out that although maximum-likelihood estimates can usually be obtained by differentiating the likelihood function, this is not always the case. Exercise 3.3.2. illustrates this point; in that problem the maximum of the likelihood occurs at a point which provides a sensible estimate of $\theta$, but this estimate cannot be obtained by differentiation.

By working through many examples in this way, we find that in all the common simple cases, maximum likelihood produces the usual estimates, or else estimates which differ from these only to the order $O(1/n)$. However, the great advantage of maximum likelihood is that it can be applied with confidence to a wide class of complicated problems, and we shall give examples of some of these in the next chapter. We shall close this section by applying the method to the simple linear-regression model.

*Example* 3.12   Suppose we have paired observations $(x_1, y_1), \ldots, (x_n, y_n)$, from the model

$$E(Y) = \alpha + \beta x$$

$$V(Y) = \sigma^2$$

(3.12)

and $Y$ is independently and normally distributed. The log-likelihood is

$$L = -\frac{n}{2}\log 2\pi - n\log\sigma - \frac{1}{2\sigma^2}\sum(y_i - \alpha - \beta x_i)^2$$

and as far as estimation of $\alpha$ and $\beta$ is concerned we maximize $L$ by minimizing

$$S = \sum (y_i - \alpha - \beta x_i)^2 \qquad (3.13)$$

which means that we must choose $\alpha$ and $\beta$ in accordance with the principle of least squares. From this we see that in any problem involving the linear model such as (3.12), maximum likelihood and least squares are equivalent, the former being the more general, and this includes standard methods for linear regression and analysis of variance. For the model (3.12), the maximum-likelihood and least-squares estimates are

$$\hat{\beta} = \frac{\sum y_i (x_i - \bar{x})}{\sum (x_i - \bar{x})^2} \qquad \hat{\alpha} = \bar{y} - \hat{\beta} x \qquad \square \ \square \ \square$$

---

## Exercises 3.4

1. Obtain maximum-likelihood estimates of the unknown parameters in:

   (a) Example 3.6;    (b) Exercise 3.3.3;    (c) Exercise 3.3.4.

2. If the exponential distribution in Exercise 3.3.3 is written $e^{-x/\theta}/\theta$, what is the maximum-likelihood estimate of $\theta$? Compare this with the result in Exercise 3.4.1(b) for the exponential distribution.

3. In a certain experiment, trials may be classified as either a 'success' or a 'failure' and the probability of a success is $\theta$, independently from trial to trial. Trials continue until 3 successes are observed, and this takes 20 trials. Derive the maximum-likelihood estimate of $\theta$. Compare your result with Example 3.10 and comment.

---

### 3.5 The Cramér–Rao inequality

We have described what we mean by parametric inference and we have just discussed one general method by which estimators can be obtained. We shall now be interested in discussing the merits of alternative estimators in any problem. Clearly, one of the chief properties we shall be considering is the variance of an estimator.

Now if we consider any problem, such as the linear-regression problem of Example 2.1, it seems clear that there must be a limited amount of information about $\beta$ in the given data. That is to say, we cannot obtain more and more accurate estimates of $\beta$ by calculating more and more complicated functions of the observations: the random variation present in the observations must reveal itself in some lack of precision of our knowledge of $\beta$.

We can therefore pose the following question. Suppose we have random variables $X_1, X_2, \ldots, X_n$, drawn independently from a distribution with a p.d.f. $f(x, \theta)$, where $\theta$ is unknown, then we shall estimate $\theta$ by calculating some function $t(\mathbf{X})$ of the $X_i$'s. We can now ask if there is a lower bound to the variance $V\{t(\mathbf{X})\}$ for all possible functions $t(\mathbf{X})$. By 'estimator' here we mean some function of the observations, not including $\theta$, which will serve as an estimate of $\theta$ whatever its value. As we have argued intuitively above, the answer is that there is such a lower bound, and it is given by the Cramér–Rao inequality.

*Definition*   Provided certain regularity conditions hold, including that the range of $x$ be independent of $\theta$ and that the information

$$I = -E\left(\frac{\partial^2 L}{\partial \theta^2}\right) > 0$$

then it can be shown that

$$V\{t(\mathbf{X})\} \geq \{1 + b'(\theta)\}^2/I \qquad (3.14)$$

where $b(\theta)$ is defined as the bias of $t(\mathbf{X})$,

$$b(\theta) = E\{t(\mathbf{X})\} - \theta \qquad (3.15)$$

□ □ □

For a proof of this result see Section 3.8. The quantity $I/n$ is called the *information per sample*; this is related to the curvature of the likelihood function at $\theta$, and we might expect the minimum variance to depend on this. The larger the quantity $I$, then in general the more sharply curved is the likelihood function, and the more precise our knowledge of the unknown parameter. In fact it will be shown in Equation (3.24) of Section 3.8 that the relationship

$$I = -E\left(\frac{\partial^2 L}{\partial \theta^2}\right) = E\left\{\left(\frac{\partial L}{\partial \theta}\right)^2\right\}$$

holds so that there are two equivalent ways of calculating $I$.

The quantity on the right-hand side of the inequality (3.14) is a lower bound, and only under special circumstances is there a function $t(\mathbf{X})$ giving equality. If we have an unbiased estimator such that

$$V\{t(\mathbf{X})\} = 1/I \qquad (3.16)$$

then we say that we have a minimum-variance unbiased estimator (MVUE). The inequality applies for both discrete and continuous distributions, and there is an obvious extension for the case when successive observations are not taken from identical distributions.

*Example* 3.13   (*Exponential distribution.*) If $x_1, x_2, \ldots, x_n$ are drawn from the distribution with p.d.f. $\lambda e^{-\lambda x}$, then

$$l = \prod_{i=1}^{n} \lambda \exp(-\lambda x_i) = \lambda^n \exp(-\lambda \sum x_i)$$

$$L = +n \log \lambda - \lambda \sum x_i$$

so that

$$\frac{\partial L}{\partial \lambda} = +\frac{n}{\lambda} - \sum x_i$$

and

$$\frac{\partial^2 L}{\partial \lambda^2} = \frac{-n}{\lambda^2}$$

Thus $I = n/\lambda^2$ and the minimum variance for unbiased estimators is $\lambda^2/n$. In this case it is not possible to find any function $t(\mathbf{X})$ which has a variance quite as low as this. □ □ □

*Example* 3.14 (*Binomial distribution.*) If we observe $r$ successes, and define $\theta$ as the probability of a success, then we have from Equation (3.11)

$$\frac{\partial L}{\partial \theta} = \frac{r}{\theta} - \frac{(n-r)}{(1-\theta)}$$

$$\frac{\partial^2 L}{\partial \theta^2} = -\frac{r}{\theta^2} - \frac{(n-r)}{(1-\theta)^2}$$

$$E\left[-\frac{\partial^2 L}{\partial \theta^2}\right] = +\frac{E(r)}{\theta^2} + \frac{E(n-r)}{(1-\theta)^2} = \frac{n\theta}{\theta^2} + \frac{n(1-\theta)}{(1-\theta)^2}$$

$$= \frac{n}{\theta(1-\theta)}$$

Therefore the Cramér–Rao lower bound for the variance of unbiased estimates of $\theta$ is $\theta(1-\theta)/n$. That this is attained by the estimator $\hat{\theta} = r/n$ is left as an exercise for the reader in Exercise 3.5.4. □ □ □

If any p.d.f. depends on a finite number $k$ of unknown parameters, say $f(x, \theta_1, \theta_2, \ldots, \theta_k)$, then the lower bound given above still applies to each individual parameter. However, in this case this bound can be improved, and we shall discuss this later.

Now that we have a lower bound to the variance of possible estimators, this gives us a basis for comparison, and we can compare the variance of an estimator with this lower bound. We therefore define efficiency and asymptotic *efficiency*.

*Definition* The *efficiency* of an estimator $t(\mathbf{X})$ is equal to

$$1/\{IV(t(\mathbf{X}))\} \qquad \square\ \square\ \square$$

*Definition* The *asymptotic efficiency* is defined as

$$\lim_{n \to \infty} 1/\{IV(t(\mathbf{X}))\} \qquad \square\ \square\ \square$$

It turns out that under certain very general conditions, maximum-likelihood estimators are asymptotically efficient.

---

### Exercises 3.5

1. Suppose you are given $x_1, x_2, \ldots, x_n$, which are independently and normally distributed $N(\mu, \sigma^2)$. Find the minimum variance for estimators of (a) $\mu$, and (b) $\sigma^2$ of the normal distribution in each case, assuming the other parameter known.

2. Find the minimum variance for estimators of $\theta$ of the exponential distribution with p.d.f. defined as $\{\exp(-x/\theta)\}/\theta$. Compare with the actual variance of $\bar{x}$ as an estimator of $\theta$. Contrast with the result given in Example 3.13 and comment.

3. Find the minimum variance for estimators of $\mu$ of the Poisson distribution, $\Pr(r) = e^{-\mu}\mu^r/r\,!$

4. Show that the actual variance of the maximum-likelihood estimator of the parameter $\theta$ of the binomial distribution is equal to the Cramér–Rao lower bound.

---

### 3.6* Sufficiency

We shall briefly introduce one more theoretical concept before proceeding with the more practical matters.

Suppose we have $x_1, x_2, \ldots, x_n$ from a Poisson distribution with parameter $\mu$, then the likelihood of the observations is

$$\Pr(x_1, x_2, \ldots, x_n) = e^{-n\mu} \mu^{\Sigma x_i} \prod_{i=1}^{n} \frac{1}{x_i\,!}$$

Now we notice that where $\mu$ occurs we only have the observations through the function $\sum x_i$. Further, the random variable $\sum X_i$ has a Poisson distribution with parameter $n\mu$, so that

$$\Pr(\textstyle\sum X_i = t) = e^{-n\mu}(n\mu)^t/t\,!$$

Therefore the conditional probability of $x_1, \ldots, x_n$ given $\sum X_i = t$ is

$$\Pr\{x_1, \ldots, x_n \mid \sum X_i = t\} = \frac{\Pr\{x_1, \ldots, x_n\}}{\Pr\{\sum X_i = t\}} = \frac{t!}{x_1! \ldots x_n!} \left(\frac{1}{n}\right)^t$$

which is a multinomial distribution, and independent of $\mu$. This is a very important result, for it tells us that given $\sum x_i$, the division of this total into the individual components $x_1, \ldots, x_n$ does not depend on $\mu$. Thus all of the relevant information in the data about $\mu$ is contained in the value $\sum x_i$, and we say here that this is a *sufficient statistic* for $\mu$.

*Definition*   Given random variables $\mathbf{X}' = (X_1, \ldots, X_n)$, drawn independently from a distribution with p.d.f. $f(x, \theta)$, the statistic $t(\mathbf{X})$ is said to be *sufficient* for $\theta$ if the conditional distribution of the observations given $t(\mathbf{X})$ is independent of $\theta$.

$$f_{\mathbf{x}|t}(x_1, \ldots, x_n \mid t(\mathbf{x})) \qquad \text{independent of } \theta \qquad (3.17)$$

□ □ □

Sufficient statistics do not always exist, but when they do, we may concentrate on estimators which are functions of them. Sufficient statistics are usually recognized from the following theorem.

*Factorization theorem*   A necessary and sufficient condition that a statistic $t(\mathbf{X})$ be sufficient for a single parameter $\theta$ is that the likelihood factorizes in the form

$$l(\mathbf{X}, \theta) = g(t, \theta)h(\mathbf{X}) \qquad (3.18)$$

where $h(\mathbf{X})$ is independent of $\theta$.

□ □ □

*Meaning*   If the likelihood factorizes into the product of a function of a statistic $t$ and $\theta$, and a function of $\mathbf{X}$ independent of $\theta$, then $t$ is a sufficient statistic.

*Proof*   For a rigorous proof of the theorem we refer to standard texts on statistical theory. However, the basis of the argument is very simple. If $t(\mathbf{X})$ is a sufficient statistic, then by definition

$$f_{\mathbf{x}|t}(\mathbf{x} \mid t(\mathbf{x})) = h(\mathbf{x})$$

which is a function of $\mathbf{x}$.
Now the likelihood of $X_1, \ldots, X_n$ is

$$l(\mathbf{x}, \theta) = f_{\mathbf{x}}(\mathbf{x}, \theta)$$
$$= f_{\mathbf{x}|t}(\mathbf{x} \mid t(\mathbf{x})) f_{t(\mathbf{x})}(t(\mathbf{x}), \theta)$$

by the rules for conditional probability. The first factor on the right-hand side does not involve $\theta$, and the second factor on the right-hand side only involves **x** through $t(\mathbf{x})$. Therefore we can write this last equation

$$l(\mathbf{x}, \theta) = g(t, \theta)h(\mathbf{x}).$$

It can be established that conditions (3.17) and (3.18) are equivalent.  □ □ □

The concept of sufficiency generalizes, and if we have many unknown parameters, we may have a set of statistics which are *jointly sufficient*. An obvious extension of the factorization theorem then applies.

*Example* 3.15   If $x_1, x_2, \ldots, x_n$ are drawn independently from $N(\mu, \sigma^2)$, then

$$l(\mathbf{x}; \mu, \sigma^2) = (2\pi)^{-n/2}\sigma^{-n}\exp\left\{ -\frac{1}{2\sigma^2}\sum(x_i - \mu)^2 \right\}$$

$$= (2\pi)^{-n/2}\sigma^{-n}\exp\left\{ -\frac{(n-1)s^2}{2\sigma^2} - \frac{n(\bar{x} - \mu)^2}{2\sigma^2} \right\}$$

Now if $\sigma^2$ is known we have

$$l(\mathbf{x}; \mu, \sigma^2) = \exp\left\{ -\frac{n(\bar{x} - \theta)^2}{2\sigma^2} \right\}(2\pi)^{-n/2}\sigma^{-n}\exp\left\{ -\frac{(n-1)s^2}{2\sigma^2} \right\} = g(t, \theta)h(\mathbf{x})$$

where $t = \mathbf{x}$. Hence $\bar{x}$ is a sufficient statistic for $\mu$. Similarly, we can show that:

(a) if $\mu$ is known, $(s^2, \bar{x})$ are jointly sufficient for $\sigma^2$ ;
(b) if $\mu$ and $\sigma^2$ are both unknown, $\bar{x}$ and $s^2$ are jointly sufficient.  □ □ □

A much fuller discussion of sufficiency is given, for example, by Cox and Hinkley (1974), Cramér (1952), Kendall and Stuart (1961), or Wilks (1962).

---

### Exercises 3.6
1. Suppose $x_1, x_2, \ldots, x_n$ are drawn independently from the gamma distribution with p.d.f. $\alpha(\alpha x)^{\beta - 1}\exp(-\alpha x)/\Gamma(\beta)$. Find the sufficient statistic for $\alpha$ if $\beta$ is known, the sufficient statistic for $\beta$ if $\alpha$ is known, and the statistics which are jointly sufficient for $(\alpha, \beta)$.

2. Suppose $x_1, \ldots, x_n$ are drawn independently from the uniform distribution on the range $(0, \theta)$, where $\theta$ is unknown. Show that $(\max x_i)$ is sufficient for $\theta$.

---

### 3.7 The multivariate normal distribution
For further development of our theory we need a brief introduction to the

continuous multivariate distributions. In the univariate case, a function $f(x)$ can be a probability density function of a random variable $X$ if

$$f(x) \geq 0$$

and

$$\int_{-\infty}^{\infty} f(x)dx = 1$$

We then have the interpretation

$$\Pr(a < X < b) = \int_{a}^{b} f(x)dx$$

or in words, the probability that a random variable lies between $a$ and $b$ is the area under the p.d.f. between these points.

The requirements for multivariate distributions are similar. A function $f(x_1, \ldots, x_n)$ can be a multivariate p.d.f. if

$$f(x_1, \ldots, x_n) \geq 0$$

and

$$\int \cdots \int f(x_1, \ldots, x_n)dx_1 \cdots dx_n = 1$$

We then have a similar interpretation,

$$\Pr\{a_1 < X_1 < b_1, \ldots, a_n < X_n < b_n\} = \int_{a_1}^{b_1} \cdots \int_{a_n}^{b_n} f(x_1, \ldots, x_n)dx_1 \cdots dx_n$$

The particular multivariate distribution we shall meet is the multivariate normal. We shall illustrate this in the bivariate case first, and then give the general formula.

*The bivariate normal distribution*
The p.d.f. of the bivariate normal distribution is

$$f(x_1, x_2) = \frac{1}{2\pi\sigma_1\sigma_2\sqrt{(1 - \rho^2)}} \exp\left[ -\frac{1}{2(1 - \rho^2)}\left\{ \left(\frac{x_1 - \mu_1}{\sigma_1}\right)^2 \right.\right.$$
$$\left.\left. - 2\rho\left(\frac{x_1 - \mu_1}{\sigma_1}\right)\left(\frac{x_2 - \mu_2}{\sigma_2^2}\right) + \left(\frac{x_2 - \mu_2}{\sigma^2}\right)^2 \right\} \right] \qquad (3.19)$$

This distribution is determined by five parameters, $(\mu_1, \mu_2, \sigma_1, \sigma_2, \rho)$, and we have

$$E(X_1) = \mu_1 \qquad V(X_1) = \sigma_1^2 \qquad C(X_1, X_2) = \rho\sigma_1\sigma_2$$
$$E(X_2) = \mu_2 \qquad V(X_2) = \sigma_2^2$$

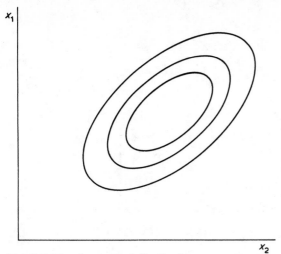

Fig. 3.7 Bivariate normal distribution

and it can be shown that

$$\int_{-\infty}^{\infty} \int_{-\infty}^{\infty} f(x_1, x_2) \, dx_1 \, dx_2 = 1$$

If the correlation $\rho$ is zero, then the density (3.19) factorizes into the product of two densities:

$f(x_1, x_2) = f_1(x_1) f_2(x_2)$ and the random variables $X_1$ and $X_2$ are statistically independent.

Contours of the bivariate normal density are shown in Fig. 3.7.

*The covariance matrix*

Before we introduce the general form of the multivariate normal distribution we must define the variance–covariance matrix, sometimes referred to simply as the covariance matrix. If $X_1, X_2, \ldots, X_k$ are random variables then the *covariance matrix* is defined as

$$\begin{bmatrix} V(X_1) & C(X_1, X_2) & C(X_1, X_3) \ldots C(X_1, X_k) \\ C(X_1, X_2) & V(X_2) & C(X_2, X_3) \ldots C(X_2, X_k) \\ C(X_1, X_3) & C(X_2, X_3) & V(X_3) \qquad \ldots \\ \vdots & & \\ C(X_1, X_k) & C(X_2, X_k) & \ldots \qquad \ldots V(X_k) \end{bmatrix}$$

where $C(X_i, X_j)$ is the covariance of $X_i$ and $X_j$.

Now

$$V(X_1) = E(X_1^2) - E^2(X_1)$$
$$C(X_1, X_2) = E(X_1 X_2) - E(X_1)E(X_2)$$

so that if

$$\mathbf{X} = \begin{bmatrix} X_1 \\ \vdots \\ X_k \end{bmatrix}$$

then

$$\mathbf{XX}' = \begin{bmatrix} X_1^2 & X_1 X_2 \cdots X_1 X_k \\ X_1 X_2 & X_2^2 & X_2 X_k \\ \vdots & \vdots & \vdots \\ X_1 X_k & X_2 X_k & X_k^2 \end{bmatrix}$$

The covariance matrix is therefore

$$\boldsymbol{\Sigma} = \{C(X_i, X_j)\} = E(\mathbf{XX}') - E(\mathbf{X})E(\mathbf{X}') \tag{3.20}$$

We notice that if the random variables $X_1, X_2, \ldots, X_k$ are independent and

$$V(X_1) = V(X_2) = \ldots = V(X_k) = \sigma^2$$

then the covariance matrix is

$$\boldsymbol{\Sigma} = \sigma^2 \mathbf{I}$$

*The multivariate normal distribution*

Given random variables $X_1, X_2, \ldots, X_k$ such that $E(X_i) = \mu_i$, $V(X_i) = \sigma_i^2$, and $C(X_i, X_j) = \rho_{ij}\sigma_i\sigma_j$ then we write

$$\mathbf{X} = \begin{bmatrix} X_1 \\ \vdots \\ X_k \end{bmatrix}, \qquad \boldsymbol{\mu} = \begin{bmatrix} \mu_1 \\ \vdots \\ \mu_k \end{bmatrix}$$

and

$$\boldsymbol{\Sigma} = \begin{bmatrix} \sigma_1^2 & \rho_{12}\sigma_1\sigma_2 \cdots \rho_{1k}\sigma_1\sigma_k \\ \rho_{12}\sigma_1\sigma_2 & \sigma_2^2 & \rho_{1k}\sigma_2\sigma_k \\ \vdots & & \\ \rho_k\sigma_1\sigma_k & \rho_{2k}\sigma_2\sigma_k & \sigma_k^2 \end{bmatrix}$$

The multivariate normal density is written as follows:

$$f(x) = \frac{1}{(2\pi)^{k/2}|\boldsymbol{\Sigma}|^{1/2}} \exp\{-\tfrac{1}{2}(\mathbf{X} - \boldsymbol{\mu})'\boldsymbol{\Sigma}^{-1}(\mathbf{X} - \boldsymbol{\mu})\} \tag{3.21}$$

For a proof that the p.d.f. (3.21) integrates to unity, we refer the reader to books on probability theory or multivariate analysis.

**Exercises 3.7**

1. Show that if $k = 2$ in Equation (3.21),

$$|\Sigma| = \sigma_1^2 \sigma_2^2 (1 - \rho_{12}^2)$$

and

$$\Sigma^{-1} = \frac{1}{\sigma_1^2 \sigma_2^2 (1 - \rho_{12}^2)} \begin{pmatrix} \sigma_2^2 & -\sigma_1 \sigma_2 \rho_{12} \\ -\sigma_1 \sigma_2 \rho_{12} & \sigma_1^2 \end{pmatrix}$$

and Equation (3.21) redues to Equation (3.19).

2. Check that Equation (3.19) integrates to unity by first using the transformation

$$u = (x_1 - \mu_1)/\sigma_1, \qquad v = (x_2 - \mu_2)/\sigma^2$$

and then integrating.

---

**3.8\* Proof of the Cramér–Rao inequality**
(*Those prepared to assume the results may omit this section*)

Before we indicate how the inequality can be established, we must derive some preliminary results. Suppose a p.d.f. $f(x, \theta)$ depends on a single unknown parameter $\theta$, where $\theta$ belongs to some well-defined interval. Suppose further that the range of $X$ is independent of $\theta$, and we must make certain further assumptions, connected with the existence of the derivative $df/d\theta$, over the range of $\theta$ defined above.

Now we have for any p.d.f.

$$\int_{-\infty}^{\infty} f(x, \theta) dx = 1$$

and by differentiating with respect to $\theta$ under the integral sign we have

$$\int_{-\infty}^{\infty} \frac{df}{d\theta} dx = 0$$

or

$$\int_{-\infty}^{\infty} \frac{1}{f} \frac{df}{d\theta} f(x, \theta) dx = \int \frac{d \log f}{d\theta} f(x, \theta) dx = 0 \qquad (3.22)$$

or

$$E\left\{ \frac{d \log f(x, \theta)}{d\theta} \right\} = 0 \qquad (3.23)$$

By differentiating Equation (3.22) again under the integral sign, we have

$$\int \left( \frac{d^2 \log f}{d\theta^2} \right) f(x, \theta) dx + \int \left( \frac{d \log f}{d\theta} \right) \frac{df(x, \theta)}{d\theta} dx = 0$$

or

$$\int\left(\frac{d^2\log f}{d\theta^2}\right)f(x,\theta)dx + \int\left(\frac{d\log f}{d\theta}\right)^2 f(x,\theta)dx = 0$$

or

$$E\left\{\left(\frac{d\log f(x,\theta)}{d\theta}\right)^2\right\} = -E\left\{\frac{d^2\log f(x,\theta)}{d\theta^2}\right\} \tag{3.24}$$

Now if we have independent observations $x_1, \ldots, x_n$, drawn from the distribution with p.d.f. $f(x,\theta)$, the log-likelihood is

$$L = \sum_{i=1}^{n}\log f(x_i,\theta)$$

One quantity related to this which we shall need is called the *information*, and is defined as

$$I = \sum E\left(\frac{d\log f(x_i,\theta)}{d\theta}\right)^2 = -\sum E\left(\frac{d^2\log f(x_i,\theta)}{d\theta^2}\right) \tag{3.25}$$

which is obtained by using Equation (3.24).

Now for any unbiased estimator $t(\mathbf{x})$ based on observation $\mathbf{x} = (x_1, \ldots, x_n)$,

$$E\{t(\mathbf{x})\} = \theta = \int \ldots \int t(\mathbf{x})f(x_1,\theta)\ldots f(x_n,\theta)dx_1 \ldots dx_n$$

so that by differentiation with respect to $\theta$ we obtain

$$1 = \int \ldots \int t(\mathbf{x})\left\{\sum\frac{d\log f(x_i,\theta)}{d\theta}\right\}f(x_1,\theta)\ldots f(x_n,\theta)dx_1 \ldots dx_n \tag{3.26}$$

Now put

$$\psi(\mathbf{x}) = \sum_{i=1}^{n}\frac{d\log f(x_i,\theta)}{d\theta}$$

then

$$E\{\psi(\mathbf{x})\} = \sum_{i=1}^{n}E\left\{\frac{d\log f(x_i,\theta)}{d\theta}\right\} = 0 \tag{3.27}$$

by Equation (3.23), and

$$V\{\psi(\mathbf{x})\} = \sum_{i=1}^{n}E\left\{\frac{d\log f(x_i,\theta)}{d\theta}\right\}^2$$

$$\tag{3.28}$$

$$= I$$

as defined in Equation (3.25).

By using Equation (3.27), we see that Equation (3.26) implies that the covariance

$$C\{t(\mathbf{x}), \quad \psi(\mathbf{x})\} = 1$$

Now by Schwarz's inequality, for any random variables $y$ and $z$,

$$\{C(y, z)\}^2 \le V(y) V(z)$$

so that by putting

$$y = t(\mathbf{x})$$
$$z = \psi(\mathbf{x})$$

we have

$$1 \le V\{t(\mathbf{x})\} V\{\psi(\mathbf{x})\}$$

and by using Equation (3.28) we have

$$V\{t(\mathbf{x})\} \ge 1/I$$

**Further reading**
For further discussion of the concepts of statistical inference and likelihood, see Silvey (1975), Cox and Hinkley (1974), or Kendall and Stuart (Vol. 2, 1973).

# CHAPTER FOUR

# Properties of the method of maximum likelihood

## 4.1 Introduction

In Chapter 3 we drew attention to the need for a general method of point estimation, and the method of maximum likelihood was introduced to fill this role. In this chapter we discuss some properties of the method, and we also discuss some practical problems which arise when using it. The following example illustrates the usefulness of the method.

*Example* 4.1    In the treatment of certain chronic diseases, the main criterion of success is length of survival after treatment. For some diseases the distribution of survival times is well approximated by an exponential distribution with a p.d.f.

$$\exp(-t/\theta)/\theta \tag{4.1}$$

for $\theta > 0, 0 < t < \infty$. Patients are treated as they become available, but at a time when it may be necessary to make a comparison between one treatment and other treatments, there may be a substantial number of patients still surviving. Let there be $n$ patients treated by some particular treatment, and at some point in time let there be $d$ deaths at times $t_j$ after treatment, $j = 1, 2, \ldots, d$, and $s$ survivors, having been observed for times $T_i$, $i = 1, 2, \ldots, s$, after treatment, where $d + s = n$. The problem is now to estimate the parameter of the exponential distribution from this data, and also obtain an estimate of the standard deviation of this estimate.

To begin with, let us suppose that there were no survivors, then the likelihood of the deaths is

$$l = \prod_{i=1}^{d} \exp\{-t_i/\theta\}/\theta$$

$$= \exp\{-\sum_{i=1}^{d} t_i/\theta\}/\theta^d \tag{4.2}$$

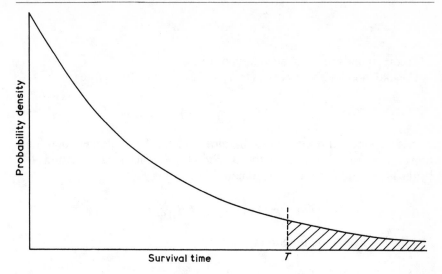

Fig. 4.1 Distribution of survival times.

The log-likelihood is

$$L = -\,\mathrm{d}\log\theta - \sum_{i=1}^{d} t_i/\theta$$

so that by differentiating we have

$$\frac{\mathrm{d}L}{\mathrm{d}\theta} = -\frac{d}{\theta} - \frac{1}{\theta^2}\sum_{i=1}^{d} t_i$$

and the maximum-likelihood estimate is

$$\hat{\theta} = \sum_{i=1}^{d} t_i/d \qquad\qquad (4.3)$$

Now when we do have survivors, it is easy to see that it is not satisfactory to ignore the survivors and to use Equation (4.3) to estimate $\theta$. Firstly, by doing this we are favouring shorter survival times, and this could lead to a badly biased estimate. (Other information on bias is given in Exercise 4.1.2). Secondly, this procedure can be very inefficient, if there are a substantial number of survivors. Clearly therefore this intuitive approach to estimation will not do, and this is an example of many situations where we need to apply a reliable method of estimation. The application of maximum likelihood to this problem is very simple.

The probability that a patient survives to a time $T_i$ after treatment is the probability that his survival time is greater than $T_i$, which is

$$\frac{1}{\theta}\int_{T_i}^{\infty} e^{-x/\theta} dx = e^{-T_i/\theta}$$

This corresponds to the shaded area of Fig. 4.1.

The likelihood of the survivors to $T_i$, $i = 1, 2, \ldots, s$, is therefore

$$\exp\left(-\frac{1}{\theta}\sum_1^s T_i\right) \tag{4.4}$$

Therefore, since all observations are independent, the total likelihood function is obtained by multiplying the likelihood of the survivors, Equation (4.4), by the likelihood of the deaths, Equation (4.2). This yields

$$l(\theta) = \frac{1}{\theta^d}\exp\left(-\frac{1}{\theta}\sum_1^d t_j - \frac{1}{\theta}\sum_1^s T_i\right)$$

and the log-likelihood function is

$$L(\theta) = -d\log\theta - \frac{1}{\theta}\sum_1^d t_j - \frac{1}{\theta}\sum_1^s T_i \tag{4.5}$$

We can now obtain the maximum-likelihood estimate $\hat{\theta}$ by differentiating $L$ with respect to $\theta$, and equating to zero. We have

$$\frac{dL}{d\theta} = -\frac{d}{\theta} + \frac{1}{\theta^2}\left(\sum_1^d t_j + \sum_1^s T_i\right)$$

from which the estimator is seen to be

$$\hat{\theta} = \left(\sum_1^d t_j + \sum_1^s T_i\right)\bigg/ d \tag{4.6}$$

$$= \text{(Total living times during testing)/(No. of deaths)}$$

If we recollect that the exponential distribution is that of intervals between events randomly distributed in time, this estimator looks very reasonable.

□ □ □

This analysis of Example 4.1 shows us a little of the value of maximum likelihood, and the method can be applied to very much more complicated situations; we shall give further examples later. The important problem now is to examine the general properties of such estimates. Let us consider some of the simple problems to which we have already applied the method. We shall see that in most cases we get estimators which are either efficient or asymptotically efficient. (For definitions of these terms see the end of Section 3.5.)

*Example* 4.2   If we observe $r$ successes from a binomial distribution with parameters $(n, \theta)$, then from Example 3.10 the maximum-likelihood estimate is

$$\hat{\theta} = r/n$$

and from first principles the variance of this

$$V(\hat{\theta}) = \theta(1 - \theta)/n$$

Example 3.14 shows that this variance is the lower bound for estimators of $\theta$, so that the maximum-likelihood estimate is efficient in this case.  □ □ □

*Example* 4.3 For estimating $\theta$ for the exponential distribution with a p.d.f. $\exp(-x/\theta)/\theta$, we have seen in Example 4.1 that the maximum-likelihood estimate is

$$\hat{\theta} = \bar{x}$$

and the results of Exercise 3.5.2 show that this is efficient. If the distribution is parameterized as in Example 3.13, it can be shown that the maximum-likelihood estimate is asymptotically efficient.  □ □ □

*Example* 4.4 If observations $r_1, \ldots, r_n$ are observed from a Poisson distribution $e^{-\mu}\mu^r/r!$, the likelihood is

$$l = \sum e^{-\mu}\mu^{r_i}/r_i!$$

The log-likelihood is

$$L = -n\mu + (\log \mu)\sum r_i + c$$

and

$$\frac{dL}{d\mu} = -n + \frac{\sum r_i}{\mu} \tag{4.7}$$

so that

$$\hat{\mu} = \sum r_i/n$$

From first principles, $V(\hat{\mu}) = \mu/n$, and by comparing this with the result of Exercise 3.5.3, we find that the maximum-likelihood estimate is efficient.  □ □ □

Other examples could be given, and in all of these simple cases maximum-likelihood estimates are exactly or asymptotically unbiased, and also exactly or asymptotically efficient. In fact it can be proved that maximum-likelihood estimates are both asymptotically unbiased and efficient under very general conditions. We shall state these properties formally in the next section.

## Exercises 4.1

1. Observations $r_1, \ldots, r_n$ are taken successively in an experiment where the $i$th observation is known to have an independent Poisson distribution with mean $(i\mu)$, so that

$$\Pr(r_i) = e^{-i\mu}(i\mu)^{r_i}/r_i!$$

Show that the maximum-likelihood estimate of $\mu$ is

$$\hat{\mu} = 2\sum r_i/\{n(n+1)\}$$

2.* The exponential distribution with p.d.f. $\{\exp(-x/\theta)\}/\theta$ is observed for a time $T$, and the $d$ deaths observed occur at times $t_j, j = 1, 2, \ldots, d$, while there are $s$ survivors at time $T$. Show that conditionally on $t < T$,

$$E(t_j) = \frac{\theta(1 - e^{-T/\theta}) - Te^{-T/\theta}}{1 - e^{-T/\theta}} < \theta$$

3.* Random variables $X_1, X_2, \ldots, X_n$ are independently distributed with a p.d.f. $\lambda e^{-\lambda x}$. Show that the maximum-likelihood estimator is $\hat{\lambda} = 1/\bar{X}$.

Also show that the p.d.f. of $Y = \sum_1^n X_i$ is

$$\lambda(\lambda y)^{n-1}e^{-\lambda y}/(n-1)! \qquad 0 < y < \infty$$

and that the p.d.f. of $Z = 1/Y$ is

$$\frac{1}{(n-1)!}\left(\frac{\lambda}{z}\right)^n\frac{e^{-\lambda/z}}{z} \qquad 0 < z < \infty$$

Hence show that

$$E\left(\frac{1}{\bar{X}}\right) = \frac{n}{(n-1)}\lambda \quad \text{and} \quad V\left(\frac{1}{\bar{X}}\right) = \frac{n^2}{(n-1)^2(n-2)}\lambda^2$$

and therefore show that the maximum-likelihood estimate is asymptotically unbiased and efficient.

## 4.2 Formal statements of main properties

When there is only one unknown parameter the main properties are summed up in the following two theorems.

*Theorem 4.1* Let $X_1, X_2, \ldots, X_n$ be mutually independent, identically distributed random variables whose p.d.f. $f(x, \theta)$ is determined except for the value of a single unknown parameter $\theta$ which takes values in a set $\Omega$. Then under certain very general conditions the maximum-likelihood estimator of $\theta$ is consistent. □ □ □

This is a very general result, and a proof is given by Wald (1949); see also Wolfowitz (1949). A simplified form of proof is given in Kendall and Stuart (Vol. 2, 1973). This result is important, but a rather more useful result is given by the next theorem. In Theorem 4.2 the maximum-likelihood estimator is denoted $\hat{\theta}$, and this is a function of the random variables $X_1, X_2, \dots, X_n$; the theorem is concerned with the probability distribution of $\hat{\theta}$.

*Theorem* 4.2    Let the random variables be as defined in Theorem 4.1, and let the true value of $\theta$ be $\theta_0$, then provided:

(i)   the information $I(\theta) = E\left\{ \left( \dfrac{\mathrm{d}L}{\mathrm{d}\theta} \right)^2 \right\} = - E\left\{ \dfrac{\mathrm{d}^2 L}{\mathrm{d}\theta^2} \right\}$          (4.8)

is finite and positive for all $\theta$ in the set considered;
(ii)  the range of $X$ does not depend on $\theta$;
(iii) certain other regularity conditions are satisfied;
The following results hold:
(a)  $\hat{\theta}$ is asymptotically normally distributed;
(b)  $\hat{\theta}$ is asymptotically unbiased, so that $\lim\limits_{n \to \infty} E(\hat{\theta}) = \theta_0$;
(c)  $\hat{\theta}$ is asymptotically efficient, so that $V(\hat{\theta}) \to 1/I$ as $n \to \infty$.    □ □ □

In fact, result (b) of Theorem 4.2 is contained in the result of Theorem 4.1, when a property such as that $V(\hat{\theta}) \to 0$ from (c) is taken into account, but it is stated again there for completeness. A proof of the theorem, for those who want it, can be found in Kendall and Stuart (Vol. 2, 1973) or in Cramér (1952). The result concerning the information is proved in Equation (3.25). A very careful and advanced discussion of both of these theorems is given in Cox and Hinkley (1974). An excellent survey and bibliography on theoretical aspects of maximum likelihood is given by Norden (1972, 1973).

The practical meaning of Theorems 4.1 and 4.2 is that under very general conditions, maximum-likelihood estimators exist, are consistent, asymptotically efficient and can be treated as approximately normally distributed about the true value as mean and with a variance

$$V(\hat{\theta}) = \left\{ - E\left( \frac{\mathrm{d}^2 L}{\mathrm{d}\theta^2} \right) \right\}_{\theta = \hat{\theta}}^{-1}$$          (4.9)

The result is asymptotic, and some qualifications upon it are given in the discussion below; it enables us to obtain confidence intervals and significance tests for $\theta$, by using methods for normally distributed variables. Occasionally we find problems for which one of the conditions of Theorem 4.2 fails (typically condition (ii)), and this theorem does not apply, but Theorem 4.1 nearly always holds.

*Example* 4.5    In an experiment where the outcome is known to be a Poisson random variable with parameter $\mu$, for $0 < \mu < \infty$, only the numbers of zero and non-zero outcomes are recorded. What is the maximum-likelihood estimate of $\mu$ if there are $r$ zero responses in $n$ trials? Also find an approximate 95 per cent confidence interval for $\mu$.

The probability distribution of the outcome is

$$\Pr(x) = e^{-\mu}\mu^x/x! \qquad x = 0, 1, 2, \ldots$$

so that the probability of a zero response is $\exp(-\mu)$. If there are $n$ trials, in each of which there are two possible outcomes (zero, non-zero), with a constant probability of the zero response, then the binomial distribution applies. The probability of $r$ zero responses in $n$ trials is therefore

$$\Pr(r) = \binom{n}{r}\theta^r(1-\theta)^{n-r}$$

where $\theta = \exp(-\mu)$. Therefore the likelihood function is

$$l = ke^{-r\mu}(1 - e^{-\mu})^{n-r} \tag{4.10}$$

and

$$L = c - r\mu + (n-r)\log(1 - e^{-\mu})$$

We find that

$$\frac{dL}{d\mu} = -r + \frac{(n-r)e^{-\mu}}{(1 - e^{-\mu})}$$

which leads to

$$\hat{\mu} = -\log_e(r/n). \tag{4.11}$$

We now need to evaluate Equation (4.8) at the maximum-likelihood estimate to obtain an approximate variance.
We have

$$\frac{d^2L}{d\mu^2} = -\frac{(n-r)e^{-\mu}}{(1 - e^{-\mu})^2} \tag{4.12}$$

so that

$$I = -E\left(\frac{d^2L}{d\mu^2}\right) = \frac{e^{-\mu}}{(1 - e^{-\mu})^2}E(n-r)$$

but

$$E(n-r) = n - E(r) = n - ne^{-\mu} = n(1 - e^{-\mu})$$

so that

$$I = \frac{ne^{-\mu}}{(1 - e^{-\mu})} \tag{4.13}$$

Therefore by Theorem 4.2, $\hat{\mu}$ is approximately normally distributed with a variance

$$\frac{1}{I} = \frac{e^{\mu}(1 - e^{-\mu})}{n} \qquad (4.14)$$

and approximate 95 per cent confidence intervals for $\mu$ are

$$\hat{\mu} \pm 1.96\sqrt{[\exp(\hat{\mu})\{1 - \exp(-\hat{\mu})\}/n]} \qquad (4.15)$$

□ □ □

Now we observe in the above example that when we insert Equation (4.11) into Equation (4.12), as an estimate of $\mu$, we get the same result as by putting Equation (4.11) into Equation (4.13). Thus we would get the same result here if we treated $\hat{\mu}$ as having an approximate variance

$$1 \left/ -\left(\frac{d^2L}{d\mu^2}\right)_{\mu=\hat{\mu}} \right. \qquad (4.16)$$

instead of

$$1 \left/ -E\left(\frac{d^2L}{d\mu^2}\right)_{\mu=\hat{\mu}} \right. \qquad (4.17)$$

Indeed, when there is a difference between the two results, Equation (4.16) is preferable for general use.

Example 4.5 has illustrated the use of Theorem 4.2 in a very simple case, but the value of the theorem lies in the fact that we can obtain an approximate variance for estimates in much more complicated problems, such as Example 4.1 For mathematical details of the application to Example 4.1, see Armitage (1959).

One mathematical point should be noted. Theorem 4.2 involves the asymptotic distribution of $\hat{\theta}$. In some cases the moments of the small-sample distribution do not converge to the moments of the asymptotic distribution. One such case arises when for all finite $n$, there is a positive (though usually very small) probability of $\hat{\theta}$ being infinite. When this happens, neither the expectation nor the variance exist for all $n$, yet this probability of an infinite estimate vanishes in the limit, so that the asymptotic distribution of $\hat{\theta}$ is according to the theorem; see Exercise 4.2.3.

The theorem proves only that there is a maximum near the true value $\theta_0$. However, in certain cases there are multiple maxima, and this gives rise to certain practical problems to which we refer later. If the conditions of Theorem 4.2 do not hold, we can still formally apply the method of maximum likelihood, but we cannot from this theorem say anything about what properties the estimates may possess, except that Theorem 4.1 will usually apply. In some special situations separate results exist which are less strong than Theorem 4.2, but more general in application. Consider the following example.

*Example* 4.6   Suppose we are given the following data:

$$1.12, \qquad 0.30, \qquad 0.87, \qquad 2.68, \qquad 1.98$$

which we know to have been sampled independently from the rectangular distribution with a p.d.f.

$$f(x, \theta) = \begin{cases} \theta^{-1} & 0 < x < \theta \\ 0 & \text{otherwise} \end{cases}$$

as shown in Fig. 4.2, where $\theta$, the maximum of the range of $x$, is unknown. What is the maximum-likelihood estimator of $\theta$?

Now the likelihood is zero for all $\theta$ less than the maximum of the observations, since then one or more actual observations has zero probability density. Thus the likelihood function is

$$l(\theta) = \begin{cases} \theta^{-5} & \theta \geq 2.68 \\ 0 & \theta < 2.68 \end{cases}$$

and the log-likelihood is

$$L(\theta) = \begin{cases} -5 \log \theta & \theta \geq 2.68 \\ -\infty & \theta < 2.68 \end{cases}$$

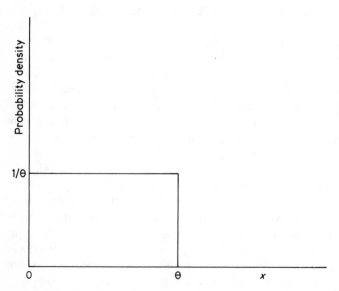

Fig. 4.2 The p.d.f. of the rectangular distribution for Example 4.6.

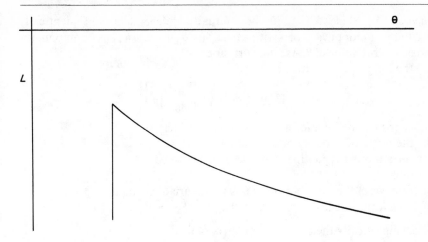

Fig. 4.3 Log-likelihood for Example 4.6.

Also for $\theta > 2.68$, we have

$$\frac{dL}{d\theta} = -\frac{5}{\theta}$$

which is negative throughout the permitted range of $\theta$. The log-likelihood function is sketched in Fig. 4.3 and we see that the maximum occurs at

$$\hat{\theta} = \max(\text{observations}) = 2.68$$

However, the conditions for Theorem 4.2 are violated; in particular the range of the random variables *does* depend on $\theta$. Therefore, although we have a maximum-likelihood estimator which is consistent, by Theorem 4.1, we cannot apply Theorem 4.2 to obtain asymptotic normality and an asymptotic variance. In such cases we are often driven back to deriving properties of the estimate ourselves. Example 4.6 also illustrates the point that we cannot always obtain maximum-likelihood estimates by solving the equation

$$\frac{dL}{d\theta} = 0,$$

which is often called the *likelihood equation*.

We shall not pursue Example 4.6 further here.                    □ □ □

We now turn to the situation in which there are a fixed number $k$ of unknown parameters $\theta_i$, which we write $\boldsymbol{\theta} = (\theta_1, \ldots, \theta_k)$. Theorem 4.1 still holds provided we insert $\boldsymbol{\theta}$, for $\theta$ as given in the theorem. The analogue to Theorem 4.2 is as follows, and we denote the maximum-likelihood estimator by $\boldsymbol{\theta}$.

*Theorem* 4.3   Let $X_1, \ldots, X_n$ be mutually independent and identically distributed random variables with a p.d.f. $f(x, \boldsymbol{\theta}_0)$, where $\boldsymbol{\theta}_0$ has $k$ components and is the true value of $\boldsymbol{\theta}$. Assume further that:

(i)   the information matrix

$$\mathbf{I}(\boldsymbol{\theta}) = \{I_{ij}(\boldsymbol{\theta})\} = -\left\{E\left[\frac{\partial^2 L}{\partial\theta_i\,\partial\theta_j}\right]\right\} \tag{4.18}$$

is positive definite for all $\boldsymbol{\theta}$ in the set considered;

(ii)   the range of $X$ does not depend on $\boldsymbol{\theta}$;

(iii)   certain other regularity conditions are satisfied;

then:

(a) the asymptotic distribution of $\hat{\boldsymbol{\theta}}$ is multivariate normal;

(b) $\hat{\boldsymbol{\theta}}$ is asymptotically unbiased, so that $\lim_{n\to\infty} E(\hat{\boldsymbol{\theta}}) = \boldsymbol{\theta}_0$;

(c) asymptotically the covariance matrix of $\hat{\boldsymbol{\theta}}$ is

$$\begin{bmatrix} V(\hat{\theta}_1) & C(\hat{\theta}_1, \hat{\theta}_2) \ldots C(\hat{\theta}_1, \hat{\theta}_k) \\ C(\hat{\theta}_1, \hat{\theta}_2) & V(\hat{\theta}_2) \\ \vdots & \\ C(\hat{\theta}_1, \hat{\theta}_k) & V(\hat{\theta}_k) \end{bmatrix} = \mathbf{I}^{-1}(\boldsymbol{\theta}_0) \tag{4.19}$$

□ □ □

The form of the covariance matrix is what we might expect as a generalization of Theorem 4.2. By inserting the estimated values $\hat{\theta}_1$ into Equation (4.19), we can obtain estimates of the variances and covariances of the $\hat{\theta}_i$. There is a generalization of the Cramér–Rao inequality to the case of $k$ unknown parameters, and the result is that lower bounds to the variances of the $\theta_i$ are in fact given by the diagonal elements of Equation (4.19). Again, therefore, the theorem shows that we have asymptotically unbiased and minimum-variance estimates, and again we can use the theorem to construct approximate methods for estimating confidence intervals, carrying out significance tests, etc. For a proof of the theorem, see the references given under Theorem 4.2.

Theorem 4.3 applies in all cases where there are a fixed number of parameters, and Example 4.7 gives a very simple illustration.

*Example* 4.7   Consider the estimation of $\mu$ and $\sigma^2$ from a random sample $x_1, \ldots, x_n$ from a normal distribution $N(\mu, \sigma^2)$. From Examples 3.8 and 3.11 we have

$$\frac{\partial L}{\partial \mu} = \frac{1}{\sigma^2}\sum(x_i - \mu)$$

$$\frac{\partial L}{\partial \sigma} = -\frac{n}{\sigma} + \frac{1}{\sigma^3}\sum(x_i - \mu)^2$$

so that

$$\frac{\partial^2 L}{\partial \mu^2} = -\frac{n}{\sigma^2}$$

$$\frac{\partial^2 L}{\partial \mu \partial \sigma} = -\frac{2}{\sigma^3}\sum(x_i - \mu)$$

$$\frac{\partial^2 L}{\partial \sigma^2} = \frac{n}{\sigma^2} - \frac{3}{\sigma^4}\sum(x_i - \mu)^2$$

By taking expectations over the $x$'s, we have

$$E\left(\frac{\partial^2 L}{\partial \mu^2}\right) = -\frac{n}{\sigma^2}$$

$$E\left(\frac{\partial^2 L}{\partial \mu \partial \sigma}\right) = 0$$

$$E\left(\frac{\partial^2 L}{\partial \sigma^2}\right) = -\frac{2n}{\sigma^2}$$

Hence, using Theorem 4.3, we obtain the covariance matrix

$$\begin{pmatrix} V(\hat{\mu}) & C(\hat{\mu}, \hat{\sigma}) \\ C(\hat{\mu}, \hat{\sigma}) & V(\hat{\sigma}) \end{pmatrix} = \begin{pmatrix} n/\sigma^2 & 0 \\ 0 & 2n/\sigma^2 \end{pmatrix}^{-1} = \begin{pmatrix} \sigma^2/n & 0 \\ 0 & \sigma^2/2n \end{pmatrix} \quad (4.20)$$

□ □ □

This last result is of some interest as it shows that the estimators of $\mu$ and $\sigma$ given in Example 3.11 are asymptotically uncorrelated. (In fact it can be shown that these estimators are statistically independent.)

It should be stressed that Theorem 4.3 only applies where there is a fixed number $k$ of parameters. The following example illustrates the situation where the number of parameters increases with the number of observations, and the theorem does not apply.

*Example* 4.8   In a certain experiment observations are taken in pairs such that

$$E(Y_{i1}) = E(Y_{i2}) = \mu_i \qquad i = 1, 2, \ldots, n$$

and all observations are subject to normally distributed error of variance $\sigma^2$. Then the likelihood is

$$l(\mu_i, \sigma) = \prod_{i=1}^{n} \frac{1}{2\pi\sigma^2} \exp\left\{-\frac{(y_{i1} - \mu_i)^2}{2\sigma^2} - \frac{(y_{i2} - \mu_i)^2}{2\sigma^2}\right\}$$

so that the log-likelihood is

$$L(\mu_i, \sigma) = -n \log 2\pi - 2n \log \sigma - \frac{1}{2\sigma^2} \sum_{i=1}^{n} \{(y_{i1} - \mu_i)^2 + (y_{i2} - \mu_i)^2\}$$

Therefore we have

$$\frac{\partial L}{\partial \mu_i} = \frac{1}{\sigma^2}\{(y_{i1} - \mu_i) + (y_{i2} - \mu_i)\}$$

and

$$\hat{\mu}_i = (y_{i1} + y_{i2})/2 \tag{4.21}$$

Similarly, we have

$$\frac{\partial L}{\partial \sigma} = -\frac{2n}{\sigma} + \frac{1}{\sigma^3} \sum_{i=1}^{n} \{(y_{i1} - \mu_i)^2 + (y_{i2} - \mu_i)^2\}$$

and by inserting Equation (4.21), this leads to

$$\hat{\sigma}^2 = \sum_{i=1}^{n} (y_{i1} - y_{i2})^2/4n \tag{4.22}$$

However, we see that

$$E(Y_{i1} - Y_{i2})^2 = V(Y_{i1} - Y_{i2})$$
$$= 2\sigma^2$$

so that

$$E(\hat{\sigma}^2) = \sigma^2/2$$

and the maximum-likelihood estimator is not consistent.   □ □ □

The trouble in Example 4.8 is that each new pair of observations brings in a new parameter, $\mu_i$, and Theorem 4.3 does not apply. We could avoid the trouble in Example 4.8 by considering that the relevant information for $\sigma^2$ is provided by the differences

$$d_i = y_{i1} - y_{i2}$$

If we take $d_i$ to be normally distributed with zero mean and variance $2\sigma^2$, we have only one unknown parameter, and maximum likelihood gives a consistent and asymptotically efficient estimator.

There is one final point on the two theorems. If the space of admissible $\theta_i$'s is defined by $\Omega$, then the true value $\theta_0$ must be inside this space, and not, for example, on the boundary, otherwise again Theorems 4.2 and 4.3 do not apply.

There are a number of other methods of estimation with the same asymptotic properties as maximum likelihood. However, maximum likelihood also has some small-sample properties to which we refer below. The difficulty of the above theory from a practical point of view is that it is asymptotic, and it is difficult in any practical case to know how large $n$ needs to be for the results to be valid. Sprott and Kalbfleisch (1969) give an interesting discussion of this point, with examples, and they emphasize the value of plotting the likelihood, rather than just relying on asymptotic theory to make inferences.

The asymptotic variance of maximum-likelihood estimators depends simply on the curvature of the likelihood function at the maximum (or more correctly, on the expectation of this). If the actual likelihood function is drawn, we may find in particular cases that it is not smooth, but skew, or else it has bumps in it, and such information will obviously affect the inference we make. Although the interpretation of likelihoods is rather difficult, it will become easier with practice. With the sort of computing machinery that we have currently available, likelihood functions are plotted fairly readily.

*Two exact properties*
The properties stated above are asymptotic, and therefore only hold approximately in any practical case. Now we give two important properties which hold exactly for all sample sizes.

*Theorem 4.4\** If sufficient statistics exist, then maximum-likelihood estimates are functions of them.                                    □ □ □

This theorem is readily proved, and the proof is left as an exercise; see Exercise 4.2.4. It is quite an important property, and distinguishes maximum likelihood from methods which are asymptotically equivalent to it. This property is not subject to the conditions of the previous theorems, and holds for all sample sizes. Coupled with the previous theorems, it is a very strong argument for considering maximum likelihood rather than any other method.

*Theorem 4.5* Maximum-likelihood estimates of functions of unknown parameters are the functions of the maximum-likelihood estimates of the parameters.                                    □ □ □

A good illustration of this is provided by the exponential distribution, see Exercises 3.4.1(*b*), 3.4.2 and Example 4.3. If we represent the p.d.f. as $\lambda \exp(-\lambda x)$, then from observations $x_1, \ldots, x_n$, the maximum-likelihood estimate of $\lambda$ is $1/\bar{x}$. However, if we represent the distribution

$$\frac{1}{\theta} e^{-x/\theta}$$

then the maximum-likelihood estimate of $\theta$ is $\bar{x}$. Thus we have

$$\hat{\theta} = 1/\hat{\lambda}$$

as we expect. If this result did not hold, we would certainly want to look hard at the method before using it, for we would not want the value of an estimate to depend on the accident of how the problem was parameterized. However, it is easy to see that Theorem 4.5 must hold in general.

Consider for simplicity the case of one unknown parameter. Let the likelihood of a set of observations $x_1, \dots, x_n$ be $l(x_1, \dots, x_n, \theta)$, and suppose this takes its maximum at a value $\hat{\theta}$. Now suppose that the p.d.f. is reparameterized to involve

$$\psi = G(\theta) \qquad \text{so that} \qquad \theta = G^{-1}(\psi)$$

The new likelihood is

$$L\{x_1, \dots, x_n, G^{-1}(\psi)\}$$

and will clearly take its maximum at a value $\hat{\psi}$ such that

$$\hat{\theta} = G(\hat{\psi})$$

This result is easily extended to the case of several unknown parameters, and it does not matter whether the observations are independent or not. We have therefore a very strong result, that the maximum-likelihood estimate of any function of unknown parameters is the function of the separate maximum-likelihood estimates (jointly estimated).

In emphasizing the importance of the result of Theorem 4.5, we should note that it cannot hold in general for sampling distribution properties. Thus, for example, unbiased estimation of $\sigma^2$ in a normal distribution is provided by

$$s^2 = \sum_1^n (x_i - \bar{x})^2/(n-1)$$

but we have from Exercise 3.2.1 that $E(s) < \sigma$. That is, the property of unbiasedness is not invariant to transformations of the parameters. Similarly, standard-error properties of estimators are not invariant to transformations. Some alternative methods of estimation, such as the method of moments, mentioned below, do not possess this property.

*Alternative method of estimating confidence intervals*
Finally in this section we quote another asymptotic result which can be used to obtain approximate confidence intervals. We can state the result in the following form:

*Theorem* 4.6    Under certain regularity conditions the distributions of

$$2\{L(\hat{\theta}) - L(\theta_0)\}$$

where $\theta_0$ is the true value and $\hat{\theta}$ is the maximum-likelihood estimate, is asymptotically $\chi_k^2$, where $k$ is the number of components of $\theta$.    □ □ □

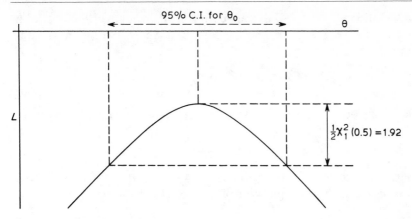

Fig. 4.4 Approximate confidence interval.

An illustration of how this result can be used to obtain approximate confidence intervals for the one-parameter case ($k = 1$) is shown in Fig. 4.4. From tables of the $\chi^2$-distribution we see that

$$\Pr\{\chi_1^2 < 3.84\} = 0.95$$

Therefore we have approximately

$$\Pr\{(L(\hat{\theta}) - L(\theta_0)) < 1.92\} = 0.95$$

and rewriting this,

$$\Pr\{L(\hat{\theta}) - 1.92 < L(\theta_0)\} = 0.95$$

Therefore we 'slice off' 1.92 from the top of the log-likelihood function and translate the boundaries into boundaries for $\theta$. This gives an approximate 95 per cent confidence interval for $\theta_0$. In general, and for $k$ parameters, we 'slice off' an amount $\frac{1}{2}\chi_k^2$ at the appropriate probability level, in order to generate the confidence region. One advantage of this method is that it is easily seen to be independent of the parameterization. Also, the method is easier for computer use, since it does not involve differentiating the likelihood.

Some other simple examples of the method are given by Box and Cox (1964, pp. 218, 220).

---

**Exercises 4.2**

1. Observations $y_1, \ldots, y_n$ are drawn independently from a normal distribution with $E(Y_r) = r\theta$ and $V(Y_r) = r^2\sigma^2$. Show that the maximum-likelihood estimates are

$$\hat{\theta} = \frac{1}{n}\sum\frac{y_r}{r} \qquad \hat{\sigma}^2 = \frac{1}{n}\sum\frac{(y_r - r\hat{\theta})^2}{r^2},$$

and find the asymptotic covariance matrix.

2. Observations $y_1, \ldots, y_n$ have independent Poisson distributions with means $\beta x_i$, for $i = 1, 2, \ldots, n$, and known $x_i > 0$. Show that the maximum-likelihood estimate of $\beta$ is

$$\hat{\beta} = \sum y_i / \sum x_i.$$

3. Show that for Example 4.5 the probability that there are no zero responses is

$$(1 - e^{-\mu})^n$$

so that with this probability, the maximum-likelihood estimate is infinite. Discuss the meaning and application of Theorem 4.2. in this problem.

4. For Example 4.8 show that $V(\hat{\sigma}^2) = \sigma^2/n$, and hence deduce that $\hat{\sigma}^2$ tends in probability to $\sigma^2/2$. Examine what happens if observations are taken $k$ at a time instead of in pairs.

5.* Prove Theorem 4.4 for the case of a single unknown parameter as follows. Write down the likelihood of Equation (3.18), form the log-likelihood, and differentiate. Hence show that the maximum-likelihood estimate is a function of the observations only through the sufficient statistic.

---

### 4.3 Practical aspects – one-parameter case

In Example 4.1 we have discussed estimation of the parameter of the exponential distribution when the data are subject to censored sampling. (Censored sampling arises in situations where some observations are given in the form $x > x_0$, or $x < x_0$.) Example 4.9 below illustrates truncated sampling, which arises when observations are not detected at all if, say $x > x_0$ (or $x < x_0$).

*Example* 4.9  In a cloud-chamber experiment, a stream of particles is passed through a chamber, of length $b$, containing some similar particles. If a collision does not occur in the chamber, the particles pass through undetected (see Fig. 4.5). Assuming that the distribution of distances to a collision is exponential, with a p.d.f. $\lambda \exp(-\lambda x)$, estimate $\lambda$ when there are $r$ recorded collisions, at distances $x_1, \ldots, x_r$.

In this problem, we are given that the p.d.f. of distances to collisions is exponential, and the only part of the distribution actually observed is the shaded area of Fig. 4.6. The probability that any particle will have a collision within the chamber is

$$\Pr(x < b) = \lambda \int_0^b e^{-x\lambda} dx = (1 - e^{-\lambda b})$$

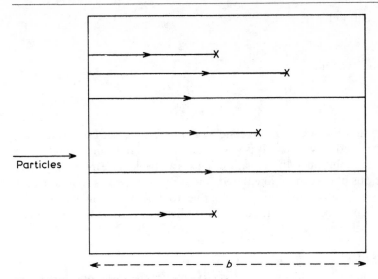

Fig. 4.5 Particle collisions in a cloud chamber.

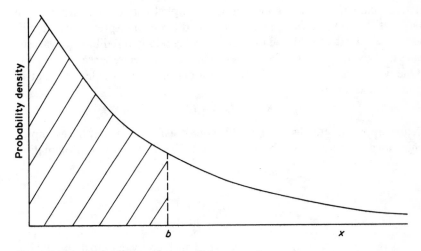

Fig. 4.6 The p.d.f. of distances to collisions.

Therefore the actual distribution of the observations is

$$\lambda e^{-x\lambda}/(1 - e^{-\lambda b}) \qquad 0 < x < b \tag{4.23}$$

so that the likelihood is

$$l = \prod_{i=1}^{r} \frac{\lambda e^{-\lambda x_i}}{(1 - e^{-\lambda b})} = \frac{\lambda^r e^{-\lambda \Sigma x_i}}{(1 - e^{-\lambda b})^r}$$

The log-likelihood is

$$L = r \log \lambda - \lambda \sum x_i - r \log(1 - e^{-\lambda b})$$

and therefore we find

$$\frac{\mathrm{d}L}{\mathrm{d}\lambda} = \frac{r}{\lambda} - \sum x_i - \frac{rb e^{-\lambda b}}{(1 - e^{-\lambda b})} \qquad (4.24)$$

In order to find the maximum-likelihood estimator, we put Equation (4.24) equal to zero and solve for $\hat{\lambda}$. It is easy to see that there is unfortunately no simple explicit solution to this equation. This is a problem we frequently meet in applications of maximum likelihood, and we nearly always have to resort to some graphical or numerical technique to find the estimate. We list three general methods below; the detailed application to Example 4.9 is left as an exercise (see Exercise 4.3.1).                    □ □ □

*Method (a) Plot the log-likelihood*
In many estimation problems it is instructive to plot the likelihood, and some statisticians contend that this should always be done (see for example Barnard *et al.*, 1962). One method would therefore be to plot the likelihood function on an appropriate scale, and then simply look at it in order to make inferences. However, some practice is needed in the interpretation of likelihood functions; the likelihood plot shows us the relative credibility of various values of $\theta$ giving rise to the data observed, in a very precise sense.

One way of summarizing a likelihood function is to consider the relative log-likelihood function

$$R(\theta) = L(\theta) - L(\hat{\theta})$$

and then approximate this by what $R(\theta)$ would be if $\hat{\theta}$ had exactly the asymptotic normal distribution given by Theorem 4.2. This leads to the approximation

$$R(\theta) \simeq -(\theta - \hat{\theta})^2 / 2\sigma^2 \qquad (4.25)$$

where

$$\sigma^2 = -1/E \left\{ \frac{\mathrm{d}^2 L}{\mathrm{d}\theta^2} \right\}_{\theta = \hat{\theta}}$$

and this is, of course, symmetrical about $\hat{\theta}$. This is equivalent to assuming the asymptotic normal distribution and standard error for $\hat{\theta}$ as indicated earlier. However Sprott and Kalbfleisch (1969) have given several examples where $R(\theta)$ differs considerably from the symmetrical approximation (4.25) and where wrong inferences could result from not plotting the actual likelihood function.

*Method (b) Plot the gradient of the likelihood*
When there is only one unknown parameter, this method is perhaps the

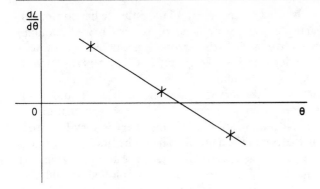

Fig. 4.7 Graphical method for maximum likelihood.

best. We simply calculate $dL/d\theta$ at various trial values of $\theta$ and plot as shown in Fig. 4.7.

Rough linear interpolation can now be used, simply by drawing a line by eye. Very rarely are more than three or four calculations necessary to obtain a good approximation to the maximum-likelihood estimate, and the slope of the line gives

$$\frac{d^2L}{d\theta^2}$$

and hence the estimate of variance of the estimator approximately.

*Method (c) Newton's iterative method*
The problem of finding the maximum-likelihood estimate is a standard one for the application of Newton's method. The equation we wish to solve is

$$\left\{\frac{dL}{d\theta}\right\}_{\theta=\theta} = 0 \tag{4.26}$$

If $\hat{\theta}_1$ is any rough estimate of $\hat{\theta}$, then the Taylor expansion of Equation (4.26) about $\hat{\theta}_1$ is

$$\left\{\frac{dL}{d\theta}\right\}_{\hat{\theta}} = \left\{\frac{dL}{d\theta}\right\}_{\hat{\theta}_1} + (\hat{\theta} - \hat{\theta}_1)\left\{\frac{d^2L}{d\theta^2}\right\}_{\hat{\theta}_1} + \ldots = 0 \tag{4.27}$$

If $\hat{\theta}_1$ is sufficiently close to $\hat{\theta}$ then we can ignore the higher terms in Equation (4.27), so that we have

$$\left\{\frac{dL}{d\theta}\right\}_{\hat{\theta}_1} + (\hat{\theta} - \hat{\theta}_1)\left\{\frac{d^2L}{d\theta^2}\right\}_{\hat{\theta}_1} \simeq 0$$

and a new estimate $\hat{\theta}_2$ can be made as

$$\hat{\theta}_2 = \hat{\theta}_1 - \left\{\frac{dL}{d\theta}\right\}_{\hat{\theta}_1} \bigg/ \left\{\frac{d^2L}{d\theta^2}\right\}_{\hat{\theta}_1} \tag{4.28}$$

Equation (4.28) can now be used recursively until the estimate has converged. (The iteration will usually converge provided $\hat{\theta}_1$ is chosen well enough.) Clearly, the method will work well if the second derivative changes very slowly, that is, if the third derivative is negligible. See Chambers (1977) for a discussion of the convergence of this and other methods of optimization.

In many cases the method works very well, and two or more iterations can be carried out without recomputing the second derivative. The main problem remaining is to find a good initial estimate $\hat{\theta}_1$, and there is very little that can be said in general on this point. Each problem has to be studied separately and the possibilities of some simple graphical device or some approximation explored. If all else fails, the log-likelihood could be plotted roughly, to get $\hat{\theta}_1$.

*Example* 4.10 (Method (*c*) for Example 4.9.) From Equation (4.24) we obtain

$$\frac{d^2 L}{d\lambda^2} = -\frac{r}{\lambda^2} + \frac{rb^2 e^{-\lambda b}}{[1 - e^{-\lambda b}]^2} \tag{4.29}$$

Given an initial estimate $\hat{\lambda}_1$, we can now substitute Equations (4.24) and (4.29) into Equation (4.28). Now it is easy to check that for sampling from the truncated exponential distribution,

$$E(\bar{x}) < 1/\lambda$$

so that

$$\lambda < 1/E(\bar{x}).$$

A suitable initial estimate $\hat{\lambda}_1$ may therefore be a value a little lower than $1/\bar{x}$.

The rate of convergence of the iteration method expressed by Equation (4.28) is usually improved slightly if the expected value of the second derivative is used:

$$\hat{\theta}_2 = \hat{\theta}_1 - \left\{\frac{dL}{d\theta}\right\}_{\hat{\theta}} \bigg/ E\left\{\frac{d^2 L}{d\theta^2}\right\}_{\theta}$$

☐ ☐ ☐

The reader should work through the numerical details of all three of these methods for himself (see Exercise 4.3.1).

*Goodness of fit, outliers*

Once a model is fitted to a set of data, two important questions can be tackled. The first problem is one of how well the model fits, and of comparing the fit with other models, or other methods of estimation. The second problem is that of detecting outliers, and clearly this is very closely related to the first problem, and somewhat difficult to disentangle from it. For example, if we have two apparently outlying observations, are these really outliers, or do they indicate that a different model should be fitted?

When the problem involves only discrete distributions, the goodness-of-fit problem can be answered by using the standard $\chi^2$ goodness-of-fit test, which is described in most elementary statistics texts, such as Wetherill (1972). For continuous distributions this method can still be used, but it is less efficient, and some simple graphical device or probability-plotting method would usually be preferred.

What is really required here are residual type methods analogous to the methods used for studying residuals in a normal-theory regression problem, and described, for example, in Section 2.4. Both the goodness-of-fit problem and discussions of outliers can be made by using generalized residuals (see Cox and Snell, 1968), but this is more suited to the multiparameter case, and will be discussed later.

**Exercises 4.3**
1. The following data are a set of fictitious observations for Example 4.9, with $b = 1.5$. Carry out all three methods of calculating $\hat{\lambda}$ and assess the goodness of fit.

| | | | | |
|---|---|---|---|---|
| 0.269 | 1.044 | 0.166 | 0.537 | 1.091 |
| 1.007 | 0.233 | 1.310 | 1.177 | 0.012 |
| 0.367 | 0.340 | 0.426 | 1.109 | 1.283 |

2. There is available a suspension $S_0$ of small particles randomly dispersed in a liquid at a concentration of $\rho$ particles per unit volume. In order to estimate $\rho$, a series of suspensions $S_0, S_1, S_2, S_3, S_4$ is prepared in which the concentrations are

$$\rho, \tfrac{1}{2}\rho, \tfrac{1}{4}\rho, \tfrac{1}{8}\rho, \tfrac{1}{16}\rho$$

From each suspension, 12 independent unit volumes are withdrawn with the intention of observing the number containing no particles. However, some of the volumes have to be rejected for technical reasons and the final observations are as shown in Table 4.1.

Table 4.1

| Suspension | $S_0$ | $S_1$ | $S_2$ | $S_3$ | $S_4$ |
|---|---|---|---|---|---|
| No. of unit volumes observed | 10 | 12 | 8 | 12 | 11 |
| No. containing no particles | 0 | 3 | 3 | 6 | 10 |

Find the maximum-likelihood estimate $\hat{\rho}$ and its asymptotic standard error. [*Hint*: Refer back to Example 4.5.]

3. In an experiment with maize, the numbers of seedlings observed in different categories were as listed in Table 4.2.

Table 4.2

| Category | Number of seedlings |
|---|---|
| I   (starchy green) | 4299 |
| II  (starchy white) | 1960 |
| III (sugary green) | 2062 |
| IV  (sugary white) | 83 |

According to genetic theory, the probabilities of a seedling belonging to each of these categories will be:

$$(I)\ \frac{2+\theta}{4}; \quad (II)\ \frac{1-\theta}{\theta}; \quad (III)\ \frac{1-\theta}{4}; \quad (IV)\ \frac{\theta}{4}$$

where $\theta$ is the recombination fraction. Show that, on this theory, the likelihood of the observations is of the form

$$c_0(2+\theta)^{c_1}(1-\theta)^{c_2}\theta^{c_3}$$

where $c_0, c_1, c_2, c_3$ are constants; and calculate the maximum-likelihood estimate of $\theta$.

## 4.4 Practical aspects – multiparameter case

In the previous section we have discussed the one-parameter case. The following example illustrates the problems which arise in the general multiparameter case. Methods analogous to (a) and (b) of Sections 4.3 can no longer be used for finding the maximum-likelihood estimate, except perhaps for two parameters, because of the difficulty of plotting; we must therefore resort to either the generalization of method (c), or else some numerical optimization algorithm for directly maximizing the likelihood.

*Example* 4.11   Feigl and Zelen (1965) discussed the analysis of survival data on patients suffering from leukaemia. When the disease is diagnosed, the white blood cell count of the patient is taken, and the log of these counts are denoted $x_i$, for patients $i = 1, 2, 3, \ldots, n$. One suitable model for the data is to consider the survival times $t_i$, as having a p.d.f.

where

$$\left. \begin{array}{l} \lambda \exp(-\lambda t) \qquad t > 0 \\[2mm] E(t_i) = 1/\lambda_i = \alpha + \beta x_i \end{array} \right\}$$

(4.30)

□ □ □

The immediate problem in Example 4.11 is to fit the model (4.30) to data $(x_i, t_i), i = 1, 2, \ldots, n$, for $n$ patients. The likelihood is

$$l(\alpha, \beta) = \prod_{i=1}^{n} \lambda_i e^{-\lambda_i t_i}$$

and the log-likelihood can be written

$$L(\alpha, \beta) = -\sum \log(\alpha + \beta x_i) - \sum t_i (\alpha + \beta x_i)^{-1} \qquad (4.31)$$

provided $(\alpha + \beta x_i) > 0$, for all $i$. From Equation (4.31) we easily obtain the equations satisfied by the maximum-likelihood estimators:

$$\frac{\partial L}{\partial \alpha} = 0 = -\sum (\hat{\alpha} + \hat{\beta} x_i)^{-1} + \sum t_i (\hat{\alpha} + \hat{\beta} x_i)^{-2} \qquad (4.32)$$

$$\frac{\partial L}{\partial \beta} = 0 = -\sum x_i (\hat{\alpha} + \hat{\beta} x_i)^{-1} + \sum t_i x_i (\hat{\alpha} + \hat{\beta} x_i)^{-2} \qquad (4.33)$$

It is easy to see that Equations (4.32) and (4.33) must be solved iteratively, and the method used is a simple generalization of method (c) of Section 4.3. We first obtain rough estimates $(\alpha_1, \beta_1)$ of the unknown parameters, and then expand in a Taylor series as follows. We have

$$\left(\frac{\partial L}{\partial \alpha}\right)_m = 0 = \left(\frac{\partial L}{\partial \alpha}\right)_1 + (\alpha - \alpha_1)\left(\frac{\partial^2 L}{\partial \alpha^2}\right)_1 + (\beta - \beta_1)\left(\frac{\partial^2 L}{\partial \alpha \partial \beta}\right)_1 + \cdots$$

$$\left(\frac{\partial L}{\partial \beta}\right)_m = 0 = \left(\frac{\partial L}{\partial \beta}\right)_1 + (\alpha - \alpha_1)\left(\frac{\partial^2 L}{\partial \alpha \partial \beta}\right)_1 + (\beta - \beta_1)\left(\frac{\partial^2 L}{\partial \beta^2}\right)_1 + \cdots$$

where the subscript $m$ to the differential coefficients indicates that they are evaluated at the actual maximum-likelihood estimates that we are trying to find, and the subscript 1 indicates that they are evaluated at the rough values $(\alpha_1, \beta_1)$. If the rough values are sufficiently accurate, we may ignore the higher terms in these expansions and we have

$$\begin{bmatrix} \alpha_1 - \alpha_2 \\ \\ \beta_1 - \beta_2 \end{bmatrix} = \begin{bmatrix} \left(\dfrac{\partial^2 L}{\partial \alpha^2}\right)_1 & \left(\dfrac{\partial^2 L}{\partial \alpha \partial \beta}\right)_1 \\ \\ \left(\dfrac{\partial^2 L}{\partial \alpha \partial \beta}\right)_1 & \left(\dfrac{\partial^2 L}{\partial \beta^2}\right)_1 \end{bmatrix}^{-1} \begin{bmatrix} \left(\dfrac{\partial L}{\partial \alpha}\right)_1 \\ \\ \left(\dfrac{\partial L}{\partial \beta}\right)_1 \end{bmatrix} \qquad (4.34)$$

where now $(\alpha_2, \beta_2)$ should be improved estimates. This procedure can now be repeated, using $(\alpha_2, \beta_2)$ as rough estimates, and so on, until our estimates have converged satisfactorily; it should be pointed out that this iterative method is not bound to converge, although it will nearly always do so if the first rough estimates $(\alpha_1, \beta_1)$ are sufficiently accurate. For working on a desk machine, it is usually not necessary to evaluate the second differential coefficient of the log-likelihood in every iteration, and this saves a considerable amount of labour since the matrix of these coefficients has to be inverted in Equation (4.34). Again, as in the one-parameter case, an alternative form of optimization is to use the expected values of the second derivatives in Equation (4.34); this is known as the 'method of scoring'.

The details of applying this method to Example 4.11 are left as an exercise for the student (see Exercise 4.4.1). In this example, initial estimates of $\alpha$ and $\beta$ are obtained by fitting a line by eye to a plot of $t_i$ against $x_i$, since from Equation (4.30),

$$E(t_i) = \alpha + \beta x_i \qquad (4.35)$$

The extension of this method to $k$ parameters is obvious; we have an equation similar to Equation (4.34) in which we have a $k \times k$ matrix of second differential coefficients of the log-likelihood to invert. The problem of finding initial estimates is in general more difficult of course, and there is a slightly greater chance of the method not converging.

*Optimization algorithms*

It has been mentioned above that one alternative to the iterative procedure just described is to use a computer algorithm for directly maximizing the likelihood. One such algorithm has been given by Nelder and Mead (1965). Their method starts by considering the function values at $(k + 1)$ vertices of a general simplex in $k$ dimensions. The vertex having the lowest value is now replaced by another point, obtained by reflecting it through the centroid. The simplex changes shape, adapting itself to the local landscape, and contracts to find the maximum. The stopping rule used is to calculate the standard error of the function values of the simplex, and stop when this falls below a pre-set figure. A FORTRAN algorithm for the procedure has been published by O'Neill (1971), with corrections noted by Chambers and Ertel (1974).

In general, optimization methods using knowledge of the derivatives are quicker than straight search routines of the Nelder–Mead type, provided the derivatives are not too difficult to calculate. See the book by Chambers (1977) for a thorough discussion of optimization methods.

*Discussion\**

It should be pointed out that occasionally likelihood surfaces arise which have a very unusual shape, or have local maxima, and the above methods do not work well. A discussion of the problems which arise in this situation, with an account of other methods of solution, is given by Barnett (1966).

Clearly, the interpretation of likelihoods with several more or less equal maxima is rather difficult. A confidence region found by using Theorem 4.6 would lead to a region which was composed of several disjointed bits, one near each of the maxima.

One practical way of avoiding the risk of missing local maxima, which also helps in the interpretation of results, is to plot the likelihood near the estimated maximum. The problem of plotting multiparameter likelihoods is discussed by Sprott and Kalbfleisch (1969). In general it is necessary to eliminate all but one, or perhaps two parameters, and one way of doing this is to plot

$$L_m(\theta_1, \theta_2) = \max_{\theta_3,\ldots,\theta_k} L(\theta_1,\ldots,\theta_k) \tag{4.36}$$

$$L_m(\theta_1) = \max_{\theta_2,\ldots,\theta_k} L(\theta_1,\ldots,\theta_k) \tag{4.37}$$

A difficulty in the use of such functions is that the parameters over which the likelihood is maximized are essentially being assumed known, and the uncertainty arising from lack of knowledge of them is not reflected in the final function. It is easy to see that a misleading picture of the plausibility of a parameter could be produced, but this will usually become serious only if a large number of parameters is eliminated in this way.

Another way to eliminate parameters is to integrate the likelihood function over a prior distribution for these parameters. This method is subject to similar dangers to the first, and may be somewhat sensitive to the prior distributions used.

One very important point of interpretation which should be clearly understood is that inferences made from the maximum likelihood are dependent on the model being assumed. In any particular case there may be a large number of simple models which would fit the data reasonably well. In a choice between different models, three different sorts of consideration play a role: external evidence, mathematical considerations, and statistical considerations.

*External evidence\** When building a model there are two basic approaches. One is to work through the process in detail, and build up a model in which the parameters have a direct meaning, or in which the form of the equation can be defended on physical grounds. The other approach is to proceed empirically, and include any terms which seem to explain observed variation.

An interesting example of a purely empirical approach is given by Coen, Gomme and Kendall (1969), who obtained a formula for predicting an economic time series by, in effect, carrying out multiple regression on a series of quantities, quite apart from any consideration of whether the inclusion of these terms could be defended on purely economic grounds.

Some serious criticism of the work was given by Box and Newbold (1971), who show that the assumptions made by Coen, Gomme and Kendall about the probability structure of error terms leads to misleading results. With the latter approach there will clearly be a lot more latitude in the choice of a model than with the former. In either case, previous examples of similar data may influence our decisions.

*Mathematical considerations** Subject to any decisions already made under the heading of external evidence, the usual rule is to include as few parameters as possible, consistent with the data. The way in which parameters are included may also be affected by mathematical considerations. For example, Cox and Snell (1968) discussed Example 4.11, but instead of using Equation (4.30) they used the model

$$t_i = \beta_1 \exp\{\beta_2(x_i - \bar{x})\} z_i \qquad (4.38)$$

where $z_i$ are independently exponentially distributed with unit mean, and $\bar{x} = \Sigma x_i / n$. The advantage of Equation (4.38) over Equation (4.30) is that $t_i > 0$ for all $x_i$, all $\beta_2$, and all $\beta_1 > 0$, whereas in (4.30), $\lambda_i > 0$ over a more difficult region, and all parameters are subject to peculiar restrictions.

*Statistical considerations** Some of the matters which arise under this heading were raised at the end of Section 4.3. Some kind of analysis of residuals is usually called for. A general definition of residuals is given by Cox and Snell (1968).

Suppose we have a model relating observed random variables $Y_i$ to a vector of unknown parameters $\boldsymbol{\beta}$, and a vector of independent and identically distributed random variables $\boldsymbol{\varepsilon}$, such that

$$Y_i = g_i(\boldsymbol{\beta}, \varepsilon_i) \qquad i = 1, 2, \ldots, n$$

where the $\varepsilon_i$ represent the unobserved error terms. If $\hat{\boldsymbol{\beta}}$ is the maximum-likelihood estimate of $\boldsymbol{\beta}$, then under general conditions an $r_i$ exists such that

$$Y_i = g_i(\hat{\boldsymbol{\beta}}, r_i) \qquad (4.39)$$

If Equation (4.39) has a unique solution for $r_i$,

$$r_i = h_i(Y_i, \hat{\boldsymbol{\beta}}) \qquad i = 1, 2, \ldots, n \qquad (4.40)$$

then the $r_i$ in Equation (4.40) are defined as the residuals.

*Example* 4.12   For Example 4.11, model (4.30), the generalized residuals are defined as follows. Equation (4.39) gives

$$t_i = (\alpha + \beta x_i)\varepsilon_i$$

where the $\varepsilon_i$ are exponentially distributed with unit mean. The residuals are then

$$r_i = t_i / (\hat{\alpha} + \hat{\beta} x_i) \qquad (4.41)$$

□ □ □

Cox and Snell (1968) showed that in certain cases the quantities in Equation (4.40) are not identically distributed, and some adjustments are necessary; see their paper for details. A study of the generalized residuals may lead to the detection of outliers, or revision of model.

It should be emphasized here that even after a study of points under these three headings, there will usually be considerable scope for alternative models to represent a set of data.

---

**Exercises 4.4**

1. The data given below are from Feigl and Zelen (1965), to which Example 4.11 refers; $y_i$ is the survival time in weeks, and $x_i$ is $\log_{10}$ (white blood cell count). Fit the model (4.30) by maximum likelihood, plot the likelihood surface for $(\alpha, \beta)$, and assess the goodness of fit of the model.

| $y_i$ | 65    | 156   | 100   | 134   | 16    | 108   | 121   |
|-------|-------|-------|-------|-------|-------|-------|-------|
| $x_i$ | 3.362 | 2.875 | 3.633 | 3.415 | 3.778 | 4.021 | 4.000 |
| $y_i$ | 4     | 39    | 143   | 56    | 26    | 22    | 1     |
| $x_i$ | 4.230 | 3.732 | 3.845 | 3.973 | 4.505 | 4.544 | 5.000 |
| $y_i$ | 1     | 5     | 65    |       |       |       |       |
| $x_i$ | 5.000 | 4.716 | 5.000 |       |       |       |       |

2. Sampford and Taylor (1959) give the logarithms of survival times (in minutes) of rats poisoned with carbon tetrachloride. One of each pair of litter mates was injected with vitamin $B_{12}$, the other received no additional drug. Observation was suspended after 16 hours.

The log (survival time) figures are as shown in Table 4.3.

By iteration, find the maximum-likelihood estimates of the mean and standard deviation of $Z = t_1 - t_2$, assuming this to be normally distributed. Discuss the effect of injection of vitamin $B_{12}$ on the rate of action of carbon tetrachloride.

3. Observations $(x_i, y_i)$, $i = 1, 2, \ldots, n$, are sampled independently from a

Table 4.3

| $B_{12}(t_1)$ | Control $(t_2)$ | $B_{12}(t_1)$ | Control $(t_2)$ |
|---------------|-----------------|---------------|-----------------|
| 2.73          | > 2.98          | 2.79          | 2.82            |
| 2.80          | > 2.98          | 2.90          | 2.79            |
| 2.01          | 2.84            | 2.78          | 2.64            |
| 2.19          | 2.76            | 2.78          | 2.48            |
| 2.34          | 2.83            | 2.97          | 2.64            |
| 2.61          | 2.73            | 2.74          | 2.31            |
| 2.51          | 2.62            | 2.96          | 2.51            |
| 2.65          | 2.70            | > 2.98        | 2.68            |

bivariate normal distribution with a p.d.f.

$$\frac{1}{2\pi\sqrt{(1-\rho^2)}} \exp\left\{-\frac{1}{2(1-\rho^2)}(x^2 - 2\rho xy + y^2)\right\} \qquad -\infty \le x, \quad y \le \infty,$$
$$-1 < \rho < 1$$

Show that the maximum-likelihood estimator of $\rho$ is a solution of the equation

$$\hat{\rho}(1-\hat{\rho}^2) + (1+\hat{\rho}^2)\frac{1}{n}\sum x_i y_i - \hat{\rho}\left(\frac{1}{n}\sum x_i^2 + \frac{1}{n}\sum y_i^2\right) = 0$$

[Do not try to solve this equation.] If this equation has three real roots, how would you choose among them for your estimator of $\rho$?

## 4.5 Other methods of estimation

If any estimator of an unknown parameter $\theta$ is asymptotically normal, unbiased and of minimum asymptotic variance, then we say that it is a BAN estimate (best asymptotically normal). We have seen that in general, maximum-likelihood estimates are BAN estimates. However, if we choose as estimates values maximizing any function sufficiently like the likelihood in the limit, then these estimates will be asymptotically equivalent, and will also be BAN estimates. One example of such a method is the minimum chi-square method (Kendall and Stuart, 1973), and another is the minimum logit chi-square method (Berkson, 1953, 1955). Such methods rarely possess the small-sample advantages of maximum likelihood, and are often applied to particular problems for special reasons.

One method which can be derived from the method of maximum likelihood is the method of least squares. This is a very important method, of very general application, and is discussed in detail in the following chapters. Least-squares theory covers a very large and important area of statistics.

One method of estimation which has the virtue of simplicity is the method of moments. In this method the first, second, etc., moments of the data are equated to the first, second, etc., moments of the population until there are enough equations to estimate all the unknown parameters. The method is well-suited to fitting distributions to observed data in cases where mathematical simplicity is required. However, the efficiency of the method may be very low, and is asymptotically zero in certain special cases.

For further numerical illustration, see the references given in Appendix B.

# CHAPTER FIVE

# The method of least squares

## 5.1 Basic model

In the next three chapters we shall discuss a particular form of statistical model, which gives rise to simple statistical methods of very wide applicability. The basic model has been mentioned in Section 2.1, Equation (2.2), see also Example 3.12, but the following example illustrates how it arises in practice.

*Example* 5.1   In the petroleum refining industry, scheduling and planning play a very important role. One phase of the process is the production of petroleum spirit from crude oil, and its subsequent processing. To enable the plant to be scheduled in an optimum manner, an estimate is required of the percentage yield of petroleum spirit from the crude oil, based upon certain rough laboratory determinations of properties of the crude oil. Table 5.1 shows values of actual per cent yields of petroleum spirit ($y$), and four properties of the crude oil, for samples from 32 different crudes. The properties recorded are as follows (data from Prater, 1956):

$x_1$ : specific gravity of the crude, a function of the API measurement.[†]
$x_2$ : crude oil vapour pressure, measured in pounds per square inch absolute.
$x_3$ : the ASTM 10 per cent distillation point, in °F.
$x_4$ : the petroleum fraction end point, in °F.

It is required to use this data to provide an equation for predicting $y$, from measurements of these four explanatory variables. In the subsequent text we shall denote the set of observations on the sample from the $i$th crude by $(x_{1i}, x_{2i}, x_{3i}, x_{4i}, y_i)$.   □ □ □

It is proposed to obtain an estimating equation for percentage petrol yield

[†]The relationship between the American Petroleum Institution measurement and specific gravity is:

specific gravity $= 141.5/\{131.5 + °\text{API}\}$.

Table 5.1 *Data on percentage yields of petroleum spirit*

| $x_1$ | $x_2$ | $x_3$ | $x_4$ | $y$ |
|-------|-------|-------|-------|------|
| 38.4 | 6.1 | 220 | 235 | 6.9 |
| 40.3 | 4.8 | 231 | 307 | 14.4 |
| 40.0 | 6.1 | 217 | 212 | 7.4 |
| 31.8 | 0.2 | 316 | 365 | 8.5 |
| 40.8 | 3.5 | 210 | 218 | 8.0 |
| 41.3 | 1.8 | 267 | 235 | 2.8 |
| 38.1 | 1.2 | 274 | 285 | 5.0 |
| 50.8 | 8.6 | 190 | 205 | 12.2 |
| 32.2 | 5.2 | 236 | 267 | 10.0 |
| 38.4 | 6.1 | 220 | 300 | 15.2 |
| 40.3 | 4.8 | 231 | 367 | 26.8 |
| 32.2 | 2.4 | 284 | 351 | 14.0 |
| 31.8 | 0.2 | 316 | 379 | 14.7 |
| 41.3 | 1.8 | 267 | 275 | 6.4 |
| 38.1 | 1.2 | 274 | 365 | 17.6 |
| 50.8 | 8.6 | 190 | 275 | 22.3 |
| 32.2 | 5.2 | 236 | 360 | 24.8 |
| 38.4 | 6.1 | 220 | 365 | 26.0 |
| 40.3 | 4.8 | 231 | 395 | 34.9 |
| 40.0 | 6.1 | 217 | 272 | 18.2 |
| 32.2 | 2.4 | 284 | 424 | 23.2 |
| 31.8 | 0.2 | 316 | 428 | 18.0 |
| 40.8 | 3.5 | 210 | 273 | 13.1 |
| 41.3 | 1.8 | 267 | 358 | 16.1 |
| 38.1 | 1.2 | 274 | 444 | 32.1 |
| 50.8 | 8.6 | 190 | 345 | 34.7 |
| 32.2 | 5.2 | 236 | 402 | 31.7 |
| 38.4 | 6.1 | 220 | 410 | 33.6 |
| 40.0 | 6.1 | 217 | 340 | 30.4 |
| 40.8 | 3.5 | 210 | 347 | 26.6 |
| 41.3 | 1.8 | 267 | 416 | 27.8 |
| 50.8 | 8.6 | 190 | 407 | 45.7 |

by fitting the following model. The observations $y$ are assumed to be independently and normally distributed with an expectation

$$E(Y_i) = \beta_0 + \beta_1 x_{1i} + \beta_2 x_{2i} + \beta_3 x_{3i} + \beta_4 x_{4i} \tag{5.1}$$

and constant variance

$$V(Y_i | x_{1i}, x_{2i}, x_{3i}, x_{4i}) = \sigma^2 \tag{5.2}$$

In fact it turns out to be more convenient to modify Equation (5.1) to

$$E(Y_i) = \alpha + \beta_1 (x_{1i} - \bar{x}_{1.}) + \ldots + \beta_4 (x_{4i} - \bar{x}_{4.}) \tag{5.3}$$

where $\bar{x}_{1.}$, etc., are the sample means of the $x_{1i}$, etc., for the data used in estimating $\alpha$, $\beta_1, \ldots, \beta_4$. Clearly, Equations (5.1) and (5.3) are equivalent models, and the argument is similar to that used in Section 2.3 for simple linear regression. There are several important points about this model, and its use in problems such as Example 5.1:

(i) We have to assume that if there is an error in the measurement of the explanatory variables $x_{ij}$, then this is negligible compared with the random variation in $y_i$.

(ii) The expectation in Equations (5.1) or (5.3) is assumed to be a good fit over the range covered by the observations. This can be examined by an appropriate analysis of residuals, or by statistical methods to be described below. Such analyses may lead to a revision of the model; see Chapter 6.

(iii) We must examine whether Equation (5.2) fits reasonably well, by an analysis of residuals. Also, by analysis of residuals, we must examine the assumptions of independence and normality of errors.

In these analyses, it is important to bear in mind the use to which the model is being put. We do not need an exact fit to the model, but merely a fit which will give us satisfactory results for our purpose. It is important therefore to know which parts of the model are critical, and which parts are not.

The statistical inferences to be made concern the individual parameters or predictions of the response variable and take the form of confidence intervals or significance tests. Most of this chapter is devoted to the theory behind the method of fitting, and the theory of these methods of inference. When making predictions it is important to realize that the model is not validated in parts of the $x$-space where observations are scarce, and extrapolation in particular can be very misleading. Some of these important practical aspects will be discussed in following chapters, but for the present we concentrate on the theory.

We shall deal with a slightly more general form of model than that given above. Let there be a vector of response variables $\mathbf{Y}$ where

$$\mathbf{Y}' = (Y_1, Y_2, \ldots, Y_n),$$

and a vector of unknown parameters $\boldsymbol{\theta}$,

$$\boldsymbol{\theta}' = (\theta_1, \theta_2, \ldots, \theta_p)$$

related by the equation

$$E(\mathbf{Y}) = \mathbf{a}\boldsymbol{\theta} \tag{5.4}$$

where $\mathbf{a}$ is an $n \times p$ matrix of known constants. The vector of response variables $\mathbf{Y}$ is assumed to be composed of uncorrelated random variables with equal variance, so that the variance–covariance matrix of $\mathbf{Y}$ is

$$V(\mathbf{Y}) = \sigma^2 \mathbf{I} \tag{5.5}$$

where $\mathbf{I}$ is the $n \times n$ identity matrix. If we add the assumption that the distribution of $Y$ is multivariate normal, then together with Equation (5.5), this implies that the $Y_i$ are independent.

*Example* 5.2 　If $\boldsymbol{\theta}' = (\beta_0, \beta_1, \dots, \beta_4)$ and the matrix $\mathbf{a}$ is

$$\mathbf{a} = \begin{bmatrix} 1 & x_{1,1} & \cdots & x_{4,1} \\ 1 & x_{1,2} & \cdots & x_{4,2} \\ \vdots & & & \\ 1 & x_{1,32} & \cdots & x_{4,32} \end{bmatrix}$$

where $x_{ij}$ is measurement $j$ on the $i$th explanatory variable, then we have the model (5.1) and (5.2) for fitting to data $(x_{1i}, x_{2i}, x_{3i}, x_{4i}, y_i)$, $i = 1, 2, \dots, 32$, for Example 5.1.　　　　□ □ □

Now from Chapter 4 we see that in order to estimate the unknown parameters $\boldsymbol{\theta}$ and $\sigma^2$ we can use maximum likelihood. If $\mathbf{y}$ is drawn from a multivariate normal distribution specified by Equation (5.4) and (5.5) then from Equation (3.21) the likelihood is

$$l = (2\pi)^{-n/2} \sigma^{-n} \exp\{ - (\mathbf{y} - \mathbf{a}\boldsymbol{\theta})'(\mathbf{y} - \mathbf{a}\boldsymbol{\theta})/2\sigma^2 \}. \tag{5.6}$$

Another way of obtaining the likelihood is as follows. The random variables $Y_i$, $i = 1, 2, \dots, n$, are independently and normally distributed with $V(Y_i) = \sigma^2$, and

$$E(Y_i) = \sum_{j=1}^{p} a_{ij}\theta_j.$$

(This is a general form for an expression like Equation (5.3).)
Therefore the likelihood for $n$ observations $y_i$ is

$$l = \prod_{i=1}^{n} \frac{1}{\sqrt{(2\pi)}\sigma} \exp\left\{ -\frac{1}{2\sigma^2}\left( y_i - \sum_{j=1}^{p} a_{ij}\theta_j \right)^2 \right\}$$

$$= (2\pi)^{-n/2} \sigma^{-n} \exp\left\{ -\frac{1}{2\sigma^2} \sum_{i=1}^{n}\left( y_i - \sum_{j=1}^{p} a_{ij}\theta_j \right)^2 \right\}$$

However, the vector of differences between the observations and their expected values is

$$(\mathbf{y} - \mathbf{a}\boldsymbol{\theta}) = \begin{bmatrix} y_1 - \sum\limits_{j=1}^{p} a_{1j}\theta_j \\ y_2 - \sum\limits_{j=1}^{p} a_{2j}\theta_j \\ \vdots \\ y_n - \sum\limits_{j=1}^{p} a_{nj}\theta_j \end{bmatrix}$$

so that the sum of squares of these differences is

$$\sum_{i=1}^{n} \left( y_i - \sum_{j=1}^{p} a_{ij}\theta_j \right)^2 = (\mathbf{y} - \mathbf{a}\boldsymbol{\theta})'(\mathbf{y} - \mathbf{a}\boldsymbol{\theta})$$

and the likelihood is given by Equation (5.6).

The log-likelihood is therefore

$$L = -\frac{n}{2}\log 2\pi - n \log \sigma - \frac{1}{2\sigma^2}(\mathbf{y} - \mathbf{a}\boldsymbol{\theta})'(\mathbf{y} - \mathbf{a}\boldsymbol{\theta})$$

and in order to maximize the likelihood for choice of $\boldsymbol{\theta}$ we must minimize

$$S = (\mathbf{y} - \mathbf{a}\boldsymbol{\theta})'(\mathbf{y} - \mathbf{a}\boldsymbol{\theta}) \qquad (5.7)$$

That is, we must choose $\boldsymbol{\theta}$ to minimize the sums of squared differences of the observations from their expected values. With this model, therefore, maximizing the likelihood with respect to choice of estimates of $\boldsymbol{\theta}$ is equivalent to the ordinary method of least squares used in linear regression, which was described for a simple case in Chapter 2.

If we multiply out Equation (5.7), we obtain

$$S = \mathbf{y}'\mathbf{y} - \mathbf{y}'\mathbf{a}\boldsymbol{\theta} - \boldsymbol{\theta}'\mathbf{a}'\mathbf{y} + \boldsymbol{\theta}'\mathbf{a}'\mathbf{a}\boldsymbol{\theta} \qquad (5.8)$$

where each term on the right-hand side is a scalar. For example, $\mathbf{y}'\mathbf{a}\boldsymbol{\theta}$ is $(1 \times n)(n \times p)(p \times 1)$ or $(1 \times 1)$.

We must now minimize $S$ with respect to choice of each $\theta_i$, and to do this we shall differentiate with respect to each $\theta_i$, for $i = 1, 2, \ldots, p$. Now if we differentiate $\boldsymbol{\theta}$ with respect to $\theta_j$, say, we have

$$\frac{\partial \boldsymbol{\theta}'}{\partial \theta_j} = \boldsymbol{\Delta}'_j = (0, \ldots, 0, 1, 0, \ldots, 0)$$

with unity in the $j$th position. We can now differentiate Equation (5.8) directly, by using this $\boldsymbol{\Delta}_j$ notation, and using the ordinary rules for differentiating a product, to obtain

$$\frac{\partial S}{\partial \theta_i} = -\mathbf{y}'\mathbf{a}\boldsymbol{\Delta}_i - \boldsymbol{\Delta}'_i\mathbf{a}'\mathbf{y} + \boldsymbol{\Delta}'_i\mathbf{a}'\mathbf{a}\boldsymbol{\theta} + \boldsymbol{\theta}'\mathbf{a}'\mathbf{a}\boldsymbol{\Delta}_i$$

where again each term on the right-hand side is a scalar. We can rewrite this equation as

$$\frac{\partial S}{\partial \theta_i} = 2\boldsymbol{\Delta}'_i(\mathbf{a}'\mathbf{a}\boldsymbol{\theta} - \mathbf{a}'\mathbf{y})$$

and since the term in the brackets is independent of $i$, we see that the minimum of $S$ is attained at the value $\hat{\boldsymbol{\theta}}$ of $\boldsymbol{\theta}$ satisfying the equations

$$\mathbf{a}'\mathbf{a}\hat{\boldsymbol{\theta}} = \mathbf{a}'\mathbf{y} \qquad (5.9)$$

These equations are called the *normal equations of least squares*. If $(\mathbf{a}'\mathbf{a})$ is non-singular, the solution is

$$\hat{\boldsymbol{\theta}} = (\mathbf{a}'\mathbf{a})^{-1}\mathbf{a}'\mathbf{y} \qquad (5.10)$$

It will be noted that we have not yet considered the estimation of $\sigma^2$, and we shall do this later. By analogy with Chapter 2 we may expect this to depend on the minimized sum of squares. The minimized value of $S$ is readily obtained by substituting Equation (5.9) into Equation (5.8), and on doing this we have

$$S_{\min} = \mathbf{y}'\mathbf{y} - \hat{\boldsymbol{\theta}}'\mathbf{a}'\mathbf{y} \qquad (5.11)$$

*Example* 5.3    Carrying Example 5.2 further, we have

$$\mathbf{a}'\mathbf{a} = \begin{bmatrix} 32 & \sum x_{1i} & \sum x_{2i} & \cdots & \sum x_{4i} \\ \sum x_{1i} & \sum x_{1i}^2 & \sum x_{1i}x_{2i} & & \sum x_{1i}x_{4i} \\ \vdots & & & & \\ \sum x_{4i} & \sum x_{4i}x_{1i} & & & \sum x_{4i}^2 \end{bmatrix}$$

and

$$\mathbf{y}'\mathbf{a} = [\sum y_i, \quad \sum y_i x_{1i}, \quad \cdots, \quad \sum y_i x_{4i}]$$

Substitution of these into Equation (5.10) yields the least-squares estimates of $\beta_0, \ldots, \beta_4$.    □ □ □

---

## Exercises 5.1

1. Write out in full $\mathbf{y}$, $\mathbf{a}$ and $\boldsymbol{\theta}$ for fitting the model (2.33) to Example 2.3 data, using the actual results obtained, and then compute $\hat{\boldsymbol{\theta}}$ and $S_{\min}$.

2. Set out the algebra for fitting Example 5.1 data, using the model in the form (5.3) rather than the form (5.1). Verify that if only $\alpha$ and one of the $x$'s is included in the model, the simple linear-regression formulae result.

3. $Y_1, \ldots, Y_n$ are independently and normally distributed random variables following linear regression on $x_1, \ldots, x_n$ through the origin and with a variance that depends on $x$, i.e.,

$$E(Y_i) = \beta x_i \qquad V(Y_i) = \sigma^2 x_i^{2\alpha} \qquad x_i \geq 0$$

Show that in this case the maximum-likelihood estimator of $\beta$ is obtained by minimizing a weighted sum of squares. Also find the standard error of $\hat{\beta}$, and examine the special cases $\alpha = 0, \frac{1}{2}, 1$.

4. Show that with the assumptions set out in the section above, the statistic $\mathbf{a'y}$ is sufficient for $\theta$.

## 5.2  Properties of the method

One important property of the method of least squares is that it can be used in a much more general situation than that described in the previous section. Given a vector of observations $\mathbf{y}$ drawn from a population for which Equations (5.4) and (5.5) hold, but not assuming normality, we can still choose estimates of the unknown parameters $\theta_i$ to minimize the sum of squared differences of the observations from their expectations. That is, we minimize Equation (5.7) and use estimates (5.10). This method of estimation can be shown to have some very desirable properties even without assuming normality. In order to examine these properties we shall assume that all sums of squares, etc., are functions of random variables $\mathbf{Y}$ rather than observations $\mathbf{y}$. We shall assume that Equations (5.4) and (5.5) hold, and that $(\mathbf{a'a})$ is nonsingular. The assumption of normality will be brought in when required. Before stating the results in the form of a series of theorems, we must deal with a property of the covariance matrix.

*The covariance matrix*
In the subsequent theorems we shall be using the variance–covariance matrix, defined in Equation (3.20) as

$$V(\mathbf{Y}) = E(\mathbf{YY'}) - E(\mathbf{Y})E(\mathbf{Y'})$$

and we see that if $\mathbf{Y}$ has $n$ components, then this is an $n \times n$ matrix. If we consider a linear function of $\mathbf{Y}$, say

$$\mathbf{Z} = \mathbf{bY}$$

where $\mathbf{b}$ is an $m \times n$ matrix of constants, then clearly

$$E(\mathbf{Z}) = \mathbf{b}E(\mathbf{Y})$$

Also, by definition we have

$$\begin{aligned}
V(\mathbf{Z}) &= E(\mathbf{ZZ'}) - E(\mathbf{Z})E(\mathbf{Z'}) \\
&= E(\mathbf{bYY'b'}) - E(\mathbf{bY})E(\mathbf{Y'b'}) \\
&= \mathbf{b}E(\mathbf{YY'})\mathbf{b'} - \mathbf{b}E(\mathbf{Y})E(\mathbf{Y'})\mathbf{b'} \\
&= \mathbf{b}V(\mathbf{Y})\mathbf{b'}
\end{aligned} \tag{5.12}$$

The particular importance of this result here is that the least-squares estimator $\hat{\theta}$ given by Equation (5.10) is simply a linear function of $\mathbf{Y}$. (See Exercise 5.2.1. for a simple application of this result.)

*Main theorems*

*Theorem* 5.1    Least-squares estimators are unbiased, so that $E(\hat{\theta}) = \theta$, and have a variance–covariance matrix

$$V(\hat{\theta}) = (\mathbf{a}'\mathbf{a})^{-1}\sigma^2 \tag{5.13}$$

□ □ □

*Meaning.* It is clearly very useful that least-squares estimators of a linear model (5.4) are unbiased. Further, the same matrix that we have to invert to obtain the least-squares estimators (5.10) also gives us the variance–covariance matrix of the estimator $\hat{\theta}$.

*Proof.* The property of unbiasedness is easily proved:

$$\begin{aligned} E(\hat{\theta}) &= E\{(\mathbf{a}'\mathbf{a})^{-1}\mathbf{a}'\mathbf{Y}\} \\ &= (\mathbf{a}'\mathbf{a})^{-1}\mathbf{a}'E(\mathbf{Y}) = (\mathbf{a}'\mathbf{a})^{-1}\mathbf{a}'\mathbf{a}\theta \\ &= \theta \end{aligned}$$

On applying Equation (5.12) to our case we have, since $V(\mathbf{Y}) = \mathbf{I}\sigma^2$

$$\begin{aligned} V(\hat{\theta}) &= V[(\mathbf{a}'\mathbf{a})^{-1}\mathbf{a}'\mathbf{Y}] \\ &= (\mathbf{a}'\mathbf{a})^{-1}\mathbf{a}'\,V(\mathbf{Y})\{(\mathbf{a}'\mathbf{a})^{-1}\mathbf{a}'\}' \\ &= (\mathbf{a}'\mathbf{a})^{-1}\sigma^2 \end{aligned}$$

□ □ □

*Example* 5.4 (*Simple linear regression.*)    Suppose we wish to fit the model

$$E(Y) = \alpha + \beta(x - \bar{x})$$
$$V(Y) = \sigma^2$$

to data $(y_1, x_1), \dots, (y_n, x_n)$. Then we have

$$\mathbf{y} = \begin{bmatrix} y_1 \\ \vdots \\ y_n \end{bmatrix} \quad \theta = \begin{bmatrix} \alpha \\ \beta \end{bmatrix} \quad \mathbf{a} = \begin{bmatrix} 1 & (x_1 - \bar{x}) \\ \vdots & \\ 1 & (x_n - \bar{x}) \end{bmatrix}$$

and

$$\mathbf{a}'\mathbf{y} = \begin{bmatrix} \sum y_i \\ \sum y_i(x_i - \bar{x}) \end{bmatrix} = \begin{bmatrix} n\bar{y} \\ CS(x, y) \end{bmatrix}$$

$$\mathbf{a}'\mathbf{a} = \begin{bmatrix} n & 0 \\ 0 & CS(x, x) \end{bmatrix} \quad (\mathbf{a}'\mathbf{a})^{-1} = \begin{bmatrix} \dfrac{1}{n} & 0 \\ 0 & \dfrac{1}{CS(x, x)} \end{bmatrix}$$

The normal equations (5.9) are

$$\begin{bmatrix} n & 0 \\ 0 & CS(x, x) \end{bmatrix} \begin{bmatrix} \hat{\alpha} \\ \hat{\beta} \end{bmatrix} = \begin{bmatrix} n\bar{y} \\ CS(x, y) \end{bmatrix} \tag{5.14}$$

and the least-squares estimators (5.10) are

$$\begin{bmatrix} \hat{\alpha} \\ \hat{\beta} \end{bmatrix} = \begin{bmatrix} \dfrac{1}{n} & 0 \\ 0 & \dfrac{1}{CS(x, x)} \end{bmatrix} \begin{bmatrix} n\bar{y} \\ CS(x, y) \end{bmatrix} = \begin{bmatrix} \bar{y} \\ CS(x, y)/CS(x, x) \end{bmatrix}$$

Theorem 5.1 states that these estimators are unbiased, and

$$\begin{bmatrix} V(\hat{\alpha}) & C(\hat{\alpha}\,\hat{\beta}) \\ C(\hat{\alpha}\,\hat{\beta}) & V(\hat{\beta}) \end{bmatrix} = \begin{bmatrix} \dfrac{1}{n} & 0 \\ 0 & \dfrac{1}{CS(x, x)} \end{bmatrix} \sigma^2 \tag{5.15}$$

These were the results quoted in Chapter 2, and we see that $\hat{\alpha}$ and $\hat{\beta}$ are uncorrelated.  □ □ □

We have shown that when the data are normally distributed, the least-squares and maximum-likelihood methods of estimating $\theta$ are equivalent. However, the following theorem states an important property of least-squares estimators which holds even if the data are not normally distributed.

*Theorem* 5.2   The least-squares estimators of $\theta_j, j = 1, 2, \ldots, p$, are the best linear unbiased estimators in the sense that for any other linear unbiased estimators $\tilde{\theta}_j$,

$$V(\hat{\theta}_j) \le V(\tilde{\theta}_j) \qquad\qquad\qquad □ □ □$$

*Meaning.*   The theorem makes no statement about the variance of biased estimates or of non-linear functions of the observations, and no statement about covariances of estimators of the $\theta_j$. However, the result is surprisingly strong. There is no linear unbiased estimator with a variance smaller than the variance of the least-squares estimator. For example, in Example 5.4, the theorem shows that there is no linear unbiased estimator of $\alpha$ with a variance smaller than $\sigma^2/n$, and simultaneously there is no linear unbiased estimator of $\beta$ with a variance smaller than $\sigma^2/CS(x, x)$.

*Proof.*   Write $\mathbf{a}'\mathbf{a} = \mathbf{A}$, then we have for the least-squares estimator $\hat{\boldsymbol{\theta}}$,

$$V(\hat{\boldsymbol{\theta}}) = \mathbf{A}^{-1}\sigma^2$$

Now if we have any other unbiased estimator $\tilde{\theta}$, we have

$$E(\tilde{\theta}) = \theta \qquad (5.16)$$

and if this estimator is a linear function of the observations then

$$\tilde{\theta} = \mathbf{d}\mathbf{Y} \qquad (5.17)$$

where $\mathbf{d}$ is a matrix of constants.

By inserting Equation (5.17) into Equation (5.16), we obtain

$$E(\tilde{\theta}) = \mathbf{d}E(\mathbf{Y})$$

$$= \mathbf{d}\mathbf{a}\theta \qquad \text{since } E(\mathbf{Y}) = \mathbf{a}\theta$$

Hence from Equation (5.16), and since these equations hold for all $\theta$, we have

$$\mathbf{d}\mathbf{a} = \mathbf{I} \qquad (5.18)$$

Now the variance–covariance matrix of $\tilde{\theta}$ is

$$V(\tilde{\theta}) = V(\mathbf{d}\mathbf{Y}) = \mathbf{d}C(\mathbf{Y}, \mathbf{Y})\mathbf{d}' \qquad (5.19)$$

But the observations are independent, so that

$$C(\mathbf{Y}, \mathbf{Y}) = \mathbf{I}\sigma^2$$

Hence from Equation (5.19),

$$V(\tilde{\theta}) = \mathbf{d}\mathbf{d}'\sigma^2 \qquad (5.20)$$

We now seek a relationship between Equation (5.20) and the variance of the least-squares estimator. We have

$$\mathbf{d}\mathbf{d}' = \mathbf{d}\mathbf{d}' - 2A^{-1} + 2A^{-1}$$

$$= \mathbf{d}\mathbf{d}' - \mathbf{d}\mathbf{a}A^{-1} - A^{-1}\mathbf{a}'\mathbf{d}' + 2A^{-1}$$

since $\mathbf{d}\mathbf{a} = \mathbf{I}$. Further, we have

$$\mathbf{d}\mathbf{d}' = \mathbf{d}\mathbf{d}' - \mathbf{d}\mathbf{a}A^{-1} - A^{-1}\mathbf{a}'\mathbf{d}' + A^{-1}\mathbf{a}'\mathbf{a}A^{-1} + A^{-1}$$

$$= (\mathbf{d} - A^{-1}\mathbf{a}')(\mathbf{d} - A^{-1}\mathbf{a}')' + A^{-1}$$

$$(5.21)$$

If we multiply by $\sigma^2$ and use Equations (5.20) and (5.13), we have

$$V(\tilde{\theta}) = V(\hat{\theta}) + (\mathbf{d} - A^{-1}\mathbf{a}')(\mathbf{d} - A^{-1}\mathbf{a}')'\sigma^2$$

and the diagonal elements at least of the last term are positive. Therefore the diagonal elements of $V(\tilde{\theta})$ are the diagonal elements of $V(\hat{\theta})$ plus some positive terms.                                    □ □ □

There is a simple extension of this theorem, that the best linear unbiased estimate of any linear combination of $\theta_j$'s is the linear combination of the least-squares estimates.

---

**Exercises 5.2**

1. Suppose that random variables $Y_1, \ldots, Y_5$ are such that $E(Y_i) = \mu$, $V(Y_i) = \sigma^2$, and that they are uncorrelated. Apply the result (5.12) to show that if we define the variables $Z_i$ as follows,

$$
\begin{aligned}
Z_1 &= Y_1 - Y_2 \\
Z_2 &= Y_1 + Y_2 - 2Y_3 \\
Z_3 &= Y_1 + Y_2 + Y_3 - 3Y_4 \\
Z_4 &= Y_1 + Y_2 + Y_3 + Y_4 - 4Y_5 \\
Z_5 &= Y_1 + Y_2 + Y_3 + Y_4 + Y_5
\end{aligned}
$$

then

$$
V(\mathbf{Z}) =
\begin{bmatrix}
2 & & & & 0 \\
 & 6 & & & \\
 & & 12 & & \\
 & & & 20 & \\
0 & & & & 5
\end{bmatrix}
\sigma^2
$$

2. Three independent sets of observations are taken: $x_1, \ldots, x_n$, with expectation $\theta_1$; $y_1, \ldots, y_n$, with expectation $\theta_2$; and $z_1, \ldots, z_m$, with expectation $(\theta_1 - \theta_2)$. The observations are independent and have a common variance $\sigma^2$. Obtain the least-squares estimates $\hat{\theta}_1$ and $\hat{\theta}_2$ of $\theta_1$ and $\theta_2$, and show that

$$
V(\hat{\theta}_1 - \hat{\theta}_2) = 2\sigma^2/(n + 2m)
$$

[*Hint*: Write out the matrix **a** for this problem.]

3. Continue the calculation of Exercise 5.1.1 by writing down the variance–covariance matrix of $\hat{\alpha}$, $\hat{\beta}_1$, and $\hat{\beta}_2$ in terms of the unknown variance $\sigma^2$.

4. Show that the result of Theorem 5.2 can be extended to show that for any linear combination $\mathbf{d}'\hat{\theta}$, of the $\theta_i$'s,

$$
V(\mathbf{d}'\hat{\theta}) \leq V(\mathbf{d}'\tilde{\theta})
$$

---

## 5.3 Properties of residuals

It has been made clear that an analysis of residuals is an important part of any statistical analysis using the linear model, and in this section we have three important theorems concerning properties of residuals. The vector of residuals is defined by

$$\mathbf{r} = \mathbf{y} - \mathbf{a}\hat{\theta}$$

and as an example we could take the final column of Table 2.5 to be such a vector. Now as at the end of Section 2.3, we must define random variables $R_i$ by a similar equation, but in which all observations $\mathbf{y}$ are replaced by random variables $\mathbf{Y}$.

*Theroem* 5.3    For the vector of residuals, $\mathbf{R}$, we have

$$E(\mathbf{R}) = \mathbf{0} \tag{5.22}$$

$$V(\mathbf{R}) = \mathbf{m}\sigma^2 \tag{5.23}$$

where

$$\mathbf{m} = \mathbf{I} - \mathbf{a}(\mathbf{a}'\mathbf{a})^{-1}\mathbf{a}' \tag{5.24}$$

provided the model holds.                                    □ □ □

*Meaning.*    The variance–covariance matrix of the residuals is proportional to $\mathbf{m}$. In general this will show that the residuals are correlated, and have unequal variances. Therefore a study of the matrix $\mathbf{m}$ will show when it is safe to examine the raw residuals for evidence of departure from the model.

*Proof*    The residuals are defined by

$$\mathbf{R} = \mathbf{Y} - \mathbf{a}\hat{\theta}$$
$$= \{\mathbf{I} - \mathbf{a}(\mathbf{a}'\mathbf{a})^{-1}\mathbf{a}'\}\mathbf{Y}$$
$$= \mathbf{m}\mathbf{Y}$$

Now we have

$$E(\mathbf{R}) = \mathbf{a}\theta - \mathbf{a}(\mathbf{a}'\mathbf{a})^{-1}\mathbf{a}'\mathbf{a}\theta, \qquad \text{since } E(\mathbf{Y}) = \mathbf{a}\theta$$

and hence we see

$$E(\mathbf{R}) = \mathbf{0}$$

which establishes Equation (5.22).
Now we have

$$V(\mathbf{R}) = V(\mathbf{m}\mathbf{Y})$$
$$= \mathbf{m}V(\mathbf{Y})\mathbf{m}'$$

by Equation (5.12); hence using Equation (5.5),

$$V(\mathbf{R}) = \mathbf{m}\mathbf{I}\mathbf{m}'\sigma^2$$

But we have that

$$\mathbf{mm'} = \{\mathbf{I} - \mathbf{a}(\mathbf{a'a})^{-1}\mathbf{a'}\}\{\mathbf{I} - \mathbf{a}(\mathbf{a'a})^{-1}\mathbf{a'}\}'$$
$$= \mathbf{I} - \mathbf{a}(\mathbf{a'a})^{-1}\mathbf{a'} - \mathbf{a}(\mathbf{a'a})^{-1}\mathbf{a'} + \mathbf{a}(\mathbf{a'a})^{-1}\mathbf{a'a}(\mathbf{a'a})^{-1}\mathbf{a'}$$
$$= \mathbf{I} - 2\mathbf{a}(\mathbf{a'a})^{-1}\mathbf{a'} + \mathbf{a}(\mathbf{a'a})^{-1}\mathbf{a'} \tag{5.25}$$
$$= \mathbf{I} - \mathbf{a}(\mathbf{a'a})^{-1}\mathbf{a'} = \mathbf{m}$$

so that

$$V(\mathbf{R}) = \mathbf{m}\sigma^2 \qquad\qquad \square\ \square\ \square$$

*Example* 5.5   We continue here the simple linear-regression example 5.4.

$$\mathbf{a}(\mathbf{a'a})^{-1}\mathbf{a'} = \begin{bmatrix} 1 & (x_1 - \bar{x}) \\ \vdots & \\ 1 & (x_n - \bar{x}) \end{bmatrix} \begin{bmatrix} \dfrac{1}{n} & 0 \\ 0 & \dfrac{1}{CS(x,x)} \end{bmatrix} \begin{bmatrix} 1 & \cdots & 1 \\ (x_1 - \bar{x}) & & (x_n - \bar{x}) \end{bmatrix}$$

On multiplying out, this leads to a matrix for which the $(i, j)$th term is

$$\frac{1}{n} + \frac{(x_i - \bar{x})(x_j - \bar{x})}{CS(x, x)}$$

Therefore from Equation (5.24), the diagonal terms of $\mathbf{m}$ are

$$1 - \frac{1}{n} - \frac{(x_i - \bar{x})^2}{CS(x, x)} \tag{5.26}$$

and the off-diagonal terms are

$$-\frac{1}{n} - \frac{(x_i - \bar{x})(x_j - \bar{x})}{CS(x, x)} \tag{5.27}$$

Theorem 5.3 states that $V(R_i)$ is Equation (5.26) multiplied by $\sigma^2$, and $C(R_i, R_j)$ is Equation (5.27) multiplied by $\sigma^2$. If the number of terms $n$ is large enough, and the extreme $x_i$ such that

$$\max_i \left\{ \frac{(x_i - \bar{x})^2}{CS(x, x)} \right\}$$

is reasonably small, then the $R_i$ can be treated as being approximately uncorrelated and of nearly homogeneous variance.

It is interesting at this point to return to Example 2.1, which was discussed in Section 2.4. Table 5.2 below gives the standard errors of the residuals tabulated in Table 2.5. The standard errors are obtained using the formula (5.26) above, and multiplying by the estimate of $\sigma^2$(0.000 295), and then taking the square root. The standardized residuals are simply the residuals divided

Table 5.2 *Residuals, their standard errors and standardized residuals for Example 2.1 data*

| $x_i$ | Residual | Standard error | Standardized residual |
|---|---|---|---|
| 8 | + 0.0101 | 0.015 28 | 0.66 |
| 12 | + 0.0260 | 0.015 58 | 1.67 |
| 19 | + 0.0136 | 0.015 98 | 0.85 |
| 20 | − 0.0036 | 0.016 03 | − 0.22 |
| 25 | − 0.0169 | 0.016 22 | − 1.04 |
| 32 | − 0.0166 | 0.016 36 | − 1.01 |
| 36 | − 0.0153 | 0.016 38 | − 0.93 |
| 42 | + 0.0126 | 0.016 32 | 0.77 |
| 48 | − 0.0107 | 0.016 17 | − 0.66 |
| 52 | − 0.0170 | 0.016 01 | − 1.06 |
| 100 | + 0.0180 | 0.009 13 | 1.97 |

by their standard errors. We see here that the calculation of standardized residuals is of some importance. The observation for $x = 100$ yields a standardized residual which is just about large enough to be suspect, and it emphasizes the points made in Section 2.4 about the inadequacy of the model. ☐ ☐ ☐

In the next theorem we are concerned with the covariances between the components of $\mathbf{R}$ and $\hat{\boldsymbol{\theta}}$. We could do this by working out the variance covariance matrix of a vector

$$\mathbf{Z}' = (\mathbf{R}', \hat{\boldsymbol{\theta}}')$$

However, it is more convenient to deal only with the section of $V(\mathbf{Z})$ that we want. We therefore introduce the notation

$$C(\mathbf{Y}, \mathbf{Z}) = E(\mathbf{YZ}') - E(\mathbf{Y})E(\mathbf{Z}')$$

*Theorem 5.4*   The vector of residuals is uncorrelated with the vector of estimators,

$$C(\mathbf{R}, \hat{\boldsymbol{\theta}}) = \mathbf{0}$$     ☐ ☐ ☐

*Meaning.*   This is obviously an important result in connection with the use of residuals to study departures from the model. It is also an important result regarding the development of significance tests (see the next section).

*Proof.*

$$C(\mathbf{R}, \hat{\boldsymbol{\theta}}) = C\{\mathbf{m\,Y}, (\mathbf{a}'\mathbf{a})^{-1}\mathbf{a}'\mathbf{Y}\}$$
$$= \mathbf{m}\,C(\mathbf{Y}, \mathbf{Y})\mathbf{a}(\mathbf{a}'\mathbf{a})^{-1}$$

but we have

$$C(\mathbf{Y}, \mathbf{Y}) = V(\mathbf{Y}) = \mathbf{I}\,\sigma^2$$

and hence

$$C(\mathbf{R}, \hat{\boldsymbol{\theta}}) = \mathbf{m} \, \mathbf{I} \, \mathbf{a}(\mathbf{a}'\mathbf{a})^{-1}\sigma^2$$
$$= \{\mathbf{I} - \mathbf{a}(\mathbf{a}'\mathbf{a})^{-1}\mathbf{a}'\} \, \mathbf{a}(\mathbf{a}'\mathbf{a})^{-1}\sigma^2$$
$$= \mathbf{0}$$

□ □ □

Before proceeding to the next theorem we need to establish two results. First we note that the trace of a square matrix $\mathbf{A}$ is defined as the sum of the diagonal elements. Then we note that for any two matrices $\mathbf{c}$ and $\mathbf{d}$ which are such that $\mathbf{c}\,\mathbf{d}$ and $\mathbf{d}\,\mathbf{c}$ both exist and are square, that we have

$$\text{tr}(\mathbf{c}\,\mathbf{d}) = \text{tr}(\mathbf{d}\,\mathbf{c}) \tag{5.28}$$

This is readily established by writing out what each side of this equation means.

The second result we shall need is to evaluate the trace of $\mathbf{m}$, the matrix defined in Equation (5.24). Now we have

$$\text{tr}(\mathbf{m}) = \text{tr}\{\mathbf{I} - \mathbf{a}(\mathbf{a}'\mathbf{a})^{-1}\mathbf{a}'\}$$
$$= \text{tr}\{\mathbf{I}\} - \text{tr}\{\mathbf{a}(\mathbf{a}'\mathbf{a})^{-1}\mathbf{a}'\}$$
$$= \text{tr}\{\mathbf{I}\} - \text{tr}\{(\mathbf{a}'\mathbf{a})^{-1}\mathbf{a}'\mathbf{a}\} \tag{5.29}$$

by applying Equation (5.28).

But on the right-hand side of Equation (5.29) we now have an $n \times n$ unit matrix, and a $p \times p$ unit matrix. Therefore we obtain

$$\text{tr}(\mathbf{m}) = (n - p) \tag{5.30}$$

The next theorem concerns the expectation of the minimum value of the sum of squares, defined in Equation (5.11).

*Theorem 5.5*

$$E(S_{\text{min}}) = (n - p)\sigma^2 \tag{5.31}$$

□ □ □

*Meaning.* In the theorems above we have obtained the variance–covariance matrix of the least-squares estimates, and the variance–covariance matrix of the residuals, but both of these involve the parameter $\sigma^2$. This theorem shows that an unbiased estimate of $\sigma^2$ is

$$\hat{\sigma}^2 = S_{\text{min}}/(n - p) \tag{5.32}$$

In this situation the quantity $(n - p)$ is in fact

(no. of observations) − (no. of parameters fitted)

and it is usually known as the *degrees of freedom* relating to $S_{\text{min}}$.
(In this chapter we consider only models for which the matrix $\mathbf{a}$ is of full

rank, but later we shall drop this restriction. It then turns out that the theorem holds with $p = \text{rank} \ (\mathbf{a})$.)

*Proof.*  By definition, the minimized sum of squares is the sum of squared residuals,

$$
\begin{aligned}
E(S_{\min}) &= E(\mathbf{R}'\mathbf{R}) \\
&= \text{tr} \ E(\mathbf{R}'\mathbf{R}) \quad && \text{(since } \mathbf{R}'\mathbf{R} \text{ is a scalar)} \\
&= \text{tr} \ E(\mathbf{R}\mathbf{R}') \quad && \text{(by Equation (5.28))} \\
&= \text{tr} \ V(\mathbf{R}) \quad && \text{(since } E(\mathbf{R}) = 0) \\
&= \text{tr} \ (\mathbf{m})\sigma^2 \\
&= (n - p)\sigma^2 \quad && \text{(by Equation (5.30))} \qquad \square\ \square\ \square
\end{aligned}
$$

*Example* 5.6 (*Simple linear regression.*)  Carrying on Example 5.4, we see that

$$
\begin{aligned}
S_{\min} &= \mathbf{Y}'\mathbf{Y} - \hat{\boldsymbol{\theta}}'\mathbf{a}'\mathbf{Y} \\
&= (y_1, \ldots, y_n)\begin{pmatrix} y_1 \\ \vdots \\ y_n \end{pmatrix} - (\hat{\alpha}\hat{\beta})\begin{pmatrix} n\bar{y} \\ CS(x, y) \end{pmatrix} \\
&= \sum y_i^2 - n\hat{\alpha}\bar{y} - \hat{\beta}CS(x, y) \\
&= \sum y_i^2 - n\bar{y}^2 - \{CS(x, y)\}^2/CS(x, x)
\end{aligned}
$$
(5.33)

In this example $p = 2$, so that Theorem 5.5 shows that the expected value of this quantity is $(n - 2)\sigma^2$. This result was assumed in Chapter 2. $\quad \square\ \square\ \square$

---

### Exercise 5.3
1. Find the variance–covariance matrix of the residuals for the regression model discussed in Section 2.6 [Exercises 5.1.1 and 5.2.3 refer.]

---

### 5.4 Properties of sums of squares
Sums of squares play a very important role in the analysis of the linear model by the method of least squares. We have just seen how sums of squares are used to estimate $\sigma^2$. Some other properties of sums of squares are given in this section.

Our next theorem concerns the sum of squares of the fitted values of the parameters which we denote $S_{\text{par}}$. The fitted values are

$$
\hat{\mathbf{y}} = \mathbf{a}\hat{\boldsymbol{\theta}} = \mathbf{a}(\mathbf{a}'\mathbf{a})^{-1}\mathbf{a}'\mathbf{y}
$$
(5.34)

so that the sum of squares of the fitted values is

$$
\begin{aligned}
S_{\text{par}} &= \hat{\mathbf{y}}'\hat{\mathbf{y}} = \hat{\boldsymbol{\theta}}'\mathbf{a}'\mathbf{a}\hat{\boldsymbol{\theta}} = \hat{\boldsymbol{\theta}}(\mathbf{a}'\mathbf{a})(\mathbf{a}'\mathbf{a})^{-1}\mathbf{a}'\mathbf{y} \\
&= \hat{\boldsymbol{\theta}}'\mathbf{a}'\mathbf{y}
\end{aligned}
$$
(5.35)

By comparing this result with Equation (5.11), we have

$$(\text{SS of } y) = (\text{SS of deviations}) + (\text{SS of fitted values})$$

$$= S_{min} + S_{par} \tag{5.36}$$

*Theorem 5.6*

$$E(S_{par}) = p\sigma^2 + \boldsymbol{\theta}'\mathbf{a}'\mathbf{a}\boldsymbol{\theta} \qquad \square\ \square\ \square$$

*Meaning.* The expected value of the sum of squares of the fitted constants is $p\sigma^2$ plus a quadratic function of the $\theta_i$'s. Therefore by comparing $S_{par}/p$ with $S_{min}/(n-p)$, we have some indication of whether there are $\theta_i$'s which differ from zero.

*Proof.*

$$E(S_{par}) = E(\hat{\boldsymbol{\theta}}'\mathbf{a}'\mathbf{a}\hat{\boldsymbol{\theta}}) = \text{tr } E(\hat{\boldsymbol{\theta}}'\mathbf{a}'\mathbf{a}\hat{\boldsymbol{\theta}})$$

$$= \text{tr } E(\mathbf{a}\hat{\boldsymbol{\theta}}\hat{\boldsymbol{\theta}}'\mathbf{a}') \quad \text{(by using Equation (5.28))}$$

$$= \text{tr } \mathbf{a}\{V(\hat{\boldsymbol{\theta}}) + E(\hat{\boldsymbol{\theta}})E(\hat{\boldsymbol{\theta}}')\}\mathbf{a}'$$

Hence we have

$$E(S_{par}) = \text{tr } \{\sigma^2 \mathbf{a}(\mathbf{a}'\mathbf{a})^{-1}\mathbf{a}' + \mathbf{a}(\boldsymbol{\theta}\boldsymbol{\theta}')\mathbf{a}'\}$$

$$= \text{tr } \mathbf{a}'\mathbf{a}(\mathbf{a}'\mathbf{a})^{-1}\sigma^2 + \text{tr } \boldsymbol{\theta}'\mathbf{a}'\mathbf{a}\boldsymbol{\theta} \quad \begin{array}{l}\text{(again using}\\ \text{Equation (5.28))}\end{array}$$

$$= p\sigma^2 + \boldsymbol{\theta}'\mathbf{a}'\mathbf{a}\boldsymbol{\theta} \tag{5.37}$$

since the last term in Equation (5.37) is a scalar.  $\square\ \square\ \square$

This result enables us to establish a table in which the total variation in the data is split up into meaningful components. We notice that if we calculate $S_{min}/(n-p)$, we have a quantity with an expected value of $\sigma^2$, and if we calculate $S_{par}/p$, we have a quantity with an expected value

$$\sigma^2 + (\text{quadratic in } \boldsymbol{\theta})$$

These results on sums of squares are stated concisely in Table 5.3. (An

Table 5.3 *Analysis of variance*

| Meaning | Sum of squares | d.f. | Mean square | E(Mean square) |
|---|---|---|---|---|
| Due to fitted model | $S_{par}$ | $p$ | $S_{par}/p$ | $\sigma^2 + \dfrac{1}{p}\boldsymbol{\theta}'\mathbf{a}'\mathbf{a}\boldsymbol{\theta}$ |
| Deviations from model | $S_{min}$ | $(n-p)$ | $S_{min}/(n-p)$ | $\sigma^2$ |
| | $Y'Y$ | $n$ | | |

interpretation of 'degrees of freedom' in terms of distribution theory is given below.)

It is now obvious that a measure of the evidence in the data that $\theta \neq 0$ is given by the ratio

$$\frac{S_{par}/p}{S_{min}/(n-p)} \tag{5.38}$$

We shall develop this technique below.

*Example* 5.7 (*Simple linear regression.*) Again carrying on with Example 5.4, we have from Equations (5.34) and (5.36),

$$S_{par} = n\bar{y}^2 + \{CS(x,y)\}^2/(CS(x,x))$$

Now

$$\theta'\mathbf{a}'\mathbf{a}\theta = (\alpha \ \beta)\begin{pmatrix} n & 0 \\ 0 & CS(x,x) \end{pmatrix}\begin{pmatrix} \alpha \\ \beta \end{pmatrix}$$

$$= n\alpha^2 + \beta^2 CS(x,x)$$

Therefore from Theorem 5.6,

$$E(S_{par}) = 2\sigma^2 + n\alpha^2 + \beta^2 CS(x,x) \qquad \square\,\square\,\square$$

Up to this point there has been no assumption of normality. However, we obviously look for a formal test of significance of $\theta = 0$, by comparing $S_{par}/p$ with $S_{min}/(n-p)$. Further, Theorem 5.5 provides an estimate of $\sigma^2$, but we may also require confidence intervals. We therefore add to the specification of our model (5.4) and (5.5) the following assumption:

> The observations $y_i$ are independently and normally
> distributed. (5.39)

It now follows that the estimates $\hat{\theta}$ given by Equation (5.10), being linear functions of the observations, are also normally distributed. We shall next need to know the distributions of $S_{min}$ and $S_{par}$, but instead of deriving these we shall merely quote the most important results.

*Theorem* 5.7   The distribution of $S_{min}/\sigma^2$ is a $\chi^2$ distribution on $(n-p)$ degrees of freedom. $\qquad \square\,\square\,\square$

*Meaning.*   Given our model (5.4) and (5.5) with assumption (5.39), then $S_{min}$ satisfies

$$\Pr\left\{\chi_v^2\left(\frac{\alpha}{2}\right) < S_{min}/\sigma^2 < \chi_v^2\left(1 - \frac{\alpha}{2}\right)\right\} = 1 - \alpha$$

where $v = n - p$, and $\chi_v^2(P)$ is the appropriate percentage point of the $\chi^2$-

distribution on $v$ degrees of freedom. Hence $100(1 - \alpha)$ per cent confidence intervals for $\sigma^2$ are

$$\left[ S_{\min}/\chi_v^2\left(1 - \frac{\alpha}{2}\right), S_{\min}/\chi_v^2\left(\frac{\alpha}{2}\right) \right]$$

This is simply the method used in Section 1.2(a).

We shall also find Theorem 5.7 useful in certain tests of significance. We shall not prove the theorem here, and those interested in a proof should consult Rao (1965, paragraph 3b.5), Wilks (1962), or Kendall and Stuart (1973). The same references will provide proofs of the next theorem.

*Theorem 5.8* The distribution of $S_{\text{par}}/\sigma^2$ is a $\chi^2$-distribution on $p$ degrees of freedom if $\theta = 0$. □ □ □

As stated, this theorem is not very useful, as we are nearly always sure that $\theta \neq 0$; for instance, at least for regression models the constant term $\alpha \neq 0$. However, the sum of squares due to the fitted model can often be separated into independent components, and we apply Theorem 5.8 to each component. A simple example will make the point clear. More complex examples will be given later.

*Example 5.8* (*Simple linear regression.*) From Equation (5.15), $C(\hat{\alpha}, \hat{\beta}) = 0$, so that if the observations are normally distributed $\hat{\alpha}$ and $\hat{\beta}$ are independent. Then

$$S_{\text{par}} = \hat{\theta}'a'a\hat{\theta}$$
$$= (\hat{\alpha}\ \hat{\beta})\begin{pmatrix} n & 0 \\ 0 & CS(x, x) \end{pmatrix} \begin{pmatrix} \hat{\alpha} \\ \hat{\beta} \end{pmatrix}$$
$$= n\hat{\alpha}^2 + \hat{\beta}^2 CS(x, x) \tag{5.40}$$

Therefore the two terms on the right-hand side of Equation (5.40) are independent. It also follows that if we fit the model in Example 5.4 including one or other or both of the parameters $\alpha$ and $\beta$, then the corresponding value of $S_{\text{par}}$ simply includes one or other or both of the terms on the right of Equation (5.40), correspondingly. Further, by using Equation (5.15) it is easy to verify that

$$E(n\hat{\alpha}^2) = n\{V(\hat{\alpha}) + E^2(\hat{\alpha})\} = \sigma^2 + n\alpha^2$$

and similarly,

$$E(\hat{\beta}^2 CS(x, x)) = \sigma^2 + \beta^2 CS(x, x) \tag{5.41}$$

Theorem 5.8 now applies to each component of $S_{\text{par}}$, and $n\hat{\alpha}^2/\sigma^2$ has a $\chi_1^2$-distribution if $\alpha = 0$, and $\hat{\beta}^2 CS(x, x)/\sigma^2$ has a $\chi_1^2$-distribution if $\beta = 0$.

Table 5.4 *Analysis of variance for simple linear regression*

| Source | Sum of squares | d.f. | E(Mean square) | |
|---|---|---|---|---|
| Due to constant term | $n\hat{\alpha}^2$ | 1 | $\sigma^2 + n\alpha^2$ | (A) |
| Due to regression | $\hat{\beta}^2 CS(x, x)$ | 1 | $\sigma^2 + \beta^2 CS(x, x)$ | (B) |
| Deviations | $S_{min}$ | $(n-2)$ | $\sigma^2$ | (C) |
| Total | $\sum y_i^2$ | $n$ | | |

The analysis-of-variance table can therefore be rewritten as shown in Table 5.4                                                                    □ □ □

We can now write down certain significance tests which are used in regression analysis. Theorem 5.4 showed that the residuals and estimators are uncorrelated; together with the normality assumption (5.39), this implies that, for example, in Table 5.4 the mean square (C) is independent of the mean squares (A) or (B). Therefore if $\beta = 0$, the ratio of mean square (B) to mean square (C) is the ratio of two quantities which have independent $\chi^2$-distributions, divided by their degrees of freedom; this has an $F$-distribution by Definition 7 of Appendix A. Clearly, if $\beta \neq 0$ the ratio will tend to be large. We therefore have the following tests for Example 5.4:

(i) Hypothesis $\alpha = 0$

        Calculate      {Mean square (A)}/{Mean Square (C)}

(ii) Hypothesis $\beta = 0$

        Calculate      {Mean square (B)}/{Mean square (C)}.

Both of these ratios are referred to the $F$-distribution for $(1, n-2)$ degrees of freedom. One-sided tests are required, large values being significant.

The test of $\alpha = 0$ is usually unnecessary, as we are often sure it is false, and it is frequently of very small importance. The sum of squares due to the constant $\alpha$ is usually subtracted from the total, to make the corrected sum of squares for $y$. We note that $\hat{\alpha} = \bar{y}$, and

$$\sum y_i^2 - n\hat{\alpha}^2 = \sum y_i^2 - n\bar{y}^2 = CS(y, y)$$

If we insert the value of $\hat{\beta}$, Table 5.4 can be rewritten as in Table 5.5. The heading 'CSS' of the second column makes it clear that all entries are now

Table 5.5 *Analysis of variance for simple regression*

| Source | CSS | d.f. | E(Mean square) |
|---|---|---|---|
| Due to regression | $\{CS(x, y)\}^2 / CS(x, x)$ | 1 | $\sigma^2 + \beta^2 CS(x, x)$ |
| Deviations | $S_{min}$ | $(n-2)$ | $\sigma^2$ |
| Total | $CS(y, y)$ | $(n-1)$ | |

corrected sums of squares. This is the form in which the table was used in Chapter 2.

It should be noted that single degrees of freedom can only be separated from the sum of squares due to fitting the model when the appropriate row and column of $(\mathbf{a'a})$ contains zero elements for all except the diagonal element. When the regression model is written in the form (5.3), the constant $\alpha$ can always be separated in this way.

A test of the hypothesis that $\boldsymbol{\theta} = \mathbf{0}$ in the more general regression model now follows. We have already argued that the ratio (5.38) shows up discrepancies from this hypothesis. If normality holds, the denominator of (5.38) has a distribution proportional to $\chi^2$ on $(n-p)$ degrees of freedom and the numerator has independently a distribution proportional to $\chi^2$ on $p$ degrees of freedom if $\boldsymbol{\theta} = \mathbf{0}$. Therefore the ratio (5.38) has an $F(p,(n-p))$ distribution if $\boldsymbol{\theta} = \mathbf{0}$.

*Summary*
Least squares estimator of a linear model:

(a) are unbiased;
(b) have minimum variance among linear unbiased estimators;
(c) are such that the residuals and estimators are uncorrelated;
(d) provide an unbiased estimator of the error variance from the minimized sum of squares.

A significance test of the hypothesis that the parameters $\theta_i$ are all zero is obtained by a comparison of the sum of squares of the fitted values with the minimized sum of squares, and by adding the assumption that the observations are normally distributed. The other properties are independent of normality.

---

### Exercises 5.4

1. Find the expected value of the sum of squares due to fitting the model for the regression model discussed in Section 2.6. [Exercises 5.2.3 and 5.3.1 refer.] Does the sum of squares due to $\alpha$ separate in the way illustrated in Example 5.8? Why are the sums of squares due to fitting $x$ and $w$ not independent?

2. Use Equation (A.6) of Appendix A and Theorem 5.7 to show that if we use the estimator

$$\hat{\sigma}^2 = S_{\min}/(n-p)$$

then

$$V(\hat{\sigma}^2) = 2\sigma^4/(n-p)$$

## 5.5 Application to multiple regression

We shall now apply the theory of the previous sections to Example 5.1 using a model described by Equation (5.3) and (5.5). The various vectors and matrices are:

$$\theta = \begin{bmatrix} \alpha \\ \beta_1 \\ \vdots \\ \beta_4 \end{bmatrix} \qquad a = \begin{bmatrix} 1 & x_{1,1} - \bar{x}_{1.} & \cdots & x_{4,1} - \bar{x}_{4.} \\ 1 & x_{1,2} - \bar{x}_{1.} & & x_{4,2} - \bar{x}_{4.} \\ \vdots & & & \\ 1 & x_{1,32} - \bar{x}_{1.} & & x_{4,32} - \bar{x}_{4.} \end{bmatrix}$$

$$a'y = \begin{bmatrix} \sum y_i \\ \sum y_i(x_{1,i} - \bar{x}_{1.}) \\ \vdots \\ \sum y_i(x_{4,i} - \bar{x}_{4.}) \end{bmatrix} = \begin{bmatrix} 32\bar{y} \\ CS(y, x_1) \\ \vdots \\ CS(y, x_4) \end{bmatrix}$$

$$a'a = \begin{bmatrix} 32 & 0 & \cdots & 0 \\ 0 & CS(x_1, x_1) & \cdots & CS(x_4, x_1) \\ 0 & CS(x_1, x_2) & & CS(x_4, x_2) \\ 0 & CS(x_1, x_3) & & CS(x_4, x_3) \\ 0 & CS(x_1, x_4) & & CS(x_4, x_4) \end{bmatrix}$$

where

$$CS(y, x_j) = \sum_{i=1}^{32} (y_i - \bar{y})(x_{j,i} - \bar{x}_{j.})$$

$$\sum_{i=1}^{32} y_i(x_{j,i} - \bar{x}_{j.})$$

and there are similar expressions for the corrected sums of squares and products of the $x$'s.

The inverse of $(a'a)$ will have zeros in the same places as $(a'a)$ itself, so that in fitting the model (5.3), $\hat{\alpha}$ is independent of any of the $\hat{\beta}_j$, but the $\hat{\beta}_j$ are in general correlated. The sum of squares due to regresssion (5.35) therefore comes to

$$S_{par} = 32\bar{y}^2 + \sum_{i=1}^{4} \hat{\beta}_i CS(y, x_i) \qquad (5.42)$$

and the two terms on the right-hand side are statistically independent. As the $\beta_i$ are in general correlated, the second term on the right cannot be split up any further, and the analysis-of-variance table can be set out as in Table 5.6, where $S_{min}$ is calculated by subtraction of all other contributions from the total.

Table 5.6 *Analysis of variance for Example 5.1*

| Source | Sum of squares | d.f. | Mean square |
|---|---|---|---|
| Due to fitting $\alpha$ | $32\bar{y}^2$ | 1 | $32\bar{y}^2$ |
| Due to regression on $x_i$ | $\sum_1^4 \hat{\beta}_i\, CS(y, x_i)$ | 4 | $\sum_1^4 \hat{\beta}_i\, CS(y, x_i)/4$ |
| Deviations | $S_{min}$ | 27 | $S_{min}/27$ |
| Total | $\sum y_i^2$ | 32 | |

Again we do not usually include the term due to $\alpha$, and the more common form of the table is as in Table 5.7.

Theorem 5.5 shows that $E(S_{min}/27) = \sigma^2$, and Theorem 5.6 shows that the expectation of the mean square due to regression on the $x_i$ is

$$E\left\{\sum_1^4 \hat{\beta}_i\, CS(y, x_i)\right\}\bigg/4 = \sigma^2 + \frac{1}{4}\sum_1^4 \beta_i^2\, CS(x_i, x_i) + \frac{1}{2}\sum_{i<j}\sum \beta_i\beta_j\, CS(x_i, x_j)$$

For example 5.1 data we obtain the results set out in Table 5.8.

This analysis shows that the fitted regression line accounts for the great majority of the variation in the yield. The $F$-ratio for testing the hypothesis $\beta_1 = \ldots = \beta_4 = 0$ is

$$\frac{857.32}{4.99} = 171.8$$

on $(4, 27)$ degrees of freedom, and is very highly significant. There is very little interest in this particular $F$-test. It is of greater interest to see if one of

Table 5.7 *Analysis of variance for Example 5.1*

| Source | CSS | d.f. | Mean square |
|---|---|---|---|
| Due to regression on $x_i$ | $\sum_1^4 \hat{\beta}_j\, CS(y, x_i)$ | 4 | $\sum_1^4 \hat{\beta}_i\, CS(y, x_i)/4$ |
| Deviations | $S_{min}$ | 27 | $S_{min}/27$ |
| Total | $CS(y, y)$ | 31 | |

Table 5.8 *Analysis of variance for Example 5.1*

| Source | CSS | d.f. | Mean squares |
|---|---|---|---|
| Due to regression | 3429.27 | 4 | 857.32 |
| Deviations | 134.80 | 27 | 4.99 |
| Total | 3564.07 | 31 | |

the variables $(x_1, \ldots, x_4)$ could be dropped without reducing the regression sum of squares appreciably, and we shall deal with this aspect in the next chapter. A more important result from Table 5.8 is that the mean square for deviations gives an estimate of $\sigma^2$ of 4.99.

Some of the other calculations involved in the analysis are as follows. The means of the variables are:

$$\bar{y} = 19.66 \qquad \bar{x}_1 = 39.25 \qquad \bar{x}_2 = 4.18$$
$$\bar{x}_3 = 241.50 \qquad \bar{x}_4 = 332.09$$

and the corrected sums of squares and products of the $x$'s are

$$CS(x_1, x_1) = \quad 984.50 \quad CS(x_2, x_2) = \quad 212.77 \quad CS(x_1, x_2) = \quad 284.03$$
$$CS(x_3, x_3) = 43\,690.00 \quad CS(x_1, x_3) = -4591.90 \quad CS(x_2, x_3) = -2763.00$$
$$CS(x_4, x_4) = 150\,842.72 \quad CS(x_1, x_4) = -3920.05 \quad CS(x_2, x_4) = -1688.14$$
$$CS(x_3, x_4) = \quad 33\,466.50$$

Similarly, the sums of products of the $x$'s and $y$'s are:

$$CS(x_1, y) = \quad 461.41 \qquad CS(x_2, y) = 334.46 \qquad CS(x_3, y) = -3931.05$$
$$CS(x_4, y) = 16\,497.82,$$

and these are the elements of $(\mathbf{a'y})$.

We now solve the equations (5.10) to obtain

$$\hat{\beta}_1 = \quad 0.2272 \qquad \hat{\beta}_2 = 0.5537$$
$$\hat{\beta}_3 = -0.1495 \qquad \hat{\beta}_4 = 0.1547$$

and together with $\hat{\alpha} = \bar{y}$, we have the estimated regression line (5.3) as follows:

$$E(y) = 19.66 + 0.2272(x_1 - 39.25) + 0.5537(x_2 - 4.18)$$
$$- 0.1495(x_3 - 241.50) + 0.1547(x_4 - 332.09) \qquad (5.43)$$

From the working shown above, we have that $\hat{\alpha}$ is statistically independent of the $\hat{\beta}_j$'s, and an estimate of $V(\hat{\alpha})$ is

$$\hat{\sigma}^2/32 = 4.99/32 = 0.156$$

An estimate of the variance–covariance matrix of the $\hat{\beta}_j$'s is $\hat{\sigma}^2 \mathbf{C}^{-1}$ where $\mathbf{C}$ is the matrix of corrected sums of squares and products of the explanatory variables. This leads to

$$\begin{bmatrix} 0.009\,987 & 0.001\,362 & 0.001\,115 & 0.000\,027 \\ 0.001\,362 & 0.136\,713 & 0.009\,144 & -0.000\,463 \\ 0.001\,115 & 0.009\,144 & 0.000\,854 & -0.000\,058 \\ 0.000\,027 & -0.000\,463 & -0.000\,058 & 0.000\,041 \end{bmatrix} \qquad (5.44)$$

The diagonal elements of this matrix can be used to start examining if there is one of the variables which is not contributing significantly to the regression, but we shall delay a discussion on this and other practical aspects of multiple regression analysis to the next chapter.

One important point to notice is that in general the estimates of the $\beta_j$'s are correlated. Therefore, although the regression sum of squares is computed from

$$\sum \hat{\beta}_j CS(y, x_j)$$

this cannot be separated into a term attributable to each of the $x_j$s, since $\hat{\beta}_j$ involves all the $x_i$'s through Equation (5.10). The only case in which the regression sum of squares can be decomposed is when all the off-diagonal terms in $(\mathbf{a}'\mathbf{a})^{-1}$ are zero, that is, when

$$CS(x_i, x_j) = 0 \qquad i \neq j \tag{5.45}$$

In two dimensions, one design of the $x$-points for which this holds is shown in Fig. 5.1.

When Equation (5.45) holds, then we have all estimates $\hat{\beta}_j$ independent.

$$\hat{\beta}_j = CS(y, x_j)/CS(x_j, x_j)$$
$$V(\hat{\beta}_j) = \sigma^2/CS(x_j, x_j)$$

In this case the regression sum of squares decomposes to give the analysis-of-variance table shown in Table 5.9, if only linear terms in the explanatory variables $x_i$ are allowed.

We can now test whether each variable should be included by an $F$-test of each mean square for regression to the deviations mean square, on $(1, n - k - 1)$ degrees of freedom. Each term is interpreted independently of

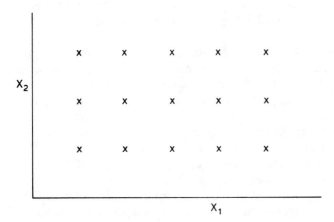

Fig. 5.1 Design for multiple regression.

Table 5.9 *Analysis of variance*

| Source | CSS | d.f. |
|--------|-----|------|
| Regression on $x_1$ | $\hat{\beta}_1 CS(y, x_1)$ | 1 |
| $\vdots$ | $\vdots$ | $\vdots$ |
| Regression on $x_k$ | $\hat{\beta}_k CS(y, x_k)$ | 1 |
| Deviations | (by subtraction) | $n - k - 1$ |
| Total | $CS(y, y)$ | $n - 1$ |

others, and this is a very great advantage, for as we shall see in the next chapter, the interpretation of the results of multiple regression analyses can be very difficult. Therefore generally, when the x-values are under the control of the experimenter, designs for which Equation (5.45) hold are preferred, firstly on grounds of ease of interpretation, and secondly on grounds of ease of calculation. In this case we say that the design is *orthogonal*.

### Exercises 5.5

1. Obtain an estimate of the variance of $\hat{\sigma}^2$ for the example given in this section [See Exercise 5.4.2.]. Also obtain a 90% confidence interval for $\sigma^2$ [See Theorem 5.7 and Example 1.4].

2. Write out the analysis-of-variance table for Example 2.3, carry out the F-test and report on the result. [Exercises 5.1.1., 5.2.3. and 5.3.1. refer.]

3. Figure 5.1 shows one set of data which is orthogonal. In fact any set of x's which are symmetrical about their centroid will be orthogonal. Hence produce some alternative sets of data points which are orthogonal.

4. The data given in Table 5.10 were reported by Ineson (1939) in connection with a study of the efficient operation of electricity power-generating stations. In a certain station the plant could be divided into two groups, one group which had old plant working on low-pressure steam, and the other group which had new plant working on high-pressure steam. The number of units of electricity supplied by each group in a month is given below. Also given is the total heat consumed by the station, denoted **H**, in millions of BTU's. This last quantity is a calculated figure, based on the quantity of fuel consumed and the calorific value of the fuel, and it is difficult to determine accurately. Find the regression equation for predicting **H** from the units of electricity produced. Over what ranges of the variables do you think your equation is satisfactory for use?

Table 5.10

| Units of electricity supplied | | Heat consumed H in BTU's × 10^6 |
| High-pressure plant | Low-pressure plant | |
| --- | --- | --- |
| 18.720 | 4.745 | 371 496 |
| 16.655 | 4.551 | 342 608 |
| 0.000 | 10.396 | 246 775 |
| 0.000 | 8.253 | 191 138 |
| 3.512 | 6.753 | 224 502 |
| 16.266 | 1.440 | 258 942 |
| 17.029 | 1.240 | 269 218 |
| 14.181 | 1.230 | 230 216 |
| 16.500 | 1.483 | 265 534 |
| 17.821 | 2.660 | 302 165 |
| 17.608 | 4.773 | 345 132 |
| 18.914 | 5.242 | 279 908 |

5.* In a regression problem, the independent variable is an angle $t$ confined to the range $0 \le t \le \pi$. Each component of $E(Y)$ may be assumed to have the form

$$E(Y) = \theta_1 \cos t + \theta_2 \cos 2t$$

for some chosen value of $t$. Errors in $Y$ are uncorrelated and have constant variance. To estimate $\theta_1$ and $\theta_2$, $m$ observations are made at $t = 0$ and a further $m$ can be made at a single further value, $t'$, say, which may be chosen by the experimenter. By finding the covariance matrix of the least-squares estimates of $\boldsymbol{\theta}$, discuss the choice of $t'$, bearing in mind that it may not be possible to make a choice which is optimal in all respects.

**Further reading**
For another account of the material in this chapter, see Draper and Smith (1966). For further numerical illustration see the references given in Appendix B.

# CHAPTER SIX

# Multiple regression: Further analysis and interpretation

## 6.1 Testing the significance of subsets of explanatory variables

In Section 5.5 we have seen how the theory of Sections 5.2 to 5.4 can be applied to multiple regression up to the point where we obtained the basic analysis-of-variance table, Table 5.7, and the $F$-test for testing the overall significance of the regression. Our next problem is to see how to test whether or not a specific explanatory variable, or a group of explanatory variables, contribute significantly to the regression. If the explanatory variables are orthogonal, as explained at the end of Chapter 5, then we can do separate $F$-tests of each variable. However, in general, the explanatory variables are not orthogonal and we must approach the problem in a different way. Any who have difficulties with the following discussion should read through Sections 6.1 and 6.2 and then come back to Section 6.1.

Let the model and notation be as in Equations (5.4) and (5.5), so that there are $p$ unknown parameters $\theta_1, \ldots, \theta_p$, to estimate. Suppose we wish to consider a model in which $r$ of these parameters are included, and $(p - r)$ are ignored. Let the $r$ parameters to be included be the first $r$, by rearrangement if necessary, then we can write

$$E(\mathbf{Y}) = \mathbf{a}\theta$$

$$= (\mathbf{a}_1, \mathbf{a}_2) \begin{bmatrix} \theta_1 \\ \theta_2 \end{bmatrix} \tag{6.1}$$

where $\mathbf{a}_1$ is $(n \times r)$, $\mathbf{a}_2$ is $n \times (p - r)$, $\theta_1$ is $(r \times 1)$ and $\theta_2$ is $(p - r) \times 1$. If the parameters $\theta_2$ are deleted, we have the model

$$E(\mathbf{Y}) = \mathbf{a}_1 \theta_1$$

so that, from Equation (5.10), the least-squares estimate is

$$\hat{\theta}_1 = (\mathbf{a}_1' \mathbf{a}_1)^{-1} \mathbf{a}_1' \mathbf{y} \tag{6.2}$$

and from Equation (5.35) the sum of squares due to regression is

$$S_{\text{par}(r)} = \hat{\theta}_1' \mathbf{a}_1' \mathbf{y} \tag{6.3}$$

*Example* 6.1   As an example we refer back to Example 5.1, discussed in Section 5.5. We might wish to see the effect of deleting, say, variables $x_1$ and $x_2$, and so we write

$$\theta' = (\theta'_1 ; \theta'_2) = (\alpha, \beta_3, \beta_4 ; \beta_1, \beta_2)$$

and then we have

$$\mathbf{a}_1 = \begin{bmatrix} 1 & (x_{3,1} - \bar{x}_{3.}) & (x_{4,1} - \bar{x}_{4.}) \\ 1 & (x_{3,2} - \bar{x}_{3.}) & (x_{4,2} - \bar{x}_{4.}) \\ \vdots & & \\ 1 & (x_{3,32} - \bar{x}_{3.}) & (x_{4,32} - \bar{x}_{4.}) \end{bmatrix}$$

$$\mathbf{a}_2 = \begin{bmatrix} (x_{1,1} - \bar{x}_{1.}) & (x_{2,1} - \bar{x}_{2.}) \\ (x_{1,2} - \bar{x}_{1.}) & (x_{2,2} - \bar{x}_{2.}) \\ \vdots & \\ (x_{1,32} - \bar{x}_{1.}) & (x_{2,32} - \bar{x}_{2.}) \end{bmatrix}$$

The sum of squares in Equation (6.3) is then the sum of squares for the model

$$E(\mathbf{Y}) = \mathbf{a}_1 \theta_1, \qquad V(\mathbf{Y}) = \mathbf{I}\sigma^2,$$

ignoring variables $x_1$ and $x_2$ entirely.                               □ □ □

Now for the full model (6.1) we have

$$\hat{\theta} = (\mathbf{a}'\mathbf{a})^{-1}\mathbf{a}'\mathbf{y}$$

and

$$S_{\text{par}(p)} = \hat{\theta}'\mathbf{a}'\mathbf{y} \tag{6.4}$$

Clearly the value of Equation (6.4) must be greater than that of Equation (6.3) since it contains all of the previous parameters and more besides. The extent to which Equation (6.4) is greater than Equation (6.3) is some measure of the value of adding the extra $(p - r)$ parameters to the model. Typically, we suspect that the true value of $\theta_2$ is zero, and we want a test of the hypothesis that this is so.

Let us suppose that $\theta_2 = 0$, so that we have exactly

$$E(\mathbf{Y}) = \mathbf{a}_1 \theta_1$$

then by applying Theorem 5.6 we obtain

$$E(S_{\text{par}(r)}) = r\sigma^2 + \theta'_1 \mathbf{a}'_1 \mathbf{a}_1 \theta_1$$

If $\theta_2 = 0$ we can also write

$$E(\mathbf{Y}) = \mathbf{a}\theta = (\mathbf{a}_1, \mathbf{a}_2) \begin{bmatrix} \theta_1 \\ 0 \end{bmatrix}$$

and then we obtain

$$E(S_{\text{par}(p)}) = p\sigma^2 + \theta' \mathbf{a}' \mathbf{a} \theta$$

$$= p\sigma^2 + (\theta_1' \quad 0) \begin{bmatrix} \mathbf{a}_1' \mathbf{a}_1 & \mathbf{a}_1' \mathbf{a}_2 \\ \mathbf{a}_2' \mathbf{a}_1 & \mathbf{a}_2' \mathbf{a}_2 \end{bmatrix} \begin{bmatrix} \theta_1 \\ 0 \end{bmatrix}$$

$$= p\sigma^2 + \theta_1' \mathbf{a}_1' \mathbf{a}_1 \theta_1$$

Therefore, if $\theta_2 = 0$ we have

$$E\{S_{\text{par}(p)} - S_{\text{par}(r)}\} = (p - r)\sigma^2 \qquad (6.5)$$

If $\theta_2 \neq 0$, then the expression on the left-hand side of Equation (6.5) must be greater than $(p - r)\sigma^2$. This immediately leads us to a method of testing the hypothesis $\theta_2 = 0$, but first we need the following theorem.

*Theorem* 6.1   If $\theta$ and $\mathbf{a}$ are partitioned as in Equation (6.1), then if $\theta_2 = 0$, the distribution of

$$\{S_{\text{par}(p)} - S_{\text{par}(r)}\}/\sigma^2$$

is $\chi^2$ on $(p - r)$ degrees of freedom.                    □ □ □

We shall not prove the theorem here, and those interested should consult the references listed under Theorem 5.7. The discussion above at least makes the result plausible.

We can now state the following principle, with notation defined above.

*Extra sum-of-squares principle*   In order to test the hypothesis that $\theta_2 = 0$, we calculate the extra sum of squares due to regression $\{S_{\text{par}(p)} - S_{\text{par}(r)}\}$ and then calculate the ratio

$$\{(S_{\text{par}(p)} - S_{\text{par}(r)})/(p - r)\}/\{S_{\text{min}}/v\}$$

which is referred to $F$-tables on $\{(p - r), v\}$ degrees of freedom, where $v$ is the degrees of freedom for $S_{\text{min}}$.                    □ □ □

This principle has very wide applications, but we apply it here to multiple regression.

**6.2 Application of the extra sum-of-squares principle to multiple regression**
Suppose we are given the $n$ sets of observations $(y_i, x_{1i}, \ldots, x_{ki})$, for $i = 1, \ldots, n$, then the analysis-of-variance table corresponding to Table 5.5 can be summarized as in Table 6.1.
The mean square for deviations in this table is used as the denominator for all subsequent $F$-tests mentioned in this section; it is calculated by dividing $S_{\text{min}}$ by the degrees of freedom. Since we shall be making reference to the

Table 6.1 *Analysis of variance for multiple regression*

| Source | d.f. | Mean square |
|---|---|---|
| Regression on $x_1, \ldots, x_k$ | $k$ | $A$ |
| Deviations | $n - k - 1$ | $B$ |
| Total | $n - 1$ | |

mean squares in this table and other mean squares it will be convenient to label them $A$, $B$, etc., as shown.

If we wish to test whether any one variable, $x_j$ say, contributes significantly to the regression, we first construct a table such as Table 6.2.

The corrected sum of squares for $C$ is obtained by simply ignoring $x_j$, and performing the usual multiple-regression calculations. The corrected sum of squares for $D$ is obtained by subtraction; this term represents the extra amount of variation in $y$ accounted for when $x_j$ is added to the regression, given that the variables $x_1, \ldots, x_k$ except $x_j$ are already included. A significance test of the hypothesis $\beta_j = 0$ is obtained by calculating $D/B$, and referring this to the $F$-tables for $(1, n - k - 1)$ degrees of freedom. Clearly, there are $k$ quantities such as $D$, one for each variable, and $k$ significance tests. However, except in the special case of orthogonal designs indicated at the end of Section 5.5, the $k$ separate adjusted sums of squares are not independent, and they do *not* add to the sum of squares for $A$. We shall discuss the general question of interpretation later; here we merely note the method of performing significance tests, and it is a direct application of the theorem and principle stated in Section 6.1.

There is an alternative and exactly equivalent method of testing the significance of the contribution of a single variable $x_j$ to the regression. The variance of $\hat{\beta}_j$ is given by the appropriate element of $\sigma^2(\mathbf{a}'\mathbf{a})^{-1}$, and an estimate of this variance is obtained by replacing $\sigma^2$ by its estimate $B$, on $(n - k - 1)$ degrees of freedom. Therefore in order to test the hypothesis $\beta_j = 0$, we calculate

$$\hat{\beta}_j / \sqrt{[\text{Est } V(\hat{\beta}_j)]} \tag{6.6}$$

and refer this to the $t$-tables on $(n - k - 1)$ degrees of freedom. Similarly,

Table 6.2 *Calculation of adjusted sum of squares for one variable*

| Source | d.f. | Mean square |
|---|---|---|
| Regression on $x_1, \ldots, x_k$ | $k$ | $A$ |
| Regression on $x_1, \ldots, x_{(j-1)}, x_{(j+1)}, x_k$ ($x_j$ ignored) | $k - 1$ | $C$ |
| Due to $x_j$ (adjusting for $x_1, \ldots, x_{(j-1)}, x_{(j+1)}, \ldots, x_k$) | 1 | $D$ |

Table 6.3 *Calculation of adjusted sum of squares for a group of variables*

| Source | d.f. | Mean square |
|---|---|---|
| Regression on $x_1, \ldots, x_k$ | $k$ | $A$ |
| Regression on $x_1, \ldots, x_s, (x_{s+1}, \ldots, x_k$ ignored$)$ | $s$ | $E$ |
| Due to $x_{s+1}, \ldots, x_k$ (adjusting for $x_1, \ldots, x_s$) | $k - s$ | $F$ |

confidence intervals for $\beta_j$ are, at the $100(1 - \alpha)$ per cent level,

$$\hat{\beta}_j \pm t_v\left(\frac{\alpha}{2}\right) \{\text{Est } V(\hat{\beta}_j)\}^{1/2}$$

where $v = n - k - 1$, the degrees of freedom, and $t_v(\alpha/2)$ is the upper $100(\alpha/2)$ per cent point of the $t_v$-distribution. There is an obvious generalization of this method for dealing with hypotheses in the form $\beta_j = \beta_{j0}$, say. When we are merely concerned with one parameter at a time, this approach is to be preferred to the $F$-test, since it leads directly to confidence intervals and generalizes more easily than the $F$-test.

Since the separate tests of $\beta_j$ are not independent, it is necessary to have a test for whether a group of variables jointly contribute to the regression significantly or not. First we calculate the table shown in Table 6.3

As before, the corrected sum of squares for $F$ is obtained by subtraction; it is the extra sum of squares representing the extra amount of variation in $y$ accounted for when $x_{s+1}, \ldots, x_k$ are added to a regression, given that $x_1, \ldots, x_s$ are already included. The appropriate significance test is carried out by calculating the ratio $F/B$, and referring this to $F$-tables for $\{(k - s), k - n - 1\}$ degrees of freedom. A series of tests of this type is usually necessary in any multiple regression analysis.

Before we illustrate these points on our example, it is appropriate to comment on the validity of the error mean square, for which we have used the deviations mean square, labelled $B$. In treating this as an estimate of the error variance $\sigma^2$, we are assuming that a linear model of the form (5.3) holds, and that all of the variables which contribute to the regression have been included. One way of investigating these assumptions is to analyse the residuals, and another is to replicate the observations on all sets of $x$, and so provide a separate estimate of $\sigma^2$ against which the deviations mean square can be tested. We shall discuss both of these points later.

*Further analysis of Example 5.1*
Some of the corrected sums of squares for Example 5.1 are set out in Table 6.4. The basic analysis of Table 5.8 has been reproduced at the top, and the terms have been labelled with $A, B, C, \ldots$, etc., as given earlier in this section in order to make comparison easier.

Table 6.4 *Analysis of variance for Example* 5.1

| Source | CSS | d.f. | Mean square | |
|---|---|---|---|---|
| *Anova Table* 1 | | | | |
| Regression on $x_1, x_2, x_3, x_4$ | 3429.27 | 4 | 857.32 | A |
| Deviations | 134.80 | 27 | 4.99 | B |
| *Total* | 3564.07 | 31 | | |
| *Anova Table* 2 | | | | |
| Regression on $x_1, x_2, x_3, x_4$ | 3429.27 | 4 | | A |
| Regression on $x_1, x_3, x_4 (x_2$ ign.$)$ | 3418.08 | 3 | | $C_1$ |
| *Due to $x_2$ (adj. for $x_1, x_3, x_4$)* | 11.19 | 1 | | $D_1$ |
| | $F = 11.19/4.99 = 2.24$ on (1, 27) d.f. | | | |
| *Anova Table* 3 | | | | |
| Regression on $x_1, x_2, x_3, x_4$ | 3429.27 | 4 | | A |
| Regression on $x_2, x_3, x_4 (x_1$ ign.$)$ | 3403.46 | 3 | | $C_2$ |
| *Due to $x_1$ (adj. for $x_2, x_3, x_4$)* | 25.81 | 1 | | $D_2$ |
| | $F = 25.81/4.99 = 5.17$ on (1, 27) d.f. | | | |
| *Anova Table* 4 | | | | |
| Regression on $x_1, x_2, x_3, x_4$ | 3429.27 | 4 | | A |
| Regression on $x_3, x_4, (x_1, x_2$ ign.$)$ | 3393.47 | 2 | | E |
| *Due to $x_1, x_2$ (adj. for $x_3, x_4$)* | 35.80 | 2 | | F |
| | $F = 17.90/4.99 = 3.58$ on (2, 27) d.f. | | | |
| *Other sums of squares* | | | | |
| Regression on $x_1, x_2, x_4 (x_3$ ign.$)$ | 3298.60 | 3 d.f. | | |
| Regression on $x_1, x_2, x_3 (x_4$ ign.$)$ | 555.32 | 3 d.f. | | |
| Regression on $x_2, x_4 (x_1, x_3$ ign.$)$ | 3194.21 | 2 d.f. | | |
| Regression on $x_1, x_4 (x_2, x_3$ ign.$)$ | 2702.13 | 2 d.f. | | |
| Regression on $x_3 (x_1, x_2, x_4$ ign.$)$ | 353.70 | 1 d.f. | | |
| Regression on $x_4 (x_1, x_2, x_3$ ign.$)$ | 1804.38 | 1 d.f. | | |

It should be emphasized that Table 6.4 has been set out primarily to illustrate the techniques introduced earlier in the section, and not to illustrate the detailed analysis of Example 5.1. Since the 5 per cent value of $F$ on (1, 27) d.f. is 4.21, we see that the contribution of $x_2$ to the regression, given that $x_1, x_3$ and $x_4$ are included, is not significant. It appears that $x_2$ can safely be dropped from the regression. The similar test for $x_1$ is just significant at 5 per cent, and $x_1$ is probably better included. A test of the joint contribution of $x_1$ and $x_2$ to the regression gives an $F$-value of 3.58 on (2, 27) degrees of freedom, and is significant at the 5 per cent level, showing that we cannot drop both $x_1$ and $x_2$.

Some other comments on the analysis are appropriate here. The variable which by itself gives the largest regression sum of squares is $x_4$, with a CSS of 1804.38; the corresponding value for the other three variables is less than one-third of this amount. Of the six sets of pairs of variables, the CSS's for those involving $x_4$ are given in Table 6.4, and they are very similar. Clearly, in such cases it is a matter of chance which pair of variables yields the largest regression sum of squares, unless there is a regression relationship which is strong enough to stand out markedly. Frequently, therefore, we find that the results of significance tests of the type described in this section are that some variables are clearly important, some variables are easily discarded, but others are in doubt, and there is no obvious choice for a best single regression equation.

As a final point on our illustration of the method of testing, the reader can check for himself that the four separate adjusted sums of squares for the $x_i$ singly, do not add up to the sum of squares $A$ of Table 6.4 (see Exercise 6.2.1).

At this point readers should make sure that they have completed the series of numerical exercises running back from Exercise 6.2.2 below to Example 2.3.

---

### Exercises 6.2

1. From the data given in Table 6.4, find the values of the four separate adjusted sums of squares for $x_1, x_2, x_3$ and $x_4$, and check that they do not add to the sum of squares for regression on $(x_1, x_2, x_3, x_4)$.

2. Continue the analysis of Example 2.3 by testing the contribution of $x$ and $w$ individually to the regression. [See Exercise 5.5.2.]

3. The data given below were reported by Mendenhall and Ott (1971), in connection with a problem in quality control in the paper-making industry. At one stage in the process, some small fibre particles are conveyed by a fluid at a high speed, and it is important to control the density of particles per unit volume in this fluid. A meter was installed, which operated by the force exerted by the particles in the fluid stream. In order to calibrate the meter, readings were taken and corresponding density measurements made for a number of samples. The density measurements were thought to have much more random variation than the meter reading, due to local variations in consistency of the fluid. Produce a curve for calibrating the meter over the range of observations used, by regressing density on meter reading. You should allow for the possibility that the regression equation includes a quadratic term in the meter reading.

| meter reading | 2.160 | 2.170 | 2.180 | 2.200 | 2.220 | 2.235 | 2.260 |
|---|---|---|---|---|---|---|---|
| density | 2.1091 | 2.1537 | 2.1783 | 2.2602 | 2.3532 | 2.3949 | 2.4292 |

| meter reading | 2.280 | 2.295 | 2.310 | 2.330 | 2.350 | 2.370 | 2.400 |
|---|---|---|---|---|---|---|---|
| density | 2.5108 | 2.5020 | 2.5885 | 2.5909 | 2.6418 | 2.6685 | 2.6591 |

| meter reading | 2.420 | 2.450 | 2.475 |
|---|---|---|---|
| density | 2.6829 | 2.6912 | 2.7532 |

4. Kuratori (1966) (see also Snee, 1971), gave the results presented in Table 6.5, which are measurements of the modulus of elasticity of a rocket propellant containing three components $A$, $B$, and $C$.

Table 6.5

| Run | Proportions of components in the mixture | | | Elasticity |
|---|---|---|---|---|
| | $A$ | $B$ | $C$ | |
| 1 | 1 | 0 | 0 | 2350 |
| 2 | 0 | 1 | 0 | 2450 |
| 3 | 0 | 0 | 1 | 2650 |
| 4 | $\frac{1}{2}$ | $\frac{1}{2}$ | 0 | 2400 |
| 5 | $\frac{1}{2}$ | 0 | $\frac{1}{2}$ | 2750 |
| 6 | 0 | $\frac{1}{2}$ | $\frac{1}{2}$ | 2950 |
| 7 | $\frac{1}{3}$ | $\frac{1}{3}$ | $\frac{1}{3}$ | 3000 |
| 8 | $\frac{2}{3}$ | $\frac{1}{6}$ | $\frac{1}{6}$ | 2690 |
| 9 | $\frac{1}{6}$ | $\frac{2}{3}$ | $\frac{1}{6}$ | 2770 |
| 10 | $\frac{1}{6}$ | $\frac{1}{6}$ | $\frac{2}{3}$ | 2980 |

It was expected that a suitable model would be of the following general form:

$$E(Y) = \alpha_0 + \alpha_1 x_1 + \alpha_2 x_2 + \alpha_3 x_3 + \alpha_4 x_1 x_2 + \alpha_5 x_1 x_3$$
$$+ \alpha_6 x_2 x_3 + \alpha_7 x_1^2 + \alpha_8 x_2^2 + \alpha_9 x_3^2$$

where $x_1, x_2$ and $x_3$ are the proportions of the components $A$, $B$ and $C$ respectively in the mixture.

(a) Show that by using the relationships

$$x_1 + x_2 + x_3 = 1$$

and

$$x_1^2 = x_1(1 - x_2 - x_3) = x_1 - x_1 x_2 - x_1 x_3$$

and the similar expressions for $x_2^2$ and $x_3^2$, that the model can be reduced to the form

$$E(Y) = \beta_1 x_1 + \beta_2 x_2 + \beta_3 x_3 + \beta_4 x_1 x_2 + \beta_5 x_1 x_3 + \beta_6 x_2 x_3$$

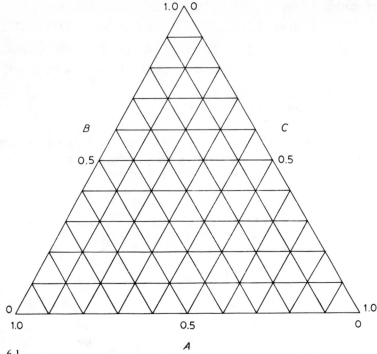

Fig. 6.1

(b) Fit the reduced form of the model to the data given above by least squares.

(c) Examine whether or not a quadratic surface is satisfactory.

(d) Plot out the combinations of components which have a response of at least 3000 on tri-linear co-ordinate paper (See Fig. 6.1).

5. Pairs of observations $x$, $y$ are available for estimating each of $k$ separate regression equations

$$E(y) = \beta_r x \qquad r = 1, 2, \ldots, k$$

where the $y$'s are independent normal with constant variance. There are $n_r$ pairs of observations for the $r$th equation. Indicate how to test the hypothesis that the regression coefficients satisfy a relation $\beta_r = r\beta$, where $\beta$ is constant.

## 6.3 Problems of interpretation

In the previous section we have outlined the technique used for testing whether or not certain regressor variables contribute significantly to a multiple regression. In practice, difficulties arise in the interpretation of the results of such analyses; before illustrating these difficulties we introduce some terminology.

*Definition*    *The total regression coefficient* is the coefficient of an explanatory (regressor) variable when it is the only such variable in a linear regression, i.e., the coefficient of $x$ in

$$E(Y) = \alpha + \beta x \qquad \qquad \square \ \square \ \square$$

*Definition*    *The partial regression coefficient* is the coefficient of an explanatory (regressor) variable when it is not the only explanatory variable, e.g., the coefficient of $x$ in

$$E(Y) = \alpha + \beta_1 x + \beta_2 z \qquad \qquad \square \ \square \ \square$$

One problem which arises is that partial regression coefficients can fluctuate widly due to the presence or absence of other variables. For example, partial and total regression coefficients can differ in sign. Suppose we have

$$E(Y) = \alpha + \beta_1 x_1 + \beta_2 x_2$$
$$\beta_1 < \beta_2, \qquad \beta_1 \text{ and } \beta_2 \text{ positive}$$

If the design of the experiment is such that $x_1$ decreases markedly as $x_2$ increases, then it can easily happen that the coefficient for the regression of $y$ on $x_1$ ignoring $x_2$, is negative. This is all shown on Fig. 6.2. Clearly, therefore, the inclusion or exclusion of a variable in a multiple regression can drastically affect the sums of squares and the other partial regression coefficients. This is especially true where the regressor variables are strongly interrelated.

As an illustration, suppose we are regressing measurements of growth of some crop against a set of variables which includes some meteorological variables, such as rainfall and temperature. If an equation is fitted which includes rainfall, then we may well find that the addition of the variable relating to temperature does not lead to a significant increase in the corrected sum of squares. This result could easily be due to the fact that rainfall and temperature are closely related, and that there is little additional effect of temperature other than the effect already accounted for in the rainfall variable. However, if the temperature variable is fitted first, we may easily find that now the rainfall variable is not worth adding, for the same reason.

*Example* 6.2    The data given in Table 6.6 on page 141 show the following quantities for the United Kingdom. Examine for a possible relationship between $y$ and $x_1$ and $x_2$.

> $y$: Fish landed, in thousands of tons
>
> $x_1$: The number of trawlers operating from the UK
>
> $x_2$: The number of drifters operating from the UK

In this example, it is natural to take $y$ as the dependent variable, and $x_1$ and $x_2$

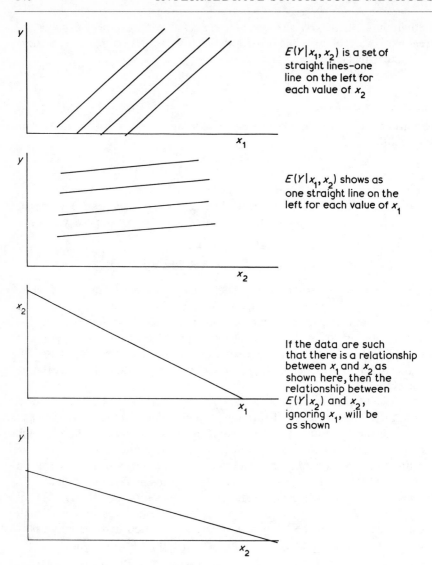

Fig. 6.2 Illustration of difficulties in multiple regression.

as the explanatory variables, and to look for a linear relationship. A graph of the data is shown in Fig. 6.3, and this gives us some evidence that it is reasonable to fit a linear relation. However, Fig. 6.4 shows sign of likely difficulties in the analysis, since $x_1$ and $x_2$ are very strongly negatively correlated.

The analysis-of-variance tables are shown in Table 6.7 on page 142, and we see that linear regression on both $x_1$ and $x_2$ is very highly significant, yielding

Table 6.6

|      | $y$   | $x_1$ | $x_2$ |
|------|-------|-------|-------|
| 1959 | 875.0 | 179   | 129   |
| 1960 | 822.8 | 160   | 131   |
| 1961 | 767.2 | 169   | 111   |
| 1962 | 807.3 | 165   | 116   |
| 1963 | 831.1 | 160   | 91    |
| 1964 | 836.1 | 271   | 79    |
| 1965 | 902.1 | 321   | 73    |
| 1966 | 928.4 | 368   | 67    |
| 1967 | 880.7 | 415   | 48    |
| 1968 | 889.3 | 432   | 34    |
| 1969 | 930.2 | 484   | 21    |

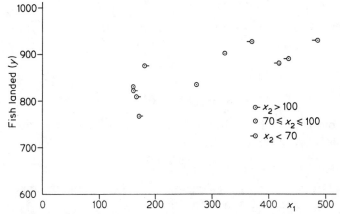

Fig. 6.3 Graph of Example 6.2 data.

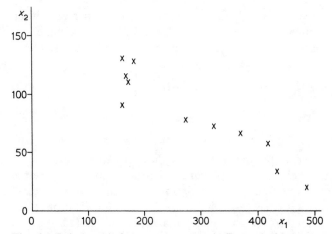

Fig. 6.4 Relationship between $x_1$ and $x_2$ in Example 6.2 data.

Table 6.7 *Analysis-of-variance tables for Example 6.2*

| Source | CSS | d.f. | Mean square |
|---|---|---|---|
| Regression on $x_1$ and $x_2$ | 18 752.77 | 2 | 9376.38 |
| Deviations | 8 308.15 | 8 | 1038.52 |
| Total | 27 060.92 | 10 | |
| Regression on $x_1$ and $x_2$ | 18 752.77 | 2 | |
| Regression on $x_1$ (ign. $x_2$) | 17 419.87 | 1 | |
| Due to $x_2$ (adj. for $x_1$) | 1 332.90 | 1 | 1332.90 |
| Regression on $x_1$ and $x_2$ | 18 752.77 | 2 | |
| Regression on $x_2$ (ign. $x_1$) | 13 010.95 | 1 | |
| Due to $x_1$ (adj. for $x_2$) | 5 741.82 | 1 | 5741.82 |

an $F$-value of

$$F = 9376.38/1038.52 = 9.03$$

on (2,8) degrees of freedom. The $F$-test for inclusion of $x_2$ in the equation, given that $x_1$ is included, gives

$$F = 1332.90/1038.52 = 1.28$$

on (1,8) degrees of freedom, and is not significant. The $F$-test for inclusion of $x_1$ in the equation given that $x_2$ is included, gives

$$F = 5741.82/1038.52 = 5.53$$

on (1,8) degrees of freedom and is just significant at the 5 per cent level. Clearly there is not much to choose between $x_1$ and $x_2$ in this example, but it is unnecessary to have both, and there is slight preference for $x_1$ (trawlers).

The three regression equations are as follows:

$$E(Y) = 605.74 + 0.61 x_1 + 0.98 x_2 \qquad (6.7)$$

$$E(Y) = 766.27 + 0.33 x_1 \qquad (6.8)$$

$$E(Y) = 939.29 \qquad\qquad - 0.96 x_2 \qquad (6.9)$$

There are two rather surprising features of these equations. Firstly, we notice the very erratic behaviour of the partial regression coefficients in the presence or absence of the other variable. Secondly, the negative coefficient of $x_2$ (drifters) clearly gives rise to some difficulty in interpretation! This arises because of the correlation between $x_1$ and $x_2$, already noted.

□ □ □

Table 6.8

|      | $x_1$   | $x_2$   | $x_3$   | $x_4$   |
|------|---------|---------|---------|---------|
| 1959 | 13.812  | 23.571  | 23.837  | 16.117  |
| 1960 | 16.995  | 23.670  | 23.369  | 16.933  |
| 1961 | 19.648  | 23.934  | 24.936  | 17.830  |
| 1962 | 23.068  | 24.644  | 28.376  | 18.910  |
| 1963 | 25.856  | 25.258  | 31.405  | 20.087  |
| 1964 | 28.951  | 25.838  | 34.162  | 21.459  |
| 1965 | 30.823  | 26.352  | 37.084  | 22.885  |
| 1966 | 33.781  | 26.590  | 38.352  | 24.232  |
| 1967 | 35.836  | 26.902  | 42.378  | 25.362  |
| 1968 | 42.092  | 26.906  | 45.036  | 27.113  |

As a final note on Example 6.2, the reader should check for himself that an analysis of residuals gives no cause for concern. The example has served to illustrate various points made earlier in the chapter. We shall discuss one more example, but deal with it rather more briefly.

*Example* 6.3   The data in Table 6.8 relate to England and Wales, for the years specified. Examine for a possible linear relationship between $x_1$ and the other variables.

$x_1$ : Expenditure on prisons, in £$10^6$

$x_2$ : Number of off licences currently held, in thousands

$x_3$ : Number of divorce decrees absolute granted, in thousands

$x_4$ : Consumer expenditure, in £$10^9$

In this example, all of the variables are very strongly positively related, since they are all strongly related to time. It is not surprising, therefore, to find that only one of these variables is worth including in the regression. A regression of $x_1$ on $x_4$ only accounts for 705.8 out of a total corrected sum of squares of 715.0. However, if time is included as a variable, $x_4$ drops out as non-significant, and the regression sum of squares goes up to 708.2. This shows the sort of spurious result which can arise if a very important variable is omitted.

However, much more serious questions arise in Example 6.3, concerned with the validity of the model. There is no reason why $x_1$ should be supposed to contain random variation and the other variables not, and the selection of the particular variables listed would also be difficult to defend. If the explanatory variables are not selected in a logical manner, it is not to be wondered at that the results of any analysis are impossible to interpret.

**Exercises 6.3**
1. Check the analysis of Example 6.3.

2. The standard linear regression model has been applied in Example 6.2 and 6.3. Give a detailed criticism of whether this is a valid model in these cases, and if not, what alternative models might be appropriate. Comment on the usefulness of the multiple regression analysis as a technique in such cases.

---

### 6.4* Relationships between sums of squares

Now we set out to clarify the relationships between the corrected sums of squares we have been discussing. Clearly, if we have four variables, then the sums of squares

$$\text{Due to regression on } x_1, x_2, x_4 (x_3 \text{ ignored})\qquad(6.10)$$

say, must be greater than the sum of squares

$$\text{Due to regression on } x_1, x_2 (x_3, x_4 \text{ ignored})\qquad(6.11)$$

although it may not be significantly greater.

If now we consider the adjusted sum of squares

$$\text{Due to regression on } x_1, x_2 \text{ (adjusting for } x_3, x_4)\qquad(6.12)$$

then we would expect (6.12) to be less than (6.11) in general, since some of the variation in $y$ will have been accounted for by $x_3, x_4$ when calculating (6.12). The following counterexample, due to D. R. Cox, shows that this need not necessarily be so.

Suppose we have an exact relationship between $y$, $x_1$ and $x_2$ described by

$$y = \alpha + \beta_1 x_1 + \beta_2 x_2$$

and where $x_2$ takes on the values $-1, 0, 1$ only, and $y$ is restricted to a range $(a, b)$, say. This relationship is shown in Fig. 6.5(a). The relationship between $y$ and $x_2$ ignoring $x_1$ is shown in Fig. 6.5(b). Here the relationship between $y$ and $x_1$ and $x_2$ is exact, so that the sum of squares due to regression on $x_1$ and $x_2$ accounts for all of the variation. We denote this sum of squares $A$. Further, a relation between $y$ and $x_1$, ignoring $x_2$, is not exact, so that the sum of squares

$$\text{Due to regression on } x_1 \ (x_2 \text{ ignored}) < A$$

but this is clearly non-zero. Therefore by subtraction we have that the quantity

$$\text{Due to regression on } x_2 \text{ (adjusting for } x_1)$$

has a positive sum of squares: let this sum of squares be denoted $B$.

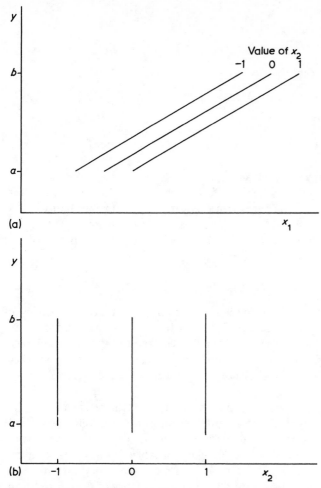

Fig. 6.5 Special example for multiple regression: (a) Relation between $y, x_1$ and $x_2$ (b) Relation between $y$ and $x_2$ ignoring $x_1$

From Fig. 6.4(*b*) we see that there is no relation between $y$ and $x_2$ when $x_1$ is ignored, so that the sum of squares

Due to regression on $x_2$ ($x_1$ ignored)

is zero and therefore less than $B$.
Therefore in certain circumstances, an adjusted sum of squares can be greater than the corresponding total regression sum of squares.

*Summary*
We always have the following relationship between corrected sums of

squares:

$$\text{Due to regression on } x_1, \dots, x_k$$
$$\geq \text{Due to regression on } x_1, \dots, x_k (x_i, x_j \text{ ign.})$$

but we can only say that the CSS

$$\text{Due to regression on } x_1, \dots, x_k \text{ (adj. for } x_i, x_j)$$

is usually less than that

$$\text{Due to regression on } x_1, \dots, x_k (x_i, x_j \text{ ignored})$$

### 6.5  Departures from assumptions

It is important to know what the effect might be on our procedures for regression analysis of departures from assumptions, and we shall discuss this subject rather briefly and give references for further reading.

A key step in this is to recognize that Theorems 5.1 to 5.6 hold provided

$$E(\mathbf{Y}) = \mathbf{a}\theta \tag{6.13}$$

$$V(\mathbf{Y}) = \mathbf{I}\sigma^2 \tag{6.14}$$

No assumption of normality is made, even down to the analysis-of-variance table and its expected mean squares, Table 5.3. It is at this stage that normality has to be introduced in order to obtain an $F$-test, and to estimate confidence intervals, etc. We may expect, therefore, that at least as far as the validity of the $F$-test is concerned, the assumption of normality is not very critical; but this is only partly true.

There have been a number of papers written on the sensitivity of the $F$-test in regression to the normality assumption (see the reference listed at the end of this section). The general conclusion is that it depends very much on the distribution of the explanatory variables, and in particular on how 'normal looking' their distribution is. If the explanatory variables are approximately normally distributed, then non-normality of the distribution of the response variable will have very little effect on the validity of the $F$-test. However, if the explanatory variables tend to be at the extremes of the region, with large gaps in between, or if the explanatory variables are clumped together with just a few which straggle a long way, then there could be some effect due to non-normality. The nature of the effects on the $F$-test are not easy to summarize, and we refer readers to the references listed for details. In general there has to be a rather unusual combination of sample size and non-normality of the response and explanatory variables for the effect to be large.

Another assumption of regression analysis is that we have chosen the correct linear form, Equation (6.13), for the expectation of the random variable. A very serious error arises if an important explanatory variable has been ignored, and this may or may not show up in an analysis of residuals. Other errors which can arise during a regression analysis are deleting too many variables, known as 'under-fitting', and including too many variables, known as 'over-fitting'.

It is obvious from the examples discussed in Section 6.3 that under-fitting can lead to seriously biased regression coefficients. It is also clear that an inflated estimate of $\sigma^2$ will tend to result. In the case of over-fitting, we do get unbiased estimates of the regression coefficients and of $\sigma^2$, but the variances of the regression coefficients are inflated. Clearly, if we have orthogonal explanatory variables (see Section 5.5) neither under-fitting nor over-fitting will lead to biased estimates of the regression coefficients.

One other feature of over-fitting has to be watched. In small sample sizes, over-fitting can take up degrees of freedom more usefully attributed to error. If, say, confidence intervals for a predicted mean are calculated, then this leads to using a $t$-distribution percentage point for fewer degrees of freedom, which consequently widens the confidence interval. This would begin to get serious if, for example, the number of degrees of freedom for estimation of $\sigma^2$ was 15 or less, and several degrees of freedom were taken up by over-fitting.

Finally, we note what happens in the case of an incorrect covariance matrix. Let the true model be

$$E(Y) = a\theta$$

$$V(Y) = V\sigma^2$$

(6.15)

where $V \neq I$,

then the ordinary least-squares estimator still gives an unbiased estimator of $\theta$, although an estimator with greater efficiency is obtained by using a procedure known as weighted least squares (see Exercise 6.5.2). Other features of the ordinary least-squares estimator are that it no longer has the usual covariance matrix (see Exercise 6.5.1), and that the estimator of $\sigma^2$ is biased. Clearly, if the matrix $V$ were known, a quite different analysis along lines of Exercise 6.5.2 would be appropriate.

For further discussion of the topics mentioned in this section see Seber (1977, Chapter 6), and the references given there.

---

### Exercises 6.5

1. Show that if ordinary least squares is used when the model (6.15) is true, then the variance of the resulting estimator is

$$V(\hat{\theta}) = \sigma^2(a'a)^{-1}a'Va(a'a)^{-1}$$

2.* Show that if $Y$ has a multivariate normal distribution with expectation and covariance matrices given by the model (6.15), then the maximum-likelihood estimator of $\theta$ is the $\hat{\theta}$ minimizing

$$S = (\mathbf{Y} - \mathbf{a}\theta)' \mathbf{V}^{-1}(\mathbf{Y} - \mathbf{a}\theta)$$

Also show that the estimator so obtained is

$$\hat{\theta} = (\mathbf{a}'\mathbf{V}^{-1}\mathbf{a})^{-1}\mathbf{a}'\mathbf{V}^{-1}\mathbf{Y}$$

This procedure is known as *weighted least squares*.

## 6.6 Predictions from regression

One of the main reasons for regression analyses is to obtain an equation to be used for prediction purposes, and this is clearly in mind in Example 5.1. It will be useful to discuss prediction in the context of that example, although the results are general.

In Section 5.5 we obtained the regression equation (5.43). Algebraically this is

$$\widehat{E(Y)} = \hat{\alpha} + \sum_{i=1}^{4} \hat{\beta}_i(x_i - \bar{x}_i) \qquad (6.16)$$

and this gives the estimated mean value of $Y$ at a new set of explanatory variables $(x_1, x_2, x_3, x_4)$. We observe that if the original random variables were normally distributed, then the expression (6.16) is also normally distributed since it is a linear function of the original random variables. The variance of (6.16) is seen to be

$$V\{\text{Estimated mean}\} = V\{\widehat{E(Y)}\} = V(\hat{\alpha}) + \sum_{i=1}^{4} (x_i - \bar{x}_i)^2 V(\hat{\beta}_i) \qquad (6.17)$$
$$+ \sum\sum_{i<j}(x_i - \bar{x}_i)(x_j - \bar{x}_j)C(\hat{\beta}_i, \hat{\beta}_j)$$

and we note that for our problem the covariance terms $C(\hat{\alpha}, \hat{\beta}_i)$ are all zero, but the other covariance terms are estimated by the terms of Equation (5.44). The first term on the right-hand side of Equation (6.17) is

$$V(\hat{\alpha}) = V(\bar{Y}) = \sigma^2/n$$

which is estimated by dividing the residual mean square by $n$. Therefore a $100(1 - \alpha)$ per cent confidence interval for the estimated mean value of $Y$ is

$$\widehat{E(Y)} \pm t_v\left(\frac{\alpha}{2}\right)\sqrt{[\text{Estimated } V\{\widehat{E(Y)}\}]} \qquad (6.18)$$

where $v$ is the number of degrees of freedom for estimating $\sigma^2$.

If we want a confidence interval for the actual response at $(x_1, x_2, x_3, x_4)$, then we note that the response variable has a variance $\sigma^2$ about its mean. Thus the difference between the response and the estimated mean, called the estimated response difference, has a variance

$$V\{\text{Estimated response}\} = V\{\widehat{E(Y)}\} + \sigma^2$$

and a formula similar to Equation (6.18) is used with Estimated $V\{\text{Estimated response}\}$ replacing Estimated $V\{\widehat{E(y)}\}$.

*Example* 6.4  Suppose that for Example 5.1 we wish to obtain a 99 per cent confidence interval for the mean response when the explanatory variables are

$$x_1 = 40, \quad x_2 = 7, \quad x_3 = 200, \quad x_4 = 380$$

We shall base our prediction on Equation (5.43) and Table 5.8, so that by substitution we have the predicted mean,

$$E(Y) = 19.66 + 0.2272(0.75) + 0.5535(2.82) - 0.1495(-41.50) + 0.1547(47.91)$$
$$= 35.00$$

By substitution in Equation (6.17), the variance of this predicted mean is 1.610, so that a 99 per cent confidence interval is

$$35.00 \pm 2.771\sqrt{(1.610)} = (31.48, \quad 38.52)$$

where 2.771 is the one-sided 0.5 percentage point for $t$ on 27 degrees of freedom. It should be noted that the results earlier in this chapter indicate that it may be better to drop $x_2$ from the prediction equation, and the calculations using this equation are left as an exercise (see Exercise 6.6.2).

□ □ □

Some comments are appropriate on the use of regression equations for prediction. Firstly, an important question is frequently how many explanatory variables to include in the final prediction equation; this is discussed in the next section. Secondly, we must take great care about the part of the $X$-space in which the prediction equation is used. It is particularly dangerous to use the equation outside the region covered by the original data, as the equation may not be valid there.

---

**Exercises 6.6**

1. Show that if, in the notation of Chapter 5, we obtain a least-squares estimator

$$\hat{\theta} = (\mathbf{a}'\mathbf{a})^{-1}\mathbf{a}'\mathbf{Y}$$

then the predicted mean at a new point

$$\mathbf{a}_0' = (a_{1,0}, \ldots, a_{p,0})$$

is

$$\widehat{E(Y)} = \mathbf{a}_0' (\mathbf{a}'\mathbf{a})^{-1} \mathbf{a}'\mathbf{Y}$$

and has a variance

$$V\{\widehat{E(Y)}\} = \mathbf{a}_0 (\mathbf{a}'\mathbf{a})^{-1} \mathbf{a}_0' \, \sigma^2$$

2. Recalculate Example 6.4 using only explanatory variables $x_1$, $x_3$ and $x_4$.

3. Recalculate Example 6.4 to obtain a 99 per cent confidence interval for the actual response.

---

### 6.7 Strategies for multiple regression analysis

The way in which a multiple regression analysis is carried out depends on the objectives of the analysis, which may fall into one of two main categories:

1. *Analyses to demonstrate or compare relationships.* Interest here lies not so much in producing a final equation, but in demonstrating whether or not certain variables $x_j$ have a significant effect on the response variable, and if there is an effect, whether this depends on the values of other explanatory variables. It may also be of interest to compare the effect of one variable with the effect of another.

2. *Prediction.* The petroleum-refining problem, Example 5.1, is a good example of this type of application. An equation is required to predict the value of $y$, from certain values of variables $x_j, j = 1, 2, \ldots, p$. An interesting but unusual application which can be counted as under this heading is given by Mansfield and Wein (1958). Operating costs of a railway goods yard are regressed on quantities such as the number of cars delivered to or picked up from other sidings. The resulting equation is used for control purposes, to see when actual operating costs are unusually high or low.

Analyses of the first type can be carried out by using a series of significance tests and estimation procedures as discussed in Section 6.1 and earlier. In such a case our conclusions may not be unambiguous if the $x$-values are strongly related. The chief difficulty with the second type of analysis is that it is necessary to decide for a single 'best' equation to use, and, as we have seen, this may be difficult to do. Various ways of making this choice of a single best equation are described below.

It may be useful to note here one type of application of multiple regression which can be counted under the prediction category, but is large enough to have a separate heading, and this is *response surface work*. In these applications the formal variables used in the multiple regression contain powers and products of other variables, such as

$$x_1, x_2, x_1^2, x_1 x_2, x_1^2 x_2, x_1 x_2^2, x_1^3$$

The methods given below are not intended to apply to response surface work since, clearly, a decision as to whether a particular term such as $x_1 x_2$ is included, may depend on what the term is.

We assume then, that we are regressing a variable $y$ on $p$ variables $x_1, \ldots, x_p$, and we wish to make a decision to use a particular equation. A simple answer which may often give a highly accurate equation would be to include all $p$ variables, but this is not usually satisfactory for the following reasons:

(a) In a prediction problem, the inclusion of variables which do not contribute to the regression inflates the error of prediction. In fact, it may be substantially better to use a subset of variables for prediction.

(b) In a case such as the petroleum-refining problem, Example 5.1, the cost of monitoring $x$-variables may be quite high, since laboratory determinations are involved. It is therefore better to exclude variables which are not contributing significantly to the regression.

(c) In any case, there is no point in including terms in an equation which have a contribution less than the error variance, unless there is some strong a priori evidence that the variables should be included.

(d) A further point which sometimes influences our decision is that an equation is usually required which is stable over a wide range of conditions. This often means having a small number of variables present.

On the other hand, in the interests of an accurate equation, all the variables which really matter should be included. Four of the most important methods of deciding for a single equation are now listed for comparison.

### (i) All possible regressions
Since each variable can be either present or absent, there are in all $2^p$ possible equations, and given access to a high-speed computer, it may be an economic proposition to compute them all. The regressions can then be divided into sets containing one, two, etc., variables, and the leading regressions in each set are considered. The final decision can be made a matter of judgement, and indeed there may be some external reason why the final equation chosen is not the leader in a set. For example, there may be fairly strong prior reasons for thinking that $x_3$ is important, but $x_3$ is not included in the leading regression of the set chosen, and including $x_3$ may lead to excluding another explanatory variable. This method can only be applied if $p$ is small, say 5 or less, or possibly up to 10 with a very efficient algorithm.

## (ii) Backward elimination

This method starts off by evaluating the regression on all $p$ variables. The variable which by itself contributes least to the regression is then deleted, and then another, etc., until all variables remaining are significant. One advantage of this procedure is that since the regression on all of $x_1, \ldots, x_p$ is carried out, we are fairly sure about obtaining a valid estimate of error against which to do tests of significance. One disadvantage is that the procedure can be very time-consuming, especially if there are, say, 25 or more variables of which about 4 or 5 are important. Another disadvantage is that when we do a regression on all of the explanatory variables we are more likely to come up with a nearly singular or singular matrix. This in turn can lead to nonsense results, as some computer routines in common use are not very accurate for this situation.

When using this method it should be noted that after each regression is fitted, the $t$-values in expression (6.6) for each explanatory variable included are easily calculated. Variables with small $t$-values, say of less than 5 per cent significance, can be tried for dropping from the model, but unless we are simply dealing with the deletion of one variable, a final decision on inclusion or exclusion should be based on the $F$-test for the extra sum of squares, as described in Section 6.1.

## (iii) Forward selection

This procedure begins at the opposite end, and starts by selecting the most important single variable, and then adding to it that variable which leads to the largest increase in the regression sum of squares. This continues until a variable is added which is non-significant. One disadvantage of the procedure is that it is difficult to know that one has a valid estimate of error. However, this procedure is very much more economic on use of computing facilities than the previous procedure.

## (iv) Stepwise procedure

This method is an improved forward-selection procedure. At every stage, all variables included in the regression are re-examined to see if they are still worth including. Thus a variable selected early in the procedure may be rejected later because of relationships with variables then included. This procedure is one of the best to date for use with a moderate or large number of regressor variables.

An illustration of how these procedures might work out can be seen from Table 6.9. The data here consist of just seven observations, with five regressor variables. Normally one would not consider fitting five regressor variables to only seven observations, but the data came from an experiment concerned with radioactive isotopes, and each observation was very expensive. The

Table 6.9 *Illustration of alternative strategies for selecting the best regression equation*

| Variables | CSS | Selection method |
|---|---|---|
| 1 | 216.96 | |
| 2 | 238.74 | X, O |
| 3 | 98.07 | |
| 4 | 13.55 | |
| 5 | 84.05 | |
| 1,2 | 468.84 | X |
| 1,3 | 411.09 | |
| 1,4 | 422.19 | |
| 1,5 | 217.27 | |
| 2,3 | 264.68 | |
| 2,4 | 416.90 | O |
| 2,5 | 322.79 | |
| 3,4 | 128.65 | |
| 3,5 | 316.27 | |
| 4,5 | 111.54 | |
| 1,2,3 | 469.76 | X |
| 1,2,4 | 469.35 | |
| 1,2,5 | 469.74 | |
| 1,3,4 | 444.75 | |
| 1,3,5 | 433.97 | |
| 1,4,5 | 451.32 | |
| 2,3,4 | 418.72 | |
| 2,3,5 | 330.50 | |
| 2,4,5 | 451.74 | O |
| 3,4,5 | 408.31 | |
| 1,2,3,4 | 470.14 | X |
| 1,2,3,5 | 469.90 | |
| 1,2,4,5 | 469.72 | |
| 1,3,4,5 | 454.76 | |
| 2,3,4,5 | 471.14 | O |
| 1,2,3,4,5 | 471.21 | X, O |

Data

| $y$ | $x_1$ | $x_2$ | $x_3$ | $x_4$ | $x_5$ |
|---|---|---|---|---|---|
| 24 | 0 | 0 | 0 | 96 | 6 |
| 2 | 0 | 32 | 110 | 0 | 6 |
| 17 | 0 | 10 | 76 | 46 | 7 |
| 14 | 0 | 12 | 90 | 75 | 13 |
| 5 | 1 | 14 | 59 | 94 | 5 |
| 0 | 1 | 14 | 59 | 94 | 5 |
| 4 | 1 | 11 | 40 | 86 | 0 |

Total corrected sum of squares = 483.71
X: forward selection
O: backward elimination

data are given here because it is a simple matter to check the calculations, and the forward-selection and backward-elimination methods are common only at the end points. The stepwise-regression procedure is difficult to illustrate with so few degrees of freedom for error (one!).

A fuller account of different methods of selecting the best regression is given in Chapter 6 of Draper and Smith (1966), or Chapter 12 of Seber (1977): see also Hocking (1976). Some important papers discussing algorithms and approaches to multiple regression analysis are Garside (1971, 1965), Beale *et al.* (1967), and a more important general discussion is given by Cox (1968). (The 1968 issue of *J.R. Statist. Soc.* Series A, contains other papers on related material beside Cox's paper, and a discussion, all of which may be of interest.)

There is also a large literature on more elaborate methods for choosing subsets of variables for prediction purposes. For example, the data could be divided at random into two subsets, and one set used to fit the model, and the other to test the validity of it. Another method is to use a criterion in which one point only of the set of data is omitted, and a prediction of this made from a regression model fitted on the rest of the data. This is then repeated for each point, and the total sum of squared prediction errors used as a criterion. We shall not discuss these techniques here, but access to the literature can be obtained from the survey paper by Hocking (1976); see also Gardner (1972), Hocking (1972), Moran and Wright (1974) and Stone (1974). There is yet another large literature on biased methods of estimation, as alternatives to least squares, which are particularly useful when there is strong multicollinearity present, see Dempster *et al.* (1977).

## 6.8 Practical details

### (*i*) *General points*

*Graphs* It is always advisable to plot some graphs, for example, of the response variable against the more important explanatory variables, or of the explanatory variables which are more strongly correlated. Some techniques which can be used were illustrated in Example 6.2, Figs. 6.3 and 6.4.

*Correlation matrix* If the correlation matrix of explanatory variables is examined, this will give us some clue to the likelihood of problems of inter-pretation arising similar to those illustrated in Section 6.3. For any pair of variables which is strongly correlated we need to do regressions with and without each, and then decide whether one or other, or both, can be dropped from the regression equation.

*Transformations* In Example 2.1 it was found that by using the transforma-tion

$$y = \log(\text{fibrin level})$$

the resulting regression is more nearly linear. Transformations can be used for either regressor variables or the dependent variable (or both), and a full discussion is given later. In simple cases, such as Example 2.1, the choice of a transformation is obvious if one or two graphs are plotted. A more detailed study of the use of transformations is given in Chapter 8.

*F-tests* The series of *F*-tests which will be required will depend on the strategy which is being adopted for deciding upon a final regression model. Unless the number of explanatory variables is very large, there is some advantage in applying the overall $F$-test for inclusion of all explanatory variables. The analysis of variance for inclusion of all explanatory variables provides a good estimate of the error variance and a mean square for use as a divisor in other $F$-tests, unless the number of degrees of freedom left for error is small.

*Residuals* It is very desirable to include some analyses of residuals with all regression analyses. The methods discussed briefly in Section 2.4 can all be used in the general case, by inserting the new definition of fitted values in appropriate places. In some cases it may be important to realize that the residuals are correlated, and calculate the variance–covariance matrix **m** of Theorem 5.3, and some of the correlations.

*Conclusions* A multiple regression analysis should always finish with a statement of conclusions written for the experimenter for whom the analysis was done. It should include a brief verbal description of the main results of the analysis, and cover the following points:
(*a*) A statement of the final regression model chosen, with some justification of how this decision was reached. This means including analysis-of-variance tables of the sort given in Section 6.2.
(*b*) The variances or standard errors of the estimated regression coefficients and also in the case of models for prediction purposes, their covariances.
(*c*) An estimate of the error variance, and possibly a confidence interval for this.
(*d*) A prediction, if required, and its associated variance.
(*e*) A statement of the main results from the analysis of residuals, together with relevant graphs.

### (ii) *Computer programs*
Most regression analysis today is carried out on ready-made package programs. These are programs available as a library, which can be tapped and used by people who do not have any knowledge of computer programming. For example, one of the most widely used packages is the biomedical programs (BMD) which were developed at the University of California (Dixon, 1972). The BMD package contains several regression programs, as well as programs for numerous other statistical techniques.

From the user's point of view, there are four main considerations about any package.

*Documentation* There should be a good written description of the package covering matters such as options available and how to put the data in.

*Statistical methods* The statistical methods should be sound, and relevant significance tests, confidence interval procedures, etc., available.

*Numerical methods*   The numerical methods used, such as matrix inversion, should be reliable and efficient.

*Output*   The output should be of a readily understood form, and all relevant graphs and tables available.

No package satisfies all these desiderata, and some criticisms of the BMD package has been given by Berk and Francis (1978), Janacek and Negus (1974), and Muxworth (1974). In particular, the multiple-regression program uses a matrix-inversion routine which may produce quite wrong results for a near-singular matrix, and there is no warning in the output of when this condition may arise. The review by Berk and Francis (1978) is followed in that journal by comments and rejoinders by a number of statisticians, and it makes interesting reading.

The BMD programs are examples of what we call the 'batch processing' type. The alternative is an *interactive program*, in which the user communicates with the computer, usually via a consol. The University of Kent interactive package MULREG has facilities which enable the user to add or delete variables at will, and its numerical algorithms are very reliable.

Multiple-regression programs can also be constructed by using one of the 'statistical systems' such as ASCOP or GENSTAT. These are multipurpose programs which are controlled by English-like statements. It is usually necessary to spend some time studying the 'language' of such systems before they can be used effectively, but the end result is much more powerful than batch or interactive programs.

There is no substitute for practical experience on using these types of computer program.

---

**Exercise 6.8**
1. Carry out an analysis of residuals for one of the sets of data discussed in this chapter.

---

**Further reading on practical points**
For further references related to the material of this section, see Anscombe (1960), Anscombe and Tukey (1963) and Tukey (1962). For an application of multiple regression which has come under rather heavy criticism see Coen, Gomme and Kendall (1969) together with the printed discussion and the paper by Box and Newbold (1971). Those interested in computer packages should turn to the 1978 (March) issue of the *Journal of the American Statistical Association*.

# CHAPTER SEVEN

# Polynomial regression

## 7.1 Introduction

We now consider a rather special application of regression, illustrated by the following examples.

*Example* 7.1   An experiment was conducted with millet in order to determine the optimum distance between plants in rows. Plots of ground were set out with rows one foot apart, and some plots were set with plants two inches apart in rows, some plots with plants four inches apart, and so on. The results below are average yields for the three central rows in the plots indicated.

| Distance of plants (inches) | 2 | 4 | 6 | 8 |
|---|---|---|---|---|
| Average yield (g) | 238 | 270 | 298 | 283 |

Fit a suitable model and estimate the planting distance for optimum yield.

□ □ □

*Example* 7.2   Measurements of the specific conductivity $K_p$ of a metal at nine temperatures gave the results listed in Table 7.1. Preliminary analysis showed that a logarithm transformation was desirable. Fit a suitable model to describe the relationship between the variables.

Table 7.1

| Temp.($^\circ$C) | $\log_e K_p$ | Temp.($^\circ$C) | $\log_e K_p$ |
|---|---|---|---|
| 5 | − 5.5103 | 30 | − 5.9820 |
| 10 | − 5.6012 | 35 | − 6.0796 |
| 15 | − 5.6954 | 40 | − 6.1774 |
| 20 | − 5.7894 | 45 | − 6.2756 |
| 25 | − 5.8847 | | |

For these problems it is reasonable to assume the statistical model:

$$E(Y|x) = \gamma_0 + \gamma_1 x + \gamma_2 x^2 + \ldots + \gamma_p x^p$$
$$V(Y|x) = \sigma^2 \tag{7.1}$$

where the $Y$'s are taken to be independently and normally distributed random variables, and $x$ is a non-stochastic variable. Now in our treatment of multiple regression we have emphasized that our methods will cover any model such as (7.1) in which the expectation of $Y$ is a linear function of unknown parameters. Since the $x$'s of the multiple regression model are non-stochastic variables, it is immaterial to the statistical model what relationships may exist between them. Therefore, least-squares estimates of $\gamma_0, \gamma_1, \ldots, \gamma_p$ in (7.1) could be obtained by a straightforward application of the multiple regression methods we have already discussed. For certain problems, this may be the most convenient way of carrying out an analysis, but we should note two points:

(i) The calculation of corrected sums of squares of powers of $x$, e.g., of $x^5$, frequently leads to very large numbers, and the matrix of corrected sums of squares will often contain elements which differ greatly in size. Under these conditions many computer programs for inverting matrices will produce sizeable errors in the inverse.

(ii) In problems such as Example 7.1 and 7.2, it is nearly always in doubt as to what degree polynomial to use. The addition of a term $\gamma_{p+1} x^{p+1}$ to $E(Y|x)$ in the model (7.1) would in general lead to modifications to all previous estimates of the parameters $\gamma_0, \gamma_1, \ldots, \gamma_p$, and would, of course, involve the inversion of a new $(p+1) \times (p+1)$ matrix.

There is an alternative method of setting out the model which is of considerable value when the set of $x$-values at which observations are taken is either an equally spaced set such as

$$-1, 0, 1, 2, \ldots, \qquad \text{or} \qquad 0, 0.75, 1.50, 2.25, \ldots, \text{etc.}$$

or else any set of $x$-values which are used repeatedly on different occasions. This alternative method leads to simpler calculations and easier interpretation.

The alternative method stems from the fact that if the off-diagonal terms of the matrix of corrected sums of squares and products of the $x$'s in multiple regression are all zero, then no matrix inversion is involved, all the sums of squares are additive, and the addition of a new term does not lead to recalculations of all previous partial regression coefficients; this is the orthogonal case, mentioned at the end of Chapter 5. Now since the expectation in the model (7.1) is written in terms of non-stochastic variables, we can rewrite it in any way we like, and in fact it is always possible to rewrite it so that it leads to an orthogonal analysis. We will consider a simple illustration.

*Example* 7.3   Let us suppose that one observation is taken at each of the values $-1, 0, 1, 4$ of an explanatory variable $x$, and denote the responses by $y_{-1}, y_0, y_1, y_4$. Find the least-squares estimates of the parameters $\gamma_0, \gamma_1$ and $\gamma_2$ of the model

$$E(Y|x) = \gamma_0 + \gamma_1 x + \gamma_2 x^2 \tag{7.2}$$
$$V(Y|x) = \sigma^2$$

In dealing with this example we shall rewrite Equation (7.2) in the following way:

$$E(Y|x) = \alpha + \beta_1 f_1(x) + \beta_2 f_2(x) \tag{7.3}$$

where the functions $f_1(x)$ and $f_2(x)$ are as follows:

$$\begin{aligned} f_1(x) &= x - 1 \\ f_2(x) &= 14x^2 - 46x - 17 \end{aligned} \tag{7.4}$$

We shall see below the reason for this rather curious choice of the form of the model, but it is easy to see that Equation (7.2) and (7.3) are equivalent. From Equation (7.3) we have

$$E(Y|x) = (\alpha - \beta_1 - 17\beta_2) + (\beta_1 - 46\beta_2)x + 14\beta_2 x^2$$

so that Equations (7.2) and (7.3) are equivalent if

$$\gamma_0 = \alpha - \beta_1 - 17\beta_2 \qquad \gamma_1 = \beta_1 - 46\beta_2 \qquad \gamma_2 = 14\beta_2$$

There are just three parameters to estimate with either expression, so we can choose the form (7.3) if it suits our purpose. In order to fit (7.3) we define new variables

$$x_1 = f_1(x) \qquad \text{and} \qquad x_2 = f_2(x)$$

so that (7.3) becomes

$$E(Y|x_1, x_2) = \alpha + \beta_1 x_1 + \beta_2 x_2 \tag{7.5}$$

and we can proceed by the methods described in the previous chapters. The values of the new variables $x_1$ and $x_2$ are shown in Table 7.2. The new variables sum to zero over the observations, and the sum of their products is also zero.

Table 7.2  *Calculations for Example 7.3*

| $x$ | $-1$ | $0$ | $1$ | $4$ | *Sum* | *Sum of squares* |
|---|---|---|---|---|---|---|
| $x_1 = f_1(x)$ | $-2$ | $-1$ | $0$ | $3$ | $0$ | $14$ |
| $x_2 = f_2(x)$ | $43$ | $-17$ | $-49$ | $23$ | $0$ | $5068$ |
| $x_1 x_2$ | $-86$ | $17$ | $0$ | $69$ | $0$ | — |

Therefore if we use the notation of Chapter 5 and put Equation (7.5) in the form $E(Y) = \mathbf{a}\,\theta$, we have $\theta$ and $\mathbf{a}$ as follows:

$$\theta' = (\alpha, \beta_1, \beta_2)$$

$$\mathbf{a} = \begin{bmatrix} 1 & -2 & 43 \\ 1 & -1 & -17 \\ 1 & 0 & -49 \\ 1 & 3 & 23 \end{bmatrix}$$

so that we find from the values in Table 7.2 that

$$(\mathbf{a}'\mathbf{a}) = \begin{bmatrix} 4 & 0 & 0 \\ 0 & 14 & 0 \\ 0 & 0 & 5068 \end{bmatrix} \tag{7.6}$$

Since this is a diagonal matrix it is now a very simple matter to obtain the least-squares estimators

$$\hat{\theta} = (\mathbf{a}'\mathbf{a})^{-1}\mathbf{a}'\mathbf{Y}$$

and the minimized sum of squares, etc., follow.

Therefore by using the form (7.3) rather than (7.2) in the model we have obtained the orthogonal arrangement described at the end of Chapter 5. If we work directly with the model (7.2) we do not have orthogonality (see Exercise 7.1.1). Clearly the model (7.3) is much simpler to fit than (7.2) and the disadvantages discussed under (i) and (ii) above do not apply. The matrix (7.6) is easy to invert and the estimators of $\alpha$, $\beta_1$ and $\beta_2$ are statistically independent. The sums of squares for fitting each of the three parameters can be listed separately in the analysis-of-variance table, and tested independently of each other; see below.                     □ □ □

It is worth while exploring Example 7.3 a little further. In order to obtain the diagonal matrix (7.6) we had to apply two conditions:

(a) the functions add to zero when summed over the observations;
(b) the sum of the products of the two functions is zero.

In the next section we examine whether these conditions can be satisfied generally.

---

### Exercise 7.1
1. Write down $\theta$, $\mathbf{a}$, and $(\mathbf{a}'\mathbf{a})$ for fitting the model (7.2) by least squares (in the notation of Chapter 5).

---

## 7.2 General theory

Let us suppose that one response is taken at each of a series of values $x_i, i = 1, 2, \ldots, n$, of an explanatory variable $x$, and let the response variables be denoted $y_i$, for $i = 1, 2, \ldots, n$. The simplifications noted in Section 7.1 arose from imposing the restrictions

$$\sum_{i=1}^{n} f_r(x_i) = 0 \qquad r = 1, 2$$

and

$$\sum_{i=1}^{n} f_1(x_i) f_2(x_i) = 0$$

where both sums run over the observations.

In order to see whether it is possible in general to find polynomials satisfying these equations, we write

$$f_1(x) = a_{11} x + a_{10}$$
$$f_2(x) = a_{22} x^2 + a_{21} x + a_{20}$$

and then we have five parameters, and only three equations (one equation for $f_1(x)$ and two for $f_2(x)$). Therefore one parameter of each polynomial can be set arbitrarily, and it is convenient to set the coefficient of the highest power of $x$ equal to unity.

Therefore, in general, instead of Equation (7.1) we write

$$E(Y|x) = \alpha + \beta_1 f_1(x) + \beta_2 f_2(x) + \ldots + \beta_p f_p(x)$$
$$V(Y|x) = \sigma^2 \tag{7.7}$$

where $f_r(x)$ is a polynomial of degree $r$ in $x$ satisfying the restrictions

$$\sum_{i=1}^{n} f_r(x_i) = 0 \qquad r = 1, 2, \ldots, p \tag{7.8}$$

and

$$\sum_{r=1}^{n} f_r(x_i) f_s(x_i) = 0 \qquad \begin{cases} s = 1, 2, \ldots, (r-1) \\ r = 2, 3, \ldots, p \end{cases} \tag{7.9}$$

Equation (7.8) sets the sums of the polynomials to zero, when summed over the observations, and Equation (7.9) sets all of the sums of products of the polynomials to zero. We can write the $r$th-degree polynomial as follows:

$$f_r(x) = a_{r,r} x^r + a_{r,r-1} x^{r-1} + \ldots + a_{r,1} x + a_{r,0}$$

and we have $(r+1)$ parameters, while there is one equation from (7.8) to satisfy and $(r-1)$ equations in (7.9) to satisfy. Again we see that there is a solution, and one parameter (or condition) can be set arbitrarily. Polynomials satisfying the restrictions (7.8) and (7.9) are said to be *orthogonal*, and we should notice that this property is a function of the particular values of $x$ used. In the next section we derive some particular polynomials as an example.

Now let us obtain the least-squares estimators in this general setting, using the model (7.7). We write the unknown parameters as follows:

$$\boldsymbol{\theta}' = (\alpha, \beta_1, \beta_2, \dots, \beta_p)$$

and the response variables are written

$$\mathbf{Y}' = (Y_1, Y_2, \dots, Y_n)$$

Then if we use the notation of Chapter 5 again, we write $E(\mathbf{Y}) = \mathbf{a}\boldsymbol{\theta}$, where

$$\mathbf{a} = \begin{bmatrix} 1 & f_1(x_1) & \cdots & f_p(x_1) \\ 1 & f_1(x_2) & & f_p(x_2) \\ \vdots & & & \\ 1 & f_1(x_n) & \cdots & f_p(x_n) \end{bmatrix}$$

so that by using Equations (7.8) and (7.9) we have

$$\mathbf{a}'\mathbf{a} = \begin{bmatrix} n & 0 & \cdots & 0 \\ 0 & \sum_{i=1}^{n} f_1^2(x_i) & 0 & 0 \\ \vdots & & & 0 \\ 0 & & 0 & \sum_{i=1}^{n} f_p^2(x_i) \end{bmatrix} \qquad (7.11)$$

Now, again from Chapter 5, the least-squares estimators are given by

$$\hat{\boldsymbol{\theta}} = (\mathbf{a}'\mathbf{a})^{-1}\mathbf{a}'\mathbf{Y}$$

so that we need the matrix $(\mathbf{a}'\mathbf{Y})$, and this is seen to be

$$\mathbf{a}'\mathbf{Y} = \begin{bmatrix} \sum_{i=1}^{n} Y_i \\ \sum_{i=1}^{n} Y_i f_1(x_i) \\ \vdots \\ \sum_{i=1}^{n} Y_i f_p(x_i) \end{bmatrix}$$

Therefore by substituting these matrices in the equation for the estimators we see that

$$\hat{\alpha} = \frac{1}{n} \sum_{i=1}^{n} Y_i \qquad (7.12)$$

and

$$\hat{\beta}_r = \frac{\sum_{i=1}^{n} Y_i f_r(x_i)}{\sum_{i=1}^{n} f_r^2(x_i)} \qquad r = 1, 2, \ldots, p \qquad (7.13)$$

Theorem (5.1) tells us that these estimators are all unbiased and have a variance–covariance matrix

$$V(\hat{\theta}) = (\mathbf{a'a})^{-1} \sigma^2$$

Since the matrix in Equation (7.11) is diagonal, this shows that all of the parameters are uncorrelated, and have variances

$$V(\hat{\alpha}) = \sigma^2/n \qquad (7.14)$$

and

$$V(\hat{\beta}_r) = \sigma^2 \left/ \left\{ \sum_{i=1}^{n} f_r^2(x_i) \right\} \right. \qquad r = 1, 2, \ldots, p \qquad (7.15)$$

If the response variables are also normally distributed, then all of the parameters are independently and normally distributed with variances given by Equations (7.14) and (7.15).

A further feature of the orthogonal case, noted at the end of Chapter 5, is that the sums of squares due to fitting the model can be separated. From Equation (5.35), the sum of squares due to regression is

$$S_{\text{par}} = \hat{\theta} \mathbf{a' Y}$$

and in this case we obtain

$$S_{\text{par}} = \frac{1}{n} \left( \sum_{i=1}^{n} Y_i \right)^2 + \left\{ \sum_{i=1}^{n} Y_i f_1(x_i) \right\}^2 \left/ \sum_{i=1}^{n} f_1^2(x_i) \right. \qquad (7.16)$$
$$+ \ldots + \left\{ \sum_{i=1}^{n} Y_i f_p(x_i) \right\}^2 \left/ \sum_{i=1}^{n} f_p^2(x_i) \right.$$

It follows from the results above that the $(p + 1)$ terms on the right-hand side of Equation (7.16) are all statistically independent. Thus the sum of squares due to fitting $\beta_r$ is

$$\left\{ \sum_{i=1}^{n} Y_i f_r(x_i) \right\}^2 \left/ \left\{ \sum_{i=1}^{n} f_r^2(x_i) \right\} \right. \qquad (7.17)$$

Table 7.3 *Analysis of variance for orthogonal polynomials*

| Source | CSS | d.f. |
|---|---|---|
| Due to $\beta_1$ | $\left\{\sum_{i=1}^{n} y_i f_1(x_i)\right\}^2 \Big/ \left\{\sum_{i=1}^{n} f_1^2(x_i)\right\}$ | 1 |
| $\vdots$ | | |
| Due to $\beta_p$ | $\left\{\sum_{i=1}^{n} y_i f_p(x_i)\right\}^2 \Big/ \left\{\sum_{i=1}^{n} f_p^2(x_i)\right\}$ | 1 |
| Residual | (by subtraction) | $(n-p-1)$ |
| Total | $\sum_{i=1}^{n} (y_i - \bar{y})^2$ | $(n-1)$ |

This involves the same terms as the estimator of Equation (7.13), but here the numerator is squared.

Again following Chapter 5, the usual form of the analysis-of-variance table is as shown in Table 7.3. This is written in terms of observations $y_i$ to emphasize that it is valid as a summary of the data.

Tests of significance can now be made for each term separately. It is normal practice to include in the final model all polynomials up to the highest power which is significant.

### 7.3 Derivation of the polynomials

In order to derive the polynomials, we merely have to solve Equations (7.8) and (7.9) for the particular set of $x$-values being considered, and for the required number of polynomials. Let us illustrate what is involved by finding the linear and quadratic polynomials for Example 7.1.

It is convenient to transform the scale of the distances (which is the explanatory variable here) so that we have

$$x = (\text{distance} - 2)/2, \quad \text{and} \quad x = 0, 1, 2, 3.$$

The polynomials are then written

$$f_1(x) = x + b$$
$$f_2(x) = x^2 + cx + d$$

with the leading coefficients unity.

Now $f_1(x)$ must satisfy Equation (7.8), which is

$$\sum_{i=1}^{4} f_1(x_i) = 0$$

or

$$6 + 4b = 0, \quad \text{so that} \quad b = -1\tfrac{1}{2}$$

and

$$f_1(x) = x - 1\tfrac{1}{2} \tag{7.18}$$

The quadratic polynomial must satisfy Equations (7.8) and (7.9),

$$\sum_{i=1}^{4} f_2(x_i) = 0 \tag{7.19}$$

and

$$\sum_{i=1}^{4} f_2(x_i)f_1(x_i) = 0 \tag{7.20}$$

Equation (7.19) is

$$14 + 6c + 4d = 0 \tag{7.21}$$

Equation (7.20) is

$$\sum_{i=1}^{4} f_2(x_i)(x - 1\tfrac{1}{2}) = 0$$

and by using Equation (7.19) this reduces to

$$\sum_{i=1}^{4} x_i f_2(x_i) = 0$$

which is

$$36 + 14c + 6d = 0 \tag{7.22}$$

The solution of Equations (7.21) and (7.22) for $c$ and $d$ yields $c = -3$, $d = 1$, so that

$$f_2(x) = x^2 - 3x + 1 \tag{7.23}$$

The values of the functions in Equations (7.18) and (7.23) are shown in Table 7.4.

At this point it is important to notice that the parameter estimates (7.13), their variances (7.15), and the sums of squares (7.17) are all obtained using the polynomial function values such as those tabulated in Table 7.4. It is *not* necessary to go back to the actual functions (7.18) and (7.23) to get any of these.

For Example 7.1 the model we are fitting is

$$E(Y \mid x) = \alpha + \beta_1 f_1(x) + \beta_2 f_2(x)$$

where $f_1(x)$ and $f_2(x)$ are given by Equations (7.18) and (7.23).

Table 7.4 *Orthogonal polynomial values for Example 7.1*

| $x$ | 0 | 1 | 2 | 3 | Sum | Sum of squares |
|---|---|---|---|---|---|---|
| $f_1(x)$ | $-1\tfrac{1}{2}$ | $-\tfrac{1}{2}$ | $\tfrac{1}{2}$ | $1\tfrac{1}{2}$ | 0 | 5 |
| $f_2(x)$ | 1 | $-1$ | $-1$ | 1 | 0 | 4 |

Therefore from Equation (7.13) we have

$$\hat{\beta}_1 = \frac{-1.5(238) - 0.5(270) + 0.5(298) + 1.5(283)}{(1.5)^2 + (0.5)^2 + (0.5)^2 + (1.5)^2}$$
$$= 81.5/5 = 16.3$$

The sum of squares for fitting the linear term is obtained using Equation (7.17), and as pointed out above, this involves the same terms as we calculated for the estimate $\hat{\beta}_1$; we get

$$\text{SS(linear)} = (81.5)^2/5 = 1328.45$$

Similarly, we get

$$\hat{\beta}_2 = (238 - 270 - 298 + 283)/4 = -11.75$$
$$\text{SS(quadratic)} = (47)^2/4 = 552.25$$

The total sum of squares is calculated in the usual way and the result is shown in Table 7.5.

At this point we would normally carry out significance tests separately for the polynomials, but there is little point in doing it here with only one degree of freedom for error. The main purpose of Example 7.1 is to illustrate the calculations.

It is important to notice that the sums of squares in Table 7.5 do not depend on the scale of the explanatory variable being used. If the $r$th polynomial was multiplied by a constant $\lambda_r$ say, Equation (7.17) would be unaffected, though the estimates (7.13) would be changed. Another important feature is that since we can always change the scale of our explanatory variable (and transform back after our model is fitted), the table of polynomial function values given in Table 7.4 could be used to fit linear and quadratic terms to any set of four observations with equally spaced $x$-values.

Now although the procedure outlined above for obtaining the polynomials is straightforward, the amount of work involved would be very large for polynomials of even a moderate degree. For example, for a sixth-degree polynomial the set of equations equivalent to (7.21) and (7.22) would be six equations in six unknowns. The orthogonal polynomial method is therefore used in two situations:

Table 7.5 *Analysis of variance for Example 7.1*

| Source | CSS | d.f. |
|---|---|---|
| Linear | 1328.45 | 1 |
| Quadratic | 552.25 | 1 |
| Deviations | 76.05 | 1 |
| Total | 1956.75 | 3 |

(i) if we are fitting a polynomial function repeatedly at the same $x$-values, we could work out our own special table equivalent to Table 7.4;

(ii) for equally spaced $x$-values, tables equivalent to Table 7.4 have been constructed, and these will be described in the next section.

**Exercises 7.3**
1. Write out the model fitted to Example 7.1 in powers of the distance, and differentiate to find the optimum.

## 7.4 Tables of orthogonal polynomials

Appendix C contains three tables relevant to orthogonal polynomials, Tables C.11, C.12 and C.13.

In order to understand Appendix C Table C.11, compare the entries for the linear and quadratic terms for $n = 4$, with Table 7.4. We see that for the linear polynomial, fractions have been avoided by multiplying through by 2, and this is indicated in the entry for $\lambda$ in Table C.11. The sums of squares of the function values are also given, ready for immediate use in Equation (7.13), (7.15) or (7.17). For $n \geq 7$ the values tabulated in Table 7.9 have been reduced by arguments based on symmetry. Only the top half of the function values is given, and the bottom half of odd polynomials differ in sign.

Appendix C, Table C.12 gives the first five polynomial functions resulting from the procedure described in the previous section, for $n$ points spaced at unit intervals and centred at zero. Table C.13 gives some of the coefficients of powers of $x$ which result from Table C.12.

Further values of some of these tables are given in *Biometrika Tables for Statisticians*, Volume 1; *Fisher and Yates' Statistical Tables*; or other advanced sets of tables.

It is important to notice that when using a polynomial in Table C.11 which has a $\lambda$ value *not* equal to unity, then this value of $\lambda$ should be used in the function given in Table C.12 when finally fitting the equation. All of the multiplying constants have been taken into account in Table C.13, and the coefficients are simply read off.

## 7.5 An illustrative example

In this section we shall discuss the analysis of Example 7.2 in order to illustrate the techniques involved. Table 7.6 below shows the data scaled by the transformation

$$y = 10^4(-\log_e K_p - 5.5) \tag{7.24}$$

together with the linear quadratic and cubic polynomials read off from Appendix C, Table C.11 for $n = 9$, and some of the calculations.

Table 7.6 *Data and calculations for Example 7.2*

| $y$ | $f_1$ | $f_2$ | $f_3$ | | |
|---|---|---|---|---|---|
| 103 | $-4$ | 28 | $-14$ | $\bar{y} =$ | $3\,884$ |
| 1012 | $-3$ | 7 | 7 | $CS(y, y) =$ | $55\,127\,038$ |
| 1954 | $-2$ | $-8$ | 13 | | |
| 2894 | $-1$ | $-17$ | 9 | $\sum yf_1 =$ | $57\,508$ |
| 3847 | 0 | $-20$ | 0 | | |
| 4820 | 1 | $-17$ | $-9$ | $\sum yf_2 =$ | $4476$ |
| 5796 | 2 | $-8$ | $-13$ | | |
| 6774 | 3 | 7 | $-7$ | $\sum yf_3 =$ | $-472$ |
| 7756 | 4 | 28 | 14 | | |
| $\sum f_i^2$ | 60 | 2772 | 990 | | |

Equations (7.13) and (7.17) now yield the estimates of the coefficients and also the amount of sums of squares attributable to the various terms.

Linear: $\quad \hat{\beta}_1 = \sum yf_1 / \sum f_1^2 = 57\,508/60 = 958.47$

$\quad\quad\quad\quad SS = (\sum yf_1)^2 / \sum f_1 = (57\,508)^2/60 = 55\,119\,501$

Quadratic: $\hat{\beta}_2 = \sum yf_2 / \sum f_2^2 = 4476/2772 = 1.6147$

$\quad\quad\quad\quad SS = (\sum yf_2)^2 / \sum f_2^2 = (4476)^2/2772 = 7227$

Cubic: $\quad \hat{\beta}_3 = \sum yf_3 / \sum f_3^2 = -472/990 = -0.4768$

$\quad\quad\quad\quad SS = (\sum yf_3)^2 / \sum f_3^2 = (-472)^2/990 = 225$

The analysis of variance can be written out as in Table 7.7, taking out the linear, linear and quadratic, etc., terms until a term is included which is not significant.

The linear and quadratic terms are both highly significant. The cubic term is significant at the 2.5 per cent level, and if the quartic term is calculated it turns out to be non-significant. The model with linear, quadratic and cubic terms is therefore the most reasonable, so that we have our estimated regression

$$\widehat{E(Y_i|x_i)} = 3884 + 958.47f_1(x) + 1.615f_2(x) - 0.4768f_3(x) \quad (7.25)$$

where the polynomials are listed in Appendix C, Table C.11 and the scale for $x$ is the scale for $f_1(x)$

$$x = (\text{temp.} - 25)/5 \quad (7.26)$$

Before transforming equation Equation (7.25) back to temperature, we can calculate the residuals. This is done simply by inserting the values of the polynomials from Table 7.6 into Equation (7.25), and calculating

$$r_i = y_i - \widehat{E(Y_i|x_i)}$$

The residuals for the linear and quadratic terms only are shown in Table 7.8.

Table 7.7 *Analysis of variance for Example 7.2*

|  | CSS | d.f. | Mean square | F |
|---|---|---|---|---|
| Linear | 55 119 501 | 1 | 55 119 501 | (large) |
| Devs | 7 537 | 7 | 1 077 | |
| Total | 55 127 038 | 8 | | |
| Linear | 55 119 501 | 1 | | |
| Quadratic | 7 227 | 1 | 7 227.0 | 140 |
| Devs | 310 | 6 | 51.7 | |
| Total | 55 127 038 | 8 | | |
| Linear | 55 119 501 | 1 | | |
| Quadratic | 7 227 | 1 | | |
| Cubic | 225 | 1 | 225 | 13.2 |
| Devs | 85 | 5 | 17 | |
| Total | 55 127 038 | 8 | | |

Table 7.8 *Residuals for Example 7.2 (linear and quadratic terms only)*

| Obs. | Residual | Obs. | Residual |
|---|---|---|---|
| 103 | 7.67 | 4820 | 4.98 |
| 1012 | − 7.89 | 5796 | 7.98 |
| 1954 | − 0.14 | 6774 | 3.30 |
| 2894 | − 4.08 | 7756 | − 7.09 |
| 3847 | − 4.71 | | |

The sum of squares of the residuals comes to 310, as shown in Table 7.7. There is a very clear pattern in these residuals, showing the need for a cubic term. The calculation of residuals including the cubic term is left to the reader, and checks for normality, heteroscedasticity (which means lack of homogeneity of variance), etc., should be done.

Finally, the fitted equation should be expressed in temperature and conductivity, and there are two stages in this. Firstly, we read off the polynomials in Appendix C:

$$f_1(x) = x, \qquad f_2(x) = 3x - 20, \qquad f_3(x) = (5x^3 - 59x)/6 \qquad (7.27)$$

Then we insert these with Equations (7.26) and (7.24) into Equation (7.25). This is left to the reader. It is important to notice that it is only at this final stage that the actual polynomials (7.27) are required; for all other purposes it is easier to use the tables of function values.

## Exercises 7.5

1. The data given below are average heights of sunflowers. It is thought

that a cubic polynomial may fit the data adequately.

| Week | 1 | 2 | 3 | 4 | 5 | 6 | 7 | 8 | 9 | 10 | 11 | 12 |
|------|---|---|---|---|---|---|---|---|---|----|----|----|
| Height (inches) | 18 | 36 | 68 | 98 | 131 | 170 | 206 | 228 | 247 | 250 | 254 | 254 |

Suggest a more suitable model for this data, and describe how the model you give could be fitted.

2. Find the orthogonal polynomials when the values of $x$ are 0, 1, 3.

3.* An experimenter wishes to fit a quadratic polynomial, and takes the following numbers of observations at the given values of the explanatory variable.

| $x$ | −2 | −1 | 0 | 1 | 2 |
|-----|----|----|---|---|---|
| No. of observations | 1 | 2 | 3 | 2 | 1 |

  (i) Work out orthogonal polynomials to use in this experiment.
 (ii) If the response variables have a variance $\sigma^2$, obtain the variance of the response function.

4.* In a plant-scale experiment on the production of a certain chemical, a batch of intermediate product was divided into six equal portions which were then processed on successive days by two different methods $P_1$ and $P_2$. The order of treatment and the yields were:

| Day | 1 | 2 | 3 | 4 | 5 | 6 |
|-----|---|---|---|---|---|---|
| Process | $P_1$ | $P_2$ | $P_1$ | $P_1$ | $P_1$ | $P_2$ |
| Yield | 5.84 | 5.73 | 7.30 | 10.46 | 9.71 | 5.91 |

It was expected that superposed on any process effect there would be a smooth, roughly parabolic trend. Experience of similar experiments showed that the standard deviation of a single observation was about 1.

(a) Estimate the process effect and the trend, and find the standard errors of your estimates.
(b) Estimate the time delay corresponding to maximum yield and give a 95 per cent confidence interval for the true delay.

# CHAPTER EIGHT

# The use of transformations

## 8.1 Introduction

When analysing any given set of data, we often seek to use multiple regression or standard analysis of variance. That is, we try for a model of the form

$$E(Y) = \alpha_0 + \alpha_1 x_1 + \alpha_2 x_2 + \dots + \alpha_k x_k$$

$$V(Y) = \sigma^2$$

(8.1)

where the distribution of the random variables can be taken to be normal. In order to try for a model of this type, we can transform the response variable, or the explanatory variables, or both. The transformations used could be designed to satisfy any of three different criteria:

(i) We would seek transformations which make the relationship between the (transformed) response and explanatory variables as simple as possible. It would be convenient if the model was linear in both the unknown parameters and the transformed explanatory variables.
(ii) We could seek transformations which make the variance homogeneous.
(iii) We could seek transformations which make the distribution of the transformed response variables more nearly normal.

Clearly, in the general case, the three methods are likely to lead to different transformations. In fact, the precise transformation used is often not too critical, and several transformations exist which approximately satisfy all three criteria. An approach by Box and Cox (1964), which we shall study below, uses maximum likelihood to derive a transformation which best satisfies all three criteria simultaneously.

In some cases there may be objections to using transformations of $y$ at all. For example, the scale used for $y$ may have a very definite meaning, such as cost, or man-hours, and it may be undesirable to transform it. A different kind of objection can arise with cases such as the use of the square-root transformation with the Poisson distribution. The square-root transformation gives Poisson variables nearly homogeneous variance (see below), and renders the distribution more nearly normal, and $\sqrt{(X + 3/8)}$ does slightly better. However, if the purpose of the analysis is to include the building of a model, people may object to using models in such an arbitrary scale.

Another factor which sometimes influences the form of a relationship is considerations of the range of $x$ over which the assumed form can be a reasonable approximation to the true relationship. For example, in experiments to determine the maximum yield of a process, the observations are yields, and the explanatory variables are process parameters. The true response is likely to be something like the full-line curve in Fig. 8.1. A quadratic polynomial (dotted-line curve in Fig. 8.1), is likely to fit well near the optimum, but is seriously in error outside the immediate neighbourhood of the optimum. Notwithstanding this deficiency, quadratic polynomial approximations are used a great deal in this type of work, where the aim is to estimate the point of maximum yield. Clearly, if a function was required which approximated to the true response over a wide range of values of $x$, then a very much more complicated response function would have to be used, which tended to the right asymptotes for extreme values of the regressor variable.

Another example of this last point occurs in bioassay work, in which the potency of drugs is estimated by the response of human or animal subjects. Several doses of known potency are given, the responses measured, and a curve fitted. Then an 'unknown' is administered, and the potency

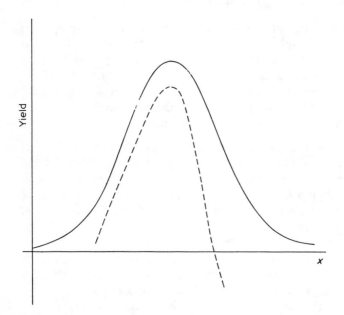

Fig. 8.1 Maximum of a response surface.

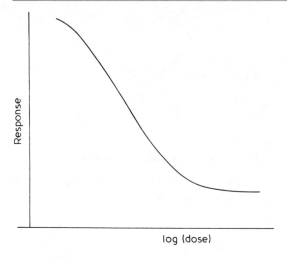

Fig. 8.2 Bioassay response curve.

estimated from the curve. The general shape of the curve is often as given in Fig. 8.2, and a simple polynomial approximation would probably be satisfactory over the important central range of $x$-values. However, the asymptotes for zero and infinite dose are often estimable, and a polynomial approximation would be unlikely to tend to the right values at these extremes. For this reason, very complicated response functions are often used in this context.

If in any particular case the use of transformations can reduce a complicated model to something of the form (8.1), which is linear in the unknown parameters, and which is also linear in the transformed variables, then our model will be easier to understand, and the statistical methods will be much simpler than if methods for fitting non-linear models have to be used. We therefore need to outline methods of finding suitable transformations.

## 8.2  One explanatory variable

If our purpose is simply to linearize the relationship between $E(Y)$ and $x$, then with one explanatory variable someone with sufficient mathematical knowledge can often devise suitable transformations. Use of the functions

$$x, 1/x, \log_e x, \sqrt{x}, x^\beta, e^{\beta x}$$

separately for $y$ and $x$ can represent a very wide variety of curves. The following list gives some of the possibilities, and Fig. 8.3 shows graphs of six cases, for positive values of all parameters except where stated:

(i)      $y = \alpha + \beta/x$

(ii)     $y = \alpha + \beta \log_e x$

(iii)    $y = 1/\{\alpha + \beta \exp(\gamma x)\}$      'logistic response curve'

(iv)    $1/y = \alpha + \beta x$

(v)     $1/y = \alpha + \beta/x$

(vi)    $\log_e y = \alpha + \beta x$

(vii)   $\log_e y = \alpha + \beta/x$

(viii)  $y = \alpha + \beta\sqrt{x}$

(ix)    $y = \alpha + \beta x^\gamma$

(x)     $y = \alpha + \beta \exp(\gamma/x)$

(xi)    $y = \alpha + \beta \exp(\gamma x)$     'exponential regression'

The relationships (i) to (viii) can be linearized in an obvious way. The remaining relationships cannot be linearized without knowledge of some of the

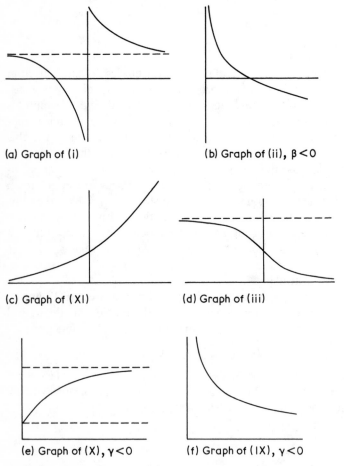

(a) Graph of (i)

(b) Graph of (ii), $\beta < 0$

(c) Graph of (XI)

(d) Graph of (iii)

(e) Graph of (X), $\gamma < 0$

(f) Graph of (IX), $\gamma < 0$

Fig. 8.3 Graphs of some transformations.

parameters, but they are relatively common. It is important to realize that while some relationships can be linearized by simple transformations, this is not always possible.

## 8.3 Transformations for homogeneity of variance

Transformations for homogeneity of variance are based upon a theorem which can be described as follows. If $T_n$ is a sequence of statistics for which the asymptotic distribution is normal with

$$E(T_n) = \theta \qquad V(T_n) = \sigma^2(\theta)/n$$

then for any function $g(\theta)$ for which $g'(\theta)$ exists and is not zero, the asymptotic distribution of $g(T_n)$ is normal with

$$E\{g(T_n)\} = g(\theta) \qquad V\{g(T_n)\} = [g'(\theta)\sigma(\theta)]^2/n$$

A demonstration of the result is as follows. Let the distribution of $X$ be $N(\theta, \sigma^2)$, and let $g(x)$ be a function which can be expanded in the form

$$g(x) = g(\theta) + (x - \theta)\left(\frac{dg}{dx}\right)_\theta + \frac{(x - \theta)^2}{2}\left(\frac{d^2g}{dx^2}\right)_\theta + \cdots$$

If we take expectations we have

$$E\{g(x)\} = g(\theta) + \frac{\sigma^2(\theta)}{2}\left(\frac{d^2g}{dx^2}\right)_\theta + \cdots$$

and similarly we obtain as an approximation

$$V\{g(x)\} = \sigma^2(\theta)\left\{\frac{dg}{dx}\right\}_\theta^2$$

In order to determine a transformation for which $V\{g(x)\} = \text{const.}$, we therefore require

$$g'(\theta)\sigma(\theta) = \text{constant}$$

Therefore we have

$$g(\theta) = \int \frac{c\,d\theta}{\sigma(\theta)} \qquad (8.2)$$

Some important applications of this formula are as follows.

*Example* 8.1   If the results of an experiment are such that the variance increases with the mean $\theta$, and $\sigma^2(\theta) = k\theta$, then we use

$$g(\theta) = c\int \frac{d\theta}{\sqrt{\theta}} = 2c\sqrt{\theta}$$

i.e., we use $g(x) = \sqrt{x}$.   □ □ □

*Example* 8.2   If we have $\sigma^2(\theta) = k\theta^2$, then we use

$$g(\theta) = c \int \frac{d\theta}{\theta} = c \log_e \theta \qquad \square\,\square\,\square$$

i.e., we use $g(x) = \log_e x$.

*Example* 8.3   If $\sigma^2(\theta) = k\theta^4$, then we use $g(x) = 1/x$. $\qquad \square\,\square\,\square$

If we are analysing data which are replicated in groups, then we should calculate the sample variance of each group, and plot the variances against the group means. Often it turns out that one of the transformations given in Examples 8.1 to 8.3 is required, and the graph of variances against means will show which one to use.

An important special case of the above theory is when we are analysing discrete data and either the binomial or Poisson distribution applies.

*Example* 8.4   If $X$ has a Poisson distribution with expectation $\theta$, then $V(X) = \theta$, and from Example 8.1 we use $\sqrt{x}$. This can be improved slightly by considering the expansion

$$\sqrt{(X + b)} = \sqrt{(\theta + b)}\sqrt{\{1 + (X - \theta)/(\theta + b)\}}$$

$$= \sqrt{(\theta + b)}\left\{1 + \frac{1}{2}\frac{(X - \theta)}{(\theta + b)} - \frac{1}{8}\frac{(X - \theta)^2}{(\theta + b)^2} + \ldots\right\}$$

By taking expectations we obtain

$$E(X + b) = \sqrt{(\theta + b)} - \frac{1}{8\sqrt{\theta}} + O(\theta^{-3/2})$$

$$V(X + b) = \frac{1}{4}\left\{1 + \frac{(3 - 8b)}{8\theta} + O(\theta^{-2})\right\}$$

From this we see that use of the transformation $\sqrt{(X + \frac{3}{8})}$ will make the variance more nearly constant. $\qquad \square\,\square\,\square$

*Example* 8.5   If $X$ has a binomial distribution with parameters $n$, $\theta$, then $V(X) = n\theta(1 - \theta)$ and we require

$$g(\theta) = \int \frac{c\,d\theta}{\sqrt{\{\theta(1 - \theta)\}}} = k \sin^{-1}\sqrt{\theta}$$

Therefore we use $g(x) = \sin^{-1}\sqrt{(x/n)}$. A slightly better transformation is

$$g(x) = \sin^{-1}\sqrt{\left\{\left(x + \frac{3}{8}\right)\Big/\left(n + \frac{3}{4}\right)\right\}} \qquad \square\,\square\,\square$$

Although the transformations given in Examples 8.4 and 8.5 are rather complex, they may be satisfactory to use in an initial analysis of a complicated set of data. It depends on the purpose of our analysis.

### 8.4 An example

The use of transformations in order to fit a model of the form (8.1) is well illustrated by the following example.

*Example* 8.6   Snoke (1956) has described experiments for the bioassay of wood preservatives. Blocks of $\frac{3}{4}$ inch cubes of wood were prepared, impregnated with a preservative, and then exposed to a particular strain of fungus for 90 days. After this the fungus growth (if any) was carefully brushed off and the weight loss noted. The amount of preservative in a block of wood is calculated on a scale called the retention level, and it was not possible accurately to attain desired retention values. Some typical results are shown in Table 8.1. A model is required to form the basis of a statistical analysis, and to facilitate comparisons between the results of separate experiments. (In Table 8.1, $R$ = retention level, and $W = \%$ weight loss.)

The data are plotted out in Fig. 8.4 and two features stand out. Firstly, the relationship between $W$ and $R$ seems to be approximately negative exponential in form. This indicates that we should try plotting $\log W$ against $R$ to see if the relationship is linear. Secondly, it is very clear from Fig. 8.4 that the variance of the observations is not homogeneous. In fact Table 8.1 shows that the range of $W$ is approximately proportional to the mean of $W$. It therefore follows from Example 8.2 above that by taking $\log W$, the sample variances should be made more homogeneous.

In this example therefore we are fortunate that the same transformation

Table 8.1  *Typical results for bioassay of wood preservatives*

| Group | 1 | | 2 | | 3 | | 4 | | |
|---|---|---|---|---|---|---|---|---|---|
| | R | W | R | W | R | W | R | W | |
| | 1.99 | 19.95 | 3.01 | 2.50 | 4.16 | 1.61 | 5.86 | 0.22 | |
| | 1.96 | 25.42 | 2.85 | 5.56 | 4.03 | 1.14 | 6.06 | 0.45 | |
| | 1.96 | 22.33 | 3.03 | 8.82 | 4.01 | 0.69 | 6.09 | 0.05 | |
| | 1.98 | 25.32 | 3.04 | 4.11 | 4.10 | 0.47 | 6.15 | 0.22 | |
| | 1.98 | 14.81 | 3.05 | 5.53 | 4.03 | 2.06 | 6.04 | 0.44 | |
| | 2.00 | 20.88 | 3.09 | 8.20 | 4.19 | 0.69 | 6.06 | 0.22 | |
| | 1.97 | 9.05 | 2.98 | 0.69 | 4.09 | 0.23 | 6.04 | 0.67 | |
| | 1.85 | 11.49 | 3.05 | 7.67 | 4.07 | 0.91 | 6.22 | 0.23 | |
| | 2.01 | 12.30 | 3.09 | 7.60 | 4.12 | 2.28 | 5.88 | 0.22 | |
| | 2.02 | 9.31 | 3.04 | 7.05 | 4.01 | 0.69 | 5.86 | 0.45 | |
| | | 17.09 | | 5.77 | | 1.08 | | 0.32 | Mean ($W$) |
| | | 16.37 | | 8.13 | | 2.05 | | 0.62 | Range ($W$) |

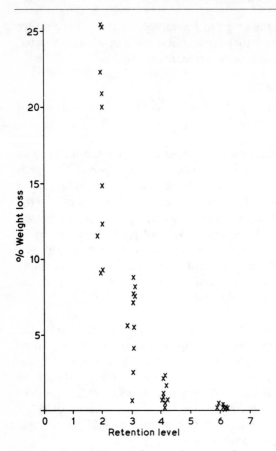

Fig. 8.4 Scatter diagram for wood preservative experiment.

($\log W$) linearizes the relationship and makes the variance more homogeneous. In order to check this the laboratory notebooks were borrowed, and $\log W$ was plotted against $R$ for a number of experiments. In every case the model

$$E(\log W) = \alpha + \beta R$$

$$V(\log W) = \sigma^2$$

(8.3)

appeared to be reasonable. A further check of the model can be made by applying the method of simple linear regression, and then plotting residuals, as indicated in Chapter 2.

The data given in Table 8.1 have been modified slightly; the original data set contained some zero and negative values of the weight loss, which cannot be accounted for by the model (8.3). The weight gains were probably due to factors such as assimilation of water from the atmosphere of the test chamber. A model which accounted for these zero and negative values would have to be more complicated than (8.3), and we shall not discuss this here.  □ □ □

It is clearly very convenient if one simple transformation enables us to fit the standard linear model, and this happens more frequently in practice than one might expect. The next section discusses a method of finding such a transformation.

---

### Exercises 8.4

1. Plot log $W$ against $R$ for the data given in Table 8.1. Under what circumstances would you be prepared to use model (8.3) for this problem? What alternative model would you suggest?

---

### 8.5 The Box–Cox transformation

So far we have not mentioned transformations to normality. We can always transform to normality, by making use of a standard result in probability theory. If a random variable $X$ has a distribution $f(x)$, and $y = H(x)$ is a strictly monotone function of $x$, then the p.d.f. of $Y = H(X)$ is

$$f(x)\left|\frac{dx}{dy}\right| \tag{8.4}$$

and we can readily check that whatever $f(x)$, we can always find a transformation $H(x)$ so that (8.4) is the normal distribution. However, we would be very fortunate if the transformation determined in this way linearized our relationship (Section 8.2), and gave a homogeneous variance (Section 8.3). Any transformation determined specially to do one of the three tasks may well fail to do either of the others.

Box and Cox (1964) developed an approach in which they sought for a transformation which simultaneously makes the model linear in the parameters, the variance homogeneous and the distribution normal. They consider a family of transformations

$$y^{(\lambda)} = \begin{cases} (y^\lambda - 1)/\lambda & \lambda \neq 0 \\ \log y & \lambda = 0 \end{cases} \tag{8.5}$$

for $y > 0$, and then assume that in terms of the transformed model

$$E(Y^{(\lambda)}) = \mathbf{a}\theta \tag{8.6}$$

$$V(Y^{(\lambda)}) = \mathbf{I}\sigma^2 \tag{8.7}$$

and that the observations are normally distributed. The method is now to estimate the value of $\lambda$ by maximum likelihood. Thus we assume that the three basic assumptions (5.4), (5.5) and normality hold simultaneously in some transformed scale, and we then estimate what that scale is.

Before going into the method we should emphasize a point made briefly in Section 8.1, that there may be objections to using a special transformation to achieve a standard linear model. This happens when the original scale is directly meaningful, such as yield, and results in any other scale would be difficult to interpret. Even in cases where transformations are permitted, there may be objections to the full range of possibilities given by Equation (8.5). In practice there is likely to be a preference for a transformation such as square-root ($\lambda = \frac{1}{2}$) or log ($\lambda = 0$) if it is reasonably close to the optimum $\lambda$ value, rather than using other values for $\lambda$. The usefulness of the Box–Cox method is that it picks out a range of $\lambda$-values from which to choose a suitable (near optimum) value.

Given the model (8.5), (8.6) and (8.7), the likelihood in relation to the original observations is

$$\frac{1}{(2\pi)^{n/2}\sigma^n} \exp\left\{-(\mathbf{y}^{(\lambda)} - \mathbf{a}\boldsymbol{\theta})'(\mathbf{y}^{(\lambda)} - \mathbf{a}\boldsymbol{\theta})/2\sigma^2\right\} J(\lambda, \mathbf{y}) \qquad (8.8)$$

where

$$J(\lambda, \mathbf{y}) = \prod_{i=1}^{n} \left|\frac{\mathrm{d}y_i^{(\lambda)}}{\mathrm{d}y_i}\right| \qquad (8.9)$$

We proceed to find the maximum-likelihood estimates of $\boldsymbol{\theta}$ and $\lambda$ in two stages. First, for given $\lambda$ we find the $\boldsymbol{\theta}$ to minimize

$$S^{(\lambda)} = (\mathbf{y}^{(\lambda)} - \mathbf{a}\boldsymbol{\theta})'(\mathbf{y}^{(\lambda)} - \mathbf{a}\boldsymbol{\theta}) \qquad (8.10)$$

yielding

$$S_{\min}^{(\lambda)} = (\mathbf{y}^{(\lambda)} - \mathbf{a}\hat{\boldsymbol{\theta}})'(\mathbf{y}^{(\lambda)} - \mathbf{a}\hat{\boldsymbol{\theta}}) \qquad (8.11)$$

Therefore for given $\lambda$, the maximised log-likelihood is, from Equation (8.8),

$$L_{\max}(\lambda) = -\frac{n}{2}\log(2\pi\hat{\sigma}^2) - \frac{1}{2\hat{\sigma}^2} S_{\min}^{(\lambda)} + \log J(\lambda, \mathbf{y})$$

but since $\hat{\sigma}^2 = S_{\min}^{(\lambda)}/n$, we have

$$L_{\max}(\lambda) = -\frac{n}{2}\log \hat{\sigma}^2 + \log J(\lambda, \mathbf{y}) + \text{constant} \qquad (8.12)$$

For the transformation (8.5),

$$\log J(\lambda, \mathbf{y}) = (\lambda - 1)\sum \log y_i$$

so that Equation (8.12) becomes

$$L_{\max}(\lambda) = -\frac{n}{2}\log S_{\min}^{(\lambda)} + (\lambda - 1)\sum \log y_i + \text{constant} \qquad (8.13)$$

The second stage of the operation is to plot Equation (8.13) for various values of $\lambda$, and so estimate $\hat{\lambda}$. Approximate confidence intervals for $\lambda$ can be obtained using the result given in Theorem 4.6 and a value such as $0, \frac{1}{2}, 1, \ldots$ within this range will usually be preferred to $\hat{\lambda}$.

Two extensions of the procedure are worth noting. One is that in the case of multiple regression the regressor variables can be transformed by a set of similar transformations. We replace $x_1, x_2, \ldots$ by $x_1^{(k_1)}, x_2^{(k_2)}, \ldots$ for the respective regressor variables. The method and formulae are similar except that Equation (8.11) becomes $S_{\min}(\lambda, k_1, k_2, \ldots)$, and that an optimization algorithm must now be employed to find the optimum values $\hat{\lambda}, \hat{k}_1, \hat{k}_2, \ldots$, see the next section for details.

Another point is that more complicated families of transformations could be used in the procedure, such as

$$y^{(\lambda)} = \begin{cases} \{(y + \lambda_2)^{\lambda_1} - 1\}/\lambda_1 & \lambda_1 \neq 0 \\ \log(y + \lambda_2) & \lambda_1 = 0 \end{cases} \tag{8.14}$$

---

**Exercises 8.5**

1. Work out the theory for using the Box–Cox technique with the family of transformations (8.14).

2. Use the transformation (8.5) on Example 5.1 data.

---

**8.6 Transformations of regressor variables**

Box and Tidwell (1962) gave an iterative procedure for finding suitable transformations of regressor variables. It is best described in terms of a special case. Suppose we are trying to estimate a relationship between $y$ and $k$ explanatory variable $x_i, i = 1, 2, \ldots, k$, and that we are considering functions of the form

$$E(Y) = f(\xi_1, \xi_2, \ldots, \xi_k)$$

where $f$ is a simple function, say linear or quadratic in the $\xi_i$, and where

$$\xi_i = \begin{cases} x_i^{\alpha_i} & \alpha_i \neq 0 \\ \log_e x_i & \alpha_i = 0 \end{cases}$$

The parameters to be estimated are the $\alpha^i$ and the coefficients in $f$. The procedure starts off with guesses of the $\alpha_i$, say $t = \alpha_i^{(0)}$, and iterates to find optimum values.

Then we have approximately, using $\xi_i^{(0)} = x_i^t$,

$$E(Y) = f(\xi_1^{(0)}, \ldots, \xi_k^{(0)}) + \sum_{i=1}^{k} (\alpha_i - t)\left(\frac{df}{d\alpha_i}\right)_t \tag{8.15}$$

But we can write

$$\left(\frac{df}{d\alpha_i}\right)_{\hat{t}} = \left(\frac{df}{d\xi_i}\right)\left(\frac{d\xi_i}{d\alpha_i}\right)_{\hat{t}} \tag{8.16}$$

and

$$\frac{d\xi_i}{d\alpha_i} = x_i^{\alpha_i} \log_e x_i \tag{8.17}$$

Hence Equation (8.15) becomes

$$E(Y) = f(\xi_1^{(0)}, \dots, \xi_k^{(0)}) + \sum_{i=1}^{k} (\alpha_i - t)\left(\frac{df}{d\xi_i}\right)x_i^t \log_e x_i \tag{8.18}$$

If we consider the case where the function $f$ is linear in the $\xi_i$, then we write

$$f(\xi_1, \dots, \xi_k) = \sum_{i=1}^{k} \beta_i \xi_i + \beta_0 \tag{8.19}$$

and Equation (8.18) becomes

$$E(Y) = f(\xi_1, \dots, \xi_k) + \sum_{i=1}^{k} (\alpha_i - t)\beta_i x_i^t \log_e x_i \tag{8.20}$$

Let us choose $t = 1$, then the procedure is as follows.

*Step* 1  Estimate the parameters of the regression

$$E(Y) = \beta_0 + \sum \beta_i x_i$$

*Step* 2  Estimate the parameters of the regression

$$E(Y) = \beta'_0 + \sum \beta'_i x_i + \sum \gamma_i x_i \log_e x_i$$

*Step* 3  Put $\hat{\alpha}_i = (\hat{\gamma}_i/\hat{\beta}_i) + 1 = w$, say

*Step* 4  Define new x-variables, equal to $x_i^w$, and repeat the procedure until it converges.

Box and Tidwell give some numerical examples in which they remark on the rapidity with which the method converges. They also say that they have had examples in which there was wild oscillation in estimates from successive iterations. Conditions under which the method may be safely applied are not known.

There are three arbitrary elements in the above procedure which can be varied to suit the particular application. These are the choice of the functions transforming $x_i$ to $\xi_i$, the choice of $f(\xi_1, \dots, \xi_k)$, and the choice of $\alpha_i^{(0)}$.

---

### Exercise 8.6

1. Show how to use the Box–Tidwell procedure for $f$ as a quadratic function of the $\xi_i$.

---

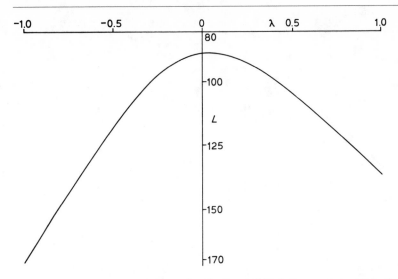

Fig. 8.5 The Box–Cox transformation for Example 8.6: linear terms only in the model.

## 8.7 Application to bioassay data

*Example* 8.7   If we apply the Box–Cox technique to Example 8.6 with the family of transformations (8.5) and with (8.6) taken to be

$$E(Y^{(\lambda)}) = \alpha + \beta R$$

then $S_{\min}^{(\lambda)}$ is the minimized sum of squares from a fitted regression line. A plot of the function (8.13) is shown in Fig. 8.5, and it is very clear from this that $\lambda = 0$ is the best value to use, confirming the logarithm transformation discussed earlier. However, if the residuals are examined they show some peculiarities. In this problem, we ought really to allow for the possibility of a quadratic term in $R$, the retention. We therefore revise our model to fit

$$E(Y^{(\lambda)}) = \alpha + \beta R + \gamma R^2$$

and the revised plot of the function (8.13) is shown in Fig. 8.6. The maximum is no longer at $\lambda = 0$, but at a value close to 0.2. We can use Theorem 4.6 to get a confidence interval for $\lambda$, and this means finding the values of $\lambda$ which have a maximized likelihood within $\frac{1}{2}(\chi_1^2)$ of the maximum. This leads to a 95 per cent confidence interval $(0.01, 0.33)$, so that $\lambda = 0$ is just excluded. In fact the maximized likelihoods are

with a quadratic term in $R$      $- 84.0$

with only a linear term in $R$      $- 88.9$

If the quadratic term were really zero, the difference between these maximized

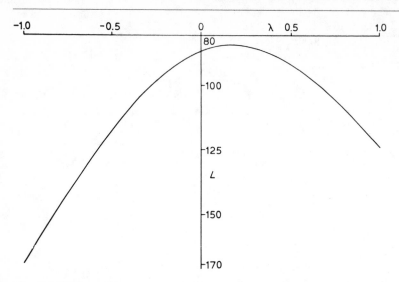

Fig. 8.6 The Box–Cox transformation for Example 8.6: quadratic terms permitted.

likelihoods would again, by Theorem 4.6, have a distribution $\frac{1}{2}(\chi_1^2)$. Since $\chi_1^2(95\%) = 3.84$, the quadratic term is seen to be significant.

This analysis has shown up more problems with modelling these data than might have been expected. The logarithm fit, though not the best, is still very good, and is very much simpler than a model involving $\lambda = 0.2$ with a quadratic term as well, and one may, therefore, be tempted to use it nevertheless.                                         □ □ □

For another numerical illustration of the technique see the source paper.

**Further reading**
For transformations for homogeneity of variance see Rao (1965, Section 6g). The papers by Box and Cox (1964), and by Box and Tidwell (1962) relate to the latter part of the chapter. Also see Bartlett (1947), Dolby (1963), and Tukey and Moore (1954).

# CHAPTER NINE

# *Correlation*

## 9.1 Definition and examples

A popular measure of the relationship between two variables is the *correlation coefficient*, which is defined in populations as

$$\rho = C(X, Y)/\sqrt{\{V(X)V(Y)\}} \tag{9.1}$$

and an estimator is given by

$$R = CS(X, Y)/\sqrt{\{CS(X, X)CS(Y, Y)\}} \tag{9.2}$$

We shall refer to a particular value of the sample coefficient by $r$.

The intuitive justification for this coefficient is clear. The covariance of two variables is a measure of linear relationship, and this is standardized by dividing by the standard deviations of the variables. Thus both population and sample values are independent of scale and location of either variable. Furthermore, it is easily shown that the coefficients lie in the range $(-1, +1)$, with $+1$ for an exact relationship in which both variables increase together, and $-1$ for an exact relationship in which one variable increases as the other decreases; see Exercise 9.2.1. For independent random variables $\rho = 0$, but unless the distribution of the random variables is bivariate normal, $\rho = 0$ does not necessarily imply independence. These points are all covered in elementary texts on statistics. A typical example of the use of the coefficient is as follows.

*Example* 9.1   Pearson and Lee (1902) collected data on the stature of brothers and sisters from 1401 families, and they found a correlation coefficient $r = 0.553$.   □ □ □

It seems more appropriate to use correlation than regression in Example 9.1, since correlation treats the variables symmetrically, whereas regression does not. (There are, however, other ways of treating variables symmetrically; see Sprent, 1969.) There are many similar applications of the correlation coefficient to the one given in Example 9.1. For instance, Galton (1888) gave some data on the height and forearm length of 348 men. Studies have recently been conducted on the correlation between skinfold thickness at different body sites. In all of these examples the underlying causal mechanism is clearly understood, but correlation is widely applied in economic and social studies, where there is sometimes very little understanding of the underlying situation.

Correlation is used a great deal in multivariate analysis. For example, in a situation where $k$ measurements are made on each of $n$ individuals, we can try to reduce the number of dimensions to rather less than $k$, by seeking for linear combinations of the original variables with maximum variance. The technique involved is called 'principal components analysis', and correlation is often used in it. There are also a number of other multivariate techniques in which correlation is used. A discussion of multivariate techniques is beyond the scope of this book, and those interested should consult the 'Further reading' list at the end of the chapter.

Sometimes correlation enters naturally into a problem. Typically, developments in genetics and inheritance are often expressed conveniently in terms of correlation. For example, some studies of inheritance using identical twins, fraternal twins of the same sex and brother-and-sister pairs, used correlation between body measurements.

In the material covered in this volume correlation is often used in an initial analysis of data. Before setting out on a multiple regression analysis, it is helpful to see the correlation matrix between the explanatory variables, so that we might see some of the linear dependencies that may exist between them.

Correlation is a measure of *linear dependence*, and it is easily shown that $\rho$ can be zero for an exact but non-linear relationship (see Wetherill, 1972, pp. 90–2). Furthermore, even if we establish a significant correlation, such as between smoking and incidence of lung cancer, this does not prove that there is a direct causal relationship. There are many well-known cases of 'nonsense' correlations to illustrate this point, such as the correlation between the price of rum in Havana and the salaries of Presbyterian ministers in Massachusetts. A good – and amusing – account of the pitfalls of correlation is given in Chapter 8 of Huff (1973). He discusses the (negative) correlation between examination grades and cigarette smoking, the correlation between cancer and milk drinking and other examples, and he shows very clearly that correlation and causality can be quite different matters.

## 9.2 Correlation or regression?

Sometimes a crucial question which arises is whether to use correlation or regression to study a relationship. One of the advantages of the correlation coefficient is that it is dimensionless. However, partly because of this it very rarely has a direct practical interpretation, whereas the regression coefficient

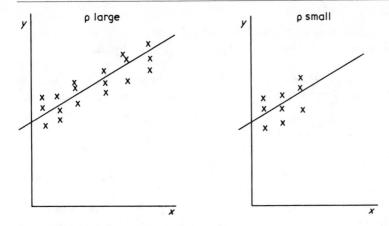

Fig. 9.1 Correlation and regression.

always has. The regression coefficient of $Y$ on $X$ is the average increase in $Y$ for a unit increase in $X$. Putting it another way, we can say that the regression coefficient estimates the relationship between $Y$ and $X$, whereas the correlation coefficient estimates the significance of the relationship. Clearly, if our units of measurement are rather arbitrary, we may prefer to use $\rho$ nevertheless.

Another important point to bear in mind is that the correlation coefficient is dependent upon the distribution of $X$, whereas the regression coefficient is not; this is illustrated in Fig. 9.1.
If we have a regression relationship given by

$$Y = \mu_Y + \beta(X - \mu_X) + \varepsilon$$

where $E(\varepsilon) = 0$, $V(\varepsilon) = \sigma_\varepsilon^2$, and $V(X) = \sigma_X^2$, then

$$C(X, Y) = \beta \sigma_X^2$$
$$V(Y) = \sigma_\varepsilon^2 + \beta^2 \sigma_X^2$$

and we have

$$\rho = \frac{\beta \sigma_X^2}{\sigma_X \{\sigma_\varepsilon^2 + \beta^2 \sigma_X^2\}^{1/2}} = \left\{ 1 + \frac{\sigma_\varepsilon^2}{\beta^2 \sigma_X^2} \right\}^{-1/2} \tag{9.3}$$

Therefore $\rho$ increases with $\sigma_X^2$.

All this points to the fact that the correlation coefficient is most frequently used when it is plausible that our data is a sample from a bivariate (or multivariate) normal population. In situations where correlation and regression are alternatives, most statisticians prefer to work in regression, because it has the more direct meaning.

**Exercises 9.2**

1. Let $X$ and $Y$ be random variables with unit variance, but not necessarily independent. By considering $V(Y + X)$ and $V(Y - X)$ show that $|\rho| \le 1$.

2. If you were analysing data on the relationship between smoking and lung cancer, say whether you would use correlation or regression, and give your reasons.

### 9.3 Estimation of $\rho$

A good discussion of maximum-likelihood estimation of $\rho$ is readily accesible in Kendall and Stuart (1973), and the results are as follows.

Suppose we are given data $x_1, \ldots, x_n, y_1, \ldots, y_n$, drawn from a bivariate normal distribution, and that it is known that the marginal distributions of $X$ and $Y$ have zero means and unit variances; then maximizing the likelihood in the usual way leads to the cubic equation

$$\rho(1 - \rho^2) + (1 + \rho^2)\frac{\sum x_i y_i}{n} - \rho\frac{(\sum x_i^2 + \sum y_i^2))}{n} = 0 \qquad (9.4)$$

(see Exercise 4.4.3).

For large $n$ this usually has only one real root. If all three roots are real, we choose the one which maximizes the likelihood function. If the marginal means and variances are also being estimated, then the maximum-likelihood estimate of $\rho$ is the sample correlation coefficient, and this is the one ordinarily used. However, the sample correlation coefficient is not unbiased; it can be shown that

$$E(R) = \rho\left\{1 - \frac{(1 - \rho^2)}{2n} + O(n^{-2})\right\}, \qquad (9.5)$$

Olkin and Pratt (1958) showed that an unbiased estimate of $\rho$ is given by

$$r_u = rF\left\{\frac{1}{2}, \frac{1}{2}, \frac{(n-2)}{2}, (1 - r^2)\right\} \qquad (9.6)$$

where

$$F(a, b, c, x) = \sum_{j=0}^{\infty} \frac{\Gamma(a + j)\Gamma(b + j)\Gamma(c)x^j}{\Gamma(a)\Gamma(b)\Gamma(c + j)\Gamma(j + 1)} \qquad (9.7)$$

is the hypergeometric function. This gives

$$r_u = r\left\{1 + \frac{(1 - r^2)}{2(n - 2)} + \frac{9(1 - r^2)^2}{8n(n - 2)} + O(n^{-3})\right\} \qquad (9.8)$$

and the authors recommended general use of this formula, but ignoring terms of $O(n^{-2})$.

The bias in $r$ is usually very small, and in any case unbiased estimation of $\rho$ is not often required. More frequently we are involved in testing hypotheses about $\rho$, for which we need the distribution of $r$, or an approximation to it.

## Exercises 9.3

1. Show that the asymptotic variance of the maximum-likelihood estimator of $\rho$ when the marginal means and variances are known is

$$(1 - \rho^2)^2 / \{n(1 + \rho^2)\}$$

## 9.4 Results on the distribution $R$

In this section we give some results on the distribution of Equation (9.2), and of transformations of it. References to readily accessible proofs are given at the end of the section. We shall assume throughout that estimation is based on a sample drawn from a bivariate normal distribution.

If $\rho = 0$, the p.d.f. of $R$ is given by

$$(1 - r^2)^{(n-4)/2} \Gamma\{(n-1)/2\}/[(\sqrt{\pi})\Gamma\{(n-2)/2\}] \tag{9.9}$$

where $n$ is the number of observations. This density is symmetric about the origin, and

$$V(R)_{\rho=0} = 1/(n-1) \tag{9.10}$$

However, the distribution tends to normality rather slowly, so that this formula is not of much practical use unless samples are large. In fact we have

$$E(R^4) = 3/\{(n-1)(n+1)\} \tag{9.11}$$

so that

$$\gamma_2 = \frac{E(R^4)}{\{E(R^2)\}^2} - 3 = -\frac{6}{(n+1)} \tag{9.12}$$

If $\rho \neq 0$, the p.d.f. of $R$ is much more complicated. One convenient form of the density is

$$\frac{(n-2)\Gamma(n-1)}{\sqrt{(2\pi)}\Gamma(n-\frac{1}{2})} (1 - \rho^2)^{(n-1)/2}(1 - r^2)^{(n-4)/2}(1 - \rho r)^{-n+3/2}$$

$$\times F\left[\frac{1}{2}, \frac{1}{2}, n - \frac{1}{2}, \frac{1+\rho r}{2}\right] \tag{9.13}$$

where $F(a, b, c, x)$ is the hypergeometric function given by Equation (9.7). This density tends to normality even slower than the distribution (9.9), especially if $\rho$ differs much from zero. The asymptotic variance of $R$ is

$$V(R) = (1 - \rho^2)^2 / (n - 1) \tag{9.14}$$

and if $\rho$ is near zero and $n$ is large, say $n \geq 400$, the distribution of $R$ can be taken as normal with variance (9.14). For sample sizes smaller than this, the actual distribution of $R$ has been tabulated by David (1938). Use of this large-scale tabulation can often be avoided by employing transformations of $R$.

If $\rho = 0$, the quantity

$$\frac{R\sqrt{(n-2)}}{\sqrt{(1-R^2)}} \qquad (9.15)$$

has exactly a $t$-distribution with $(n-2)$ degrees of freedom. Clearly (9.15) is a very convenient statistic to use to test the hypothesis $\rho = 0$, but it can only be used as a significance test.

Another very useful transformation can be derived as a variance-stabilizing function. A function $f(R)$ will have a variance independent of $\rho$ if it satisfies

$$f'(\rho) = 1/(1 - \rho^2)$$

(see Exercise 9.4.1.), and this leads to Fisher's $Z$-transformation

$$Z = \tfrac{1}{2}\log_e\{(1+R)/(1-R)\} \qquad (9.16)$$

which transforms the range $-1 \leq R \leq 1$ to $-\infty < Z < \infty$.
The first four moments of $Z$ are

$$E(Z) = \tfrac{1}{2}\log_e\frac{(1+\rho)}{(1-\rho)} + \frac{\rho}{2(n-1)}\left\{1 + \frac{5+\rho^2}{4(n-1)} + \dots\right\}$$

$$\mu_2(Z) = \frac{1}{(n-1)}\left\{1 + \frac{4-\rho^2}{2(n-1)} + \dots\right\}$$

$$\mu_3(Z) = \frac{\rho^3}{(n-1)^3} + \dots$$

$$\mu_4(Z) = \frac{1}{(n-1)^2}\left\{3 + \frac{14-3\rho^2}{(n-1)} + \dots\right\}$$

Therefore the coefficients of skewness and kurtosis are

$$\gamma_1(Z) = \rho^3/(n-1)^{3/2} + \dots$$

and

$$\gamma_2(Z) = 2/(n-1) + \dots$$

It follows that $Z$ will be very nearly normally distributed with

$$E(Z) \simeq \tfrac{1}{2}\log_e\frac{(1+\rho)}{(1-\rho)} + \frac{\rho}{2(n-1)} \qquad (9.17)$$

$$V(Z) \simeq \frac{1}{(n-1)} + \frac{4-\rho^2}{2(n-1)^2} \simeq \frac{1}{(n-3)} \qquad (9.18)$$

This approximation is satisfactory for sample sizes as small as about 25.

A summary of these results for use in carrying out significance tests and confidence intervals is given in the next section.

Most of the material in this section is covered in Anderson (1958, pp. 60–79), where proofs can be obtained. The moments of Fisher's Z-transformation are given by Gayen (1951), and David (1938) examined its small-sample accuracy.

---

**Exercises 9.4**

1. Use the results given in Equations (8.2), (9.5) and (9.14) to derive the approximate variance-stabilizing transformation for the correlation coefficient given by Equation (9.16).

---

### 9.5 Confidence intervals and hypothesis tests for $\rho$

(*i*) *Test of the hypothesis* $\rho = 0$
It follows from Equation (9.10) that significance levels for testing the hypothesis $\rho = 0$ are, for two-sided tests,

$$\left. \begin{array}{ll} 5\% \text{ level}: \ \pm 1.96/\sqrt{(n-1)} & n \geq 50 \\ 1\% \text{ level}: \ \pm 2.5758/\sqrt{(n-1)} & n \geq 100 \end{array} \right\} \qquad (9.19)$$

For sample sizes less than this, exact percentage points are available, for example, in *Biometrika Tables*.

An alternative test of the hypothesis $\rho = 0$ is to refer

$$r\sqrt{(n-2)}/\sqrt{(1-r^2)}$$

to *t*-tables on $(n-2)$ degrees of freedom.

*Example* 9.2   In order to test the hypothesis that $\rho = 0$ in Example 9.1 we can look up

$$r\sqrt{(n-1)} = 0.553\sqrt{(1400)} = 20.7$$

in standard normal tables. This correlation was obviously very highly significant, and this particular test adds very little to our knowledge! A more interesting test would be to see if the correlation differs significantly from 0.5, or to estimate confidence intervals.   □ □ □

(*ii*) *Tests and confidence intervals in the general case: Fisher's*
*Z-transformation*
The statistical methods here all follow from the results given by Equations

(9.16), (9.17) and (9.18). We put

$$\zeta(\rho) = \tfrac{1}{2}\log_e \frac{(1+\rho)}{(1-\rho)} + \frac{\rho}{2(n-1)} \tag{9.20}$$

then, for example, the two-sided 5 per cent significance levels for $Z$ are, with a hypothesis $\rho = \rho_0$,

$$\zeta(\rho_0) \pm 1.96 \frac{1}{\sqrt{(n-3)}} \tag{9.21}$$

For 1 per cent limits we use a factor 2.5758 instead of 1.96, etc.

In order to obtain, say 95 per cent confidence intervals for $\rho$ we start from the statement

$$\Pr\left\{ -\frac{1.96}{\sqrt{(n-3)}} < Z - \zeta(\rho) < \frac{1.96}{\sqrt{(n-3)}} \right\} = 0.95$$

and we now insert the observed value of $Z$ and invert the limits to obtain

$$\Pr\left\{ Z - \frac{1.96}{\sqrt{(n-3)}} < \zeta(\rho) < Z + \frac{1.96}{\sqrt{(n-3)}} \right\} = 0.95$$

Therefore confidence limits are obtained by solving for $r$ in

$$\zeta(r) = z \pm 1.96/\sqrt{(n-3)} \tag{9.22}$$

If the correction term of $O(1/n)$ in $\zeta(\rho)$ can be ignored, this is easy since the inverse $Z$-transformation is

$$r = (e^{2z} - 1)/(e^{2z} + 1) \tag{9.23}$$

and tables of (9.16) and (9.23) are given by Fisher and Yates (1963). If the correction term in Equation (9.20) cannot be ignored, some iteration or approximation is necessary. Usually it is satisfactory to use an approximate value of $r$, and correct the values of $Z$ by the term $-r/\{2(n-1)\}$ before using Equation (9.23).

*Example* 9.3    A value of $r = 0.65$ was found for 28 pairs of observations. Is this value consistent with $\rho = 0.5$? Find 95 per cent confidence intervals for $\rho$.

For a null hypothesis value $\rho = 0.5$, the transformation (9.16) gives

$$z = \tfrac{1}{2}\log_e\{1.5/0.5\} = 0.5493$$

and the correction term from Equation (9.20) is

$$\rho/\{2(n-1)\} = 0.5/(2 \times 27) = 0.0093$$

Therefore $\zeta(0.5) = 0.5493 + 0.0093 = 0.5586$, and the limits (9.21) are

$$0.5586 \pm 1.96/\sqrt{(25)} = (0.1666, 0.9506) \qquad (9.24)$$

which are rather wide.
The Z-transform for the observed correlation is

$$Z = \tfrac{1}{2}\log_e(1.65/0.35) = 0.7753$$

which is well within the limits (9.24). The observed correlation is therefore perfectly consistent with the value $\rho = 0.5$.
   Confidence intervals at the 95 per cent level for $\zeta(\rho)$ are

$$z \pm 1.96/\sqrt{(25)} = 0.7753 \pm 0.3920 = (0.3833, 1.1673)$$

and ignoring the correction term, the equivalent values of $r$ are (0.3653, 0.8234). If we now apply the correction term $\rho/\{2(n-1)\}$ to each limit, the approximate confidence interval is

$$(0.3585, \qquad 0.8082) \qquad (9.25)$$

but more exact values can be found by iteration, if this is thought worth while.

   □ □ □

   Another way of calculating the significance level for the hypothesis $\rho = 0.5$ in Example 9.3 is to refer

$$\frac{Z - \zeta(0.5)}{1/\sqrt{(n-3)}} = \frac{0.7753 - 0.5586}{1/5} = 1.083$$

to standard normal tables. *Biometrika Tables* give charts for carrying out significance test and calculating confidence intervals for a single correlation coefficient and, in effect, they follow the pattern of calculations given above. The charts give values for limited values of $n$, and they cannot be read very accurately.

*(iii) Test of the difference between two or more correlation coefficients*
In order to test the significance of the difference between two observed correlations $r_1$ and $r_2$, based on $n_1$ and $n_2$ observations respectively, we calculate the transforms $z_1$ and $z_2$, and refer

$$\frac{z_1 - z_2}{\sqrt{\{1/(n_1 - 3) + 1/(n_2 - 3)\}}} \qquad (9.26)$$

to standard normal tables.
   If we have $k$ correlations, $r_1, r_2, \ldots, r_k$, on $n_1, n_2, \ldots, n_k$ observations respectively, we calculate

$$\sum_{i=1}^{k} (n_i - 3)(z_i - \bar{z})^2 \qquad (9.27)$$

where

$$\bar{Z} = \sum(n_i - 3)Z_i / \{\sum(n_i - 3)\}$$

and refer to $\chi^2$ tables on $(k - 1)$ degrees of freedom.

Finally, we must emphasize a point made at the beginning of this section. All of the methods given here are dependent upon the assumption that the original pairs of observations were drawn from a bivariate normal distribution. Most of these methods are rather sensitive to departures from this distributional assumption.

---

### Exercises 9.5

1. Test whether the observed correlation in Example 9.1 is significantly different from 0.5. Also find 90 per cent confidence limits for the correlation coefficient.

2. Check the adequacy of the approximations given in Sections 9.5(i) and 9.5(ii) for testing a single correlation coefficient, for small values of $n$. Include a test of $\rho = 0$ using Fisher's $Z$-transformation.

3. In two repetitions of an experiment, correlations of 0.54 and 0.68 were found, each on 50 observations. Do they differ significantly?

---

### 9.6 Relationship with regression

If we carry out simple linear regression of $Y$ on $X$, then the (corrected) sum of squares due to regression is

$$\{CS(y, x)\}^2 / CS(x, x)$$

and the sum of squared deviations is

$$S = CS(y, y) - \{CS(y, x)\}^2 / CS(x, x)$$
$$= CS(y, y)\{1 - r^2\}$$

(9.28)

Therefore $r^2$ is the proportion of the variance of $y$ accounted for by regression on $x$. The analysis-of-variance table can be written as in Table 9.1.

Table 9.1

| Source | SS | d.f. |
|---|---|---|
| Due to regression | $r^2 CS(y, y)$ | 1 |
| Deviations | $(1 - r)^2 CS(y, y)$ | $n - 2$ |
| Total | $CS(y, y)$ | $n - 1$ |

It follows that the usual $F$-test for the significance of a regression coefficient is the square of the $t$-test for the correlation coefficient given in Equation (9.15).

Finally, we notice that if the regression coefficient of $Y$ on $X$ is denoted $\beta_{01}$, then

$$\beta_{01} = C(Y, X)/V(X) \qquad \beta_{10} = C(Y, X)/V(Y)$$

and

$$\rho^2 = \beta_{01}\beta_{10} \tag{9.29}$$

## 9.7 Partial correlation

The correlation coefficient just described arises when we are dealing with two variables for which the population distribution is bivariate normal; when we have more than two variables *partial correlation* arises, and it is assumed that the population distribution is multivariate normal. For example, suppose that for a group of first-year university students we take measurements of height, weight, girth, ... ; we might be interested in the strength of the association between height and weight, when the other variables are held constant. If we wished to predict one variable from the others, regression methods are appropriate, but sometimes we are not interested in prediction, but in the amount of association between variables.

Let variables $X_1, X_2, \ldots,$ be measured on a certain population, then we define the correlations

$$\rho_{i,j} = C(X_i, X_j)/\sqrt{[V(X_i)V(X_j)]} \tag{9.30}$$

and let the expectations of the $X_i$ be written $\mu_{X_i}$. Then if the $X_i$ have a multivariate normal distribution, the following regression equations apply:

$$E(X_1 | X_3) = \mu_{X_1} + \beta_{1,3}(X_3 - \mu_{X_3}) \tag{9.31}$$

$$E(X_2 | X_3) = \mu_{X_2} + \beta_{2,3}(X_3 - \mu_{X_3}) \tag{9.32}$$

where

$$\beta_{i,j} = C(X_i, X_j)/V(X_j)$$

Now define new variables

$$X_{1.3} = X_1 - \mu_{X_1} - \beta_{1,3}(X_3 - \mu_{X_3}) \tag{9.33}$$

$$X_{2.3} = X_2 - \mu_{X_2} - \beta_{2,3}(X_3 - \mu_{X_3}) \tag{9.34}$$

and we see that the variances of these variables are

$$V(X_{1.3}) = V(X_1) + \beta_{1,3}^2 V(X_3) - 2\beta_{1,3} C(X_1, X_3)$$

$$= V(X_1)\{1 - \rho_{1,3}^2\} \tag{9.35}$$

and

$$V(X_{2.3}) = V(X_2)\{1 - \rho_{2,3}^2\} \tag{9.36}$$

The correlation between these two new variables is

$$\rho_{1,2.3} = C(X_{1.3}, X_{2.3})/\sqrt{[V(X_{1.3})V(X_{2.3})]} \qquad (9.37)$$

Now the expectations of (9.33) and (9.34) over the population will be zero, so that

$$C(X_{1.3}, X_{2.3}) = E(X_{1.3}X_{2.3})$$
$$= C(X_1, X_2) - \beta_{2.3}C(X_1, X_3) - \beta_{1.3}C(X_2, X_3) + \beta_{1.3}\beta_{2.3}V(X_2)$$
$$= C(X_1, X_2) - \{C(X_1, X_3)C(X_2, X_3)\}/V(X_3)$$

by using this we obtain

$$\rho_{1,2.3} = \frac{C(X_1, X_2) - C(X_1, X_3)C(X_2, X_3)/V(X_3)}{\sqrt{\{V(X_1)V(X_2)\}}\sqrt{\{1 - \rho_{1,3}^2\}}\sqrt{\{1 - \rho_{2,3}^2\}}}$$

$$= \frac{\rho_{1,2} - \rho_{1,3}\rho_{2,3}}{\sqrt{\{1 - \rho_{1,3}^2\}}\sqrt{\{1 - \rho_{2,3}^2\}}} \qquad (9.38)$$

*Example* 9.4    From the results of a survey carried out in a certain population three variables were

$$X_1 = \text{Family size} \qquad X_2 = \text{Intelligence coefficient}$$
$$X_3 = \text{Social class}$$

and the following correlation coefficients were computed:

$$r_{1,2} = -0.4 \qquad r_{1,3} = -0.6 \qquad r_{2,3} = +0.3$$

The partial correlation between family size and intelligence, for constant social class, is

$$r_{1,2.3} = \frac{-0.4 - (-0.6)(0.3)}{\sqrt{[1 - 10.6)^2]}\sqrt{[1 - (0.3)^2]}} = -0.288 \qquad \square\ \square\ \square$$

Equation (9.38) extends in an obvious way for correlations involving four or more variables. For example, we have

$$\rho_{1,2.3,4} = \frac{\rho_{1,2.4} - \rho_{1,3.4}\rho_{2,3.4}}{\sqrt{[1 - \rho_{1,3.4}^2]}\sqrt{[1 - \rho_{2,3.4}^2]}} \qquad (9.39)$$

The distributional results are readily extended to cover partial correlation. For example, if $\rho_{1,2.3}$ is zero, then the distribution of

$$\frac{r_{1,2.3}\sqrt{(n-3)}}{\sqrt{[1 - r_{1,2.3}^2]}} \qquad (9.40)$$

is a $t$-distribution on $(n - 3)$ degrees of freedom. For $r_{1,2,3,4}$ the factor $\sqrt{(n-3)}$ in (9.40) is replaced by $\sqrt{(n-4)}$ and the distribution has $(n-4)$ degrees of freedom. Alternatively, we can use Fisher's $Z$-transformation, and treat the variable

$$Z = \tfrac{1}{2}\log_e\{(1 + r_{1.2.3.4})/(1 - r_{1.2.3.4})\} \quad (9.41)$$

as approximately normal with expectation

$$E(Z) \simeq \tfrac{1}{2}\log_e\{(1 + \rho_{1.2.3.4})/(1 - \rho_{1.2.3.4})\} \quad (9.42)$$

and

$$V(Z) = 1/(n - 3 - s) \quad (9.43)$$

where $s$ is the number of variables 'held constant' by partial correlation which is two in the above example. (For a proof of these results, see Anderson, 1958.)

Finally, Equation (9.29) also extends to partial correlation. If we define partial regression coefficients $\beta_{i.j.k}$ by

$$E(X_1) = \alpha + \beta_{1.2.3.4}x_2 + \beta_{1.3.2.4}x_3 + \beta_{1.4.2.3}x_4$$

then

$$\rho_{1.2.3.4} = \beta_{1.2.3.4}\beta_{2.1.3.4} \quad (9.44)$$

## 9.8 The multiple correlation coefficient

Suppose we carry out multiple regression of $X_1$ on explanatory variables $X_2, X_3, \ldots, X_k$, then the ordinary correlation coefficient between $X_1$ and the regression $E(X_1 | X_2, \ldots, X_k)$ is called the *multiple correlation coefficient*. In practice we usually deal with the square of this coefficient, and it is often denoted $R^2$.

Write

$$\hat{X}_1 = E(X_1 | X_2, \ldots, X_k)$$

then

$$R^2 = \{C(X_1, \hat{X}_1)/\sqrt{[V(X_1)V(\hat{X}_1)]}\}^2$$

But we have seen that for multiple regression

$$C(X_1 - \hat{X}_1, \hat{X}_1) = 0$$

since this is the covariance of the residuals with the predicted values. Hence

$$C(X_1, \hat{X}_1) = V(\hat{X}_1)$$

Table 9.2

| Source | CSS |
|---|---|
| Due to regression on $X_2, \ldots, X_k$ | $r^2 CS(X_1, X_1)$ |
| Deviations | $(1 - r^2)CS(X_1, X_1)$ |
| Total | $CS(X_1, X_1)$ |

and

$$R^2 = V(\hat{X}_1)/V(X_1)$$

That is, $R^2$ is the proportion of the variance accounted for by regression, and this is the way it is estimated from samples; see Table 9.2. It follows that a significance test for $R^2$ is provided by the usual $F$-test for regression.

---

### Exercises 9.8

1. Study the paper 'How the UGC determines allocations of recurrent grants – a curious correlation', by W. R. Cook, in *J. R. Statist. Soc.* A, 139, 374–84 (1976), and the comments on this paper given in *J. R. Statist. Soc.* A, **140**, 199–209 (1977). Write an essay based on this material, on the topic 'The use and misuse of correlation'.

---

### Further reading

References to derivations of the theory have been given in the relevant sections. Those interested in examples of applications, including applications of partial correlation, should consult Kendall and Stuart (1973), Snedecor and Cochran (1967), and Ezekiel and Fox (1959). For some applications of correlation in multivariate analysis, see Morrison (1975).

# CHAPTER TEN

# The analysis of variance

## 10.1 An example

In earlier chapters we have defined the model linear in the parameters, commonly called the linear model, and discussed the theory of estimating parameters of the model by the method of least squares. Our applications of the theory so far have been to simple linear regression, polynomial regression and multiple regression. However, there is a very wide class of models which can be analysed on the basis of the theory we have covered. In this chapter we apply our theory to data arranged in balanced arrays or cross classifications; it is part of a very big topic usually referred to as 'the analysis of variance'. The following example shows the type of problem we shall be dealing with.

*Example* 10.1 An experimenter carried out a trial to determine the effects of four treatments, *A*, *B*, *C* and *D*, to a certain crop. A field was divided into twelve plots, and three replicates of each of the four treatments were randomly assigned to the plots. The experimental layout and results were as follows (yield/plot):

| | | | |
|---|---|---|---|
| A  33.63 | D  39.62 | B  38.18 | C  41.46 |
| D  38.02 | B  35.83 | D  35.99 | A  36.58 |
| C  42.92 | A  37.80 | C  40.43 | B  37.89 |

□ □ □

A suitable linear model for this set of data might be as follows:

$$\text{Observation} = (\text{Overall average}) + \begin{pmatrix} \text{Difference from} \\ \text{average due to} \\ \text{treatment} \end{pmatrix} + (\text{Random error}) \qquad (10.1)$$

We can easily see some possible treatment effects if we rearrange the treatment results, as shown in Table 10.1.

The last two columns of this table show some evidence that treatment effects exist. We therefore proceed to fit a model to the data. In the rearranged

Table 10.1

| Treatment | Results | | | Total | Mean |
|-----------|---------|---------|---------|--------|-------|
| A | 33.63 | 37.80 | 36.58 | 108.01 | 36.00 |
| B | 35.83 | 38.18 | 37.89 | 111.90 | 37.30 |
| C | 42.92 | 40.43 | 41.46 | 124.81 | 41.60 |
| D | 38.02 | 39.62 | 35.99 | 113.63 | 37.88 |

form, we can write Equation (10.1) as

$$Y_{ij} = \mu + \gamma_i + \varepsilon_{ij} \qquad i = 1, 2, 3, 4; \qquad j = 1, 2, 3 \qquad (10.2)$$

where $i = 1, 2, 3, 4$ runs over the treatments $A$, $B$, $C$, $D$ and where $\varepsilon_{ij}$ is the random-error term, which we often take to be independently and normally distributed with expectation zero and variance $\sigma^2$. However, there are too many parameters here since there are four treatment means, and five parameters which we can write

$$\theta' = (\mu, \gamma_1, \gamma_2, \gamma_3, \gamma_4) \qquad (10.3)$$

This overparameterization leads to certain problems which we discuss below. If we write our model in the standard form

$$E(\mathbf{Y}) = \mathbf{a}\theta$$
$$V(\mathbf{Y}) = \mathbf{I}\sigma^2 \qquad (10.4)$$

then the matrix $\mathbf{a}$ has the form:

$$\mathbf{a} = \begin{bmatrix} 1 & 1 & 0 & 0 & 0 \\ 1 & 1 & 0 & 0 & 0 \\ 1 & 1 & 0 & 0 & 0 \\ 1 & 0 & 1 & 0 & 0 \\ 1 & 0 & 1 & 0 & 0 \\ 1 & 0 & 1 & 0 & 0 \\ 1 & 0 & 0 & 1 & 0 \\ 1 & 0 & 0 & 1 & 0 \\ 1 & 0 & 0 & 1 & 0 \\ 1 & 0 & 0 & 0 & 1 \\ 1 & 0 & 0 & 0 & 1 \\ 1 & 0 & 0 & 0 & 1 \end{bmatrix} \qquad (10.5)$$

In the context used in this chapter we call this matrix the *design matrix*. (This term is not used in the context of the multiple-regression applications discussed in Chapters 5, 6 and 7.)

Thus following the theory of Chapter 5, we need the matrix

$$\mathbf{a'a} = \begin{bmatrix} 12 & 3 & 3 & 3 & 3 \\ 3 & 3 & 0 & 0 & 0 \\ 3 & 0 & 3 & 0 & 0 \\ 3 & 0 & 0 & 3 & 0 \\ 3 & 0 & 0 & 0 & 3 \end{bmatrix} \qquad (10.6)$$

and it can be seen that this matrix is singular; the sum of columns two to five is equal to column one on both (10.5) and (10.6). There are now several different ways of proceeding.

*Method* 1    We could rewrite Equation (10.2) as

$$Y_{ij} = \beta_i + \varepsilon_{ij} \qquad i = 1, \ldots, 4$$

It is readily checked that least-squares estimates of $\beta_i$ can be obtained without difficulty. This approach does not extend to the more advanced types of analysis of variance, and we shall not follow it.

*Method* 2    It is easy to justify a side condition on the parameters $\gamma_i$ in Equation (10.2). For example, we can write

$$E(Y_{ij}) = (\mu + \bar{\gamma}.) + (\gamma_i - \bar{\gamma}.)$$

where

$$\bar{\gamma}. = \sum \gamma_i / 4$$

If we redefine the parameters

$$\alpha_i = \gamma_i - \bar{\gamma}. \qquad i = 1, \ldots, 4$$

we have

$$\sum \alpha_i = 0$$

This indicates that we could impose the condition

$$\sum \gamma_i = 0 \qquad (10.7)$$

on Equation (10.2) and then proceed. If we write

$$\psi = (\mu, \gamma_1, \gamma_2, \gamma_3) \qquad (10.8)$$

and use

$$\gamma_4 = -(\gamma_1 + \gamma_2 + \gamma_3) \qquad (10.9)$$

then the model

$$E(\mathbf{Y}) = \mathbf{b}\psi \qquad (10.10)$$

can be solved by the previous theory (see Exercise 10.1.1). Again we shall reject this approach, as we wish to develop a theory which will readily cover the more advanced types of experimental design frequently used.

*Method* 3   The only problem with Equation (10.2) is that it leads to the singular matrix (10.6). If we cover a few points about the theory of generalized inverses, we shall have no difficulty in dealing with equations involving singular matrices. We discuss this method in subsequent sections of this chapter.

*Discussion*
Whichever method is used, the results obtained are all equivalent. For example, by using Method 2 we obtain the estimates

$$\hat{\mu} = \bar{y}_{..} = \sum_i \sum_j y_{ij}/12 \qquad (10.11)$$

$$\hat{\gamma}_i = \bar{y}_{i.} - \bar{y}_{..} = \sum_j y_{ij}/3 - \sum_i \sum_j y_{ij}/12 \qquad (10.12)$$

The minimized sum of squares is

$$S_{\min} = \sum \sum y_{ij}^2 - \frac{1}{n}\left(\sum \sum y_{ij}\right)^2 - 3\sum_i (\bar{y}_{i.} - \bar{y}_{..})^2 \qquad (10.13)$$

and the sum of squares due to the fitted parameters $(\gamma_1, \ldots, \gamma_4)$ is

$$S_{\text{par}} = 3\sum_i (\bar{y}_{i.} - \bar{y}_{..})^2 \qquad (10.14)$$

The results can all be summarized in an analysis-of-variance table; see below for details.

---

**Exercises 10.1**
(*refer back to Method 2*)
1.  Show that by following Method 2 above we have

$$\mathbf{b'b} = \begin{bmatrix} 12 & 0 & 0 & 0 \\ 0 & 6 & 3 & 3 \\ 0 & 3 & 6 & 3 \\ 0 & 3 & 3 & 6 \end{bmatrix}$$

Show that an inverse is

$$(\mathbf{b'b})^{-1} = \begin{bmatrix} \dfrac{1}{12} & 0 & 0 & 0 \\[2mm] 0 & \dfrac{1}{4} & -\dfrac{1}{12} & -\dfrac{1}{12} \\[2mm] 0 & -\dfrac{1}{12} & \dfrac{1}{4} & -\dfrac{1}{12} \\[2mm] 0 & -\dfrac{1}{12} & -\dfrac{1}{12} & \dfrac{1}{4} \end{bmatrix}$$

leading to Equation (10.11) to (10.14).

---

## 10.2  Generalized inverses

Given any set of consistent equations

$$\mathbf{AX} = \mathbf{Y} \qquad (10.15)$$

then if $\mathbf{A}$ is non-singular there is a unique inverse such that the solution of these equations is

$$\mathbf{X} = \mathbf{A}^{-1}\mathbf{Y}$$

However, if $\mathbf{A}$ is singular we cannot do this. A singular matrix $\mathbf{A}$ means that in effect there are less restrictions than variables in $\mathbf{X}$. Thus, although solutions to Equations (10.15) exist when $\mathbf{A}$ is singular, there are many solutions, and no *unique* solution. *Generalized* inverses can be constructed if $\mathbf{A}$ is singular, and these lead us to particular solutions, and to general solutions.

*Definition*   A *generalized inverse* of an $m \times m$ matrix $\mathbf{A}$ is an $m \times m$ matrix $\mathbf{A}^-$ such that $\mathbf{AA}^-\mathbf{A} = \mathbf{A}$ $\qquad (10.16)$

□ □ □

For any $\mathbf{A}^-$ satisfying this definition,

$$\mathbf{AA}^-\mathbf{AX} = \mathbf{AX}$$

and therefore

$$\mathbf{AA}^-\mathbf{Y} = \mathbf{Y}$$

so that $\mathbf{A}^-\mathbf{Y}$ is a solution of Equation (10.15)
The general solution to Equation (10.15) is

$$\mathbf{X} = \mathbf{A}^-\mathbf{Y} + (\mathbf{A}^-\mathbf{A} - \mathbf{I})\mathbf{W} \qquad (10.17)$$

where $\mathbf{W}$ is an arbitrary vector.

Now for the equations we meet in least squares, $\mathbf{A}$ has the form

$$\mathbf{A} = \mathbf{a}'\mathbf{a}$$

where $\mathbf{a}$ is an $m \times r$ matrix.
One quantity which occurs frequently in the least-squares equation is

$$\mathbf{P} = \mathbf{a}\mathbf{A}^-\mathbf{a}' \qquad (10.18)$$

and this has two important properties:

(a) $\mathbf{P}$ does not depend on which generalized inverse is used;
(b) $\mathbf{PP} = \mathbf{P}$. $\qquad (10.19)$

Those willing to accept these results may pass on to the section headed 'Construction of a generalized inverse' or 'Summary'. The proof which follows does establish some other useful results.

First, notice that if $\mathbf{A}^-$ satisfies the definition, then so does its transpose, if $\mathbf{A}$ is symmetric.

Next, notice that for any matrix $\mathbf{W}$, say,

$$\mathbf{ZW} = 0 \quad \text{if and only if} \quad \mathbf{Z'ZW} = 0 \tag{10.20}$$

Part of this result is obvious, since

$$\mathbf{ZW} = 0 \Rightarrow \mathbf{Z'ZW} = 0$$

where the arrow means 'implies that'. The other result is easily shown since

$$\mathbf{Z'ZW} = 0 \Rightarrow \theta'\mathbf{W'Z'ZW}\theta = 0 \quad \text{for all } \theta,$$
$$\Rightarrow \mathbf{ZW}\theta = 0 \quad \text{for all } \theta,$$
$$\Rightarrow \mathbf{ZW} = 0 \quad \text{establishing Equation (10.20).}$$

Then we see that if we have $\mathbf{A}^-$ satisfying

$$\mathbf{a'a}\mathbf{A}^-\mathbf{a'a} = \mathbf{a'a}$$

this can be written

$$\mathbf{a'a}(\mathbf{A}^-\mathbf{a'a} - \mathbf{I}) = 0$$

so that by Equation (10.20)

$$\mathbf{a}(\mathbf{A}^-\mathbf{a'a} - \mathbf{I}) = 0$$

Hence

$$\mathbf{a}\mathbf{A}^-\mathbf{a'a} = \mathbf{a} \tag{10.21}$$

and by transposing,

$$\mathbf{a'a}\mathbf{A}^{-'}\mathbf{a'} = \mathbf{a'} \tag{10.22}$$

This enables us to prove Equation (10.19), for

$$\mathbf{PP} = (\mathbf{a}\mathbf{A}^-\mathbf{a'a})\mathbf{A}^-\mathbf{a'}$$
$$= \mathbf{a}\mathbf{A}^-\mathbf{a'} \quad \text{by using Equation (10.21)}$$
$$= \mathbf{P}$$

Now let $\mathbf{A}_1^-$ and $\mathbf{A}_2^-$ be any two generalized inverses satisfying Equation (10.21); then

$$\mathbf{a}(\mathbf{A}_1^- - \mathbf{A}_2^-)\mathbf{a'a} = 0$$

so that by Equation (10.20),

$$\mathbf{a}(\mathbf{A}_1^- - \mathbf{A}_2^-)\mathbf{a'} = 0$$

or

$$\mathbf{a}\mathbf{A}_1^-\mathbf{a'} = \mathbf{a}\mathbf{A}_2^-\mathbf{a'} \tag{10.23}$$

*Construction of a generalized inverse*

Given any square matrix $\mathbf{A} = \mathbf{a'a}$, we can find a non-singular $\mathbf{B}$ such that

$$\mathbf{BAB'} = \mathbf{D} \quad (\mathbf{A} = \mathbf{B}^{-1}\mathbf{DB'}^{-1})$$

where $\mathbf{D}$ is diagonal. Let $\mathbf{D}^-$ be the matrix obtained by replacing the non-zero elements of $\mathbf{D}$ by their reciprocals, then

$$\mathbf{D}\mathbf{D}^-\mathbf{D} = \mathbf{D}$$

and if we put

$$\mathbf{A}^- = \mathbf{B}'\mathbf{D}^-\mathbf{B} \tag{10.24}$$

then

$$\mathbf{A}\mathbf{A}^-\mathbf{A} = \mathbf{B}^{-1}\mathbf{D}(\mathbf{B}'^{-1}\mathbf{B}')\mathbf{D}^-(\mathbf{B}\mathbf{B}^{-1})\mathbf{D}\mathbf{B}'^{-1} = \mathbf{B}^{-1}\mathbf{D}\mathbf{D}^-\mathbf{D}\mathbf{B}'^{-1}$$
$$= \mathbf{B}^{-1}\mathbf{D}\mathbf{B}'^{-1} = \mathbf{A}$$

Hence $\mathbf{A}^- = \mathbf{B}'\mathbf{D}^-\mathbf{B}$ is a generalized inverse.
This particular generalized inverse also has the property that

$$\mathbf{A}^-\mathbf{A}\mathbf{A}^- = \mathbf{A}^- \tag{10.25}$$

which is *not* a property of all generalized inverses.

*Summary*
(Assuming $\mathbf{A} = \mathbf{a}'\mathbf{a}$)
Definition: $\qquad\qquad\qquad\qquad \mathbf{A}\mathbf{A}^-\mathbf{A} = \mathbf{A}$
General solution of

$$\mathbf{a}'\mathbf{a}\theta = \mathbf{a}'\mathbf{Y}$$

is

$$\hat{\theta} = (\mathbf{a}'\mathbf{a})^-\mathbf{a}'\mathbf{Y} + (\mathbf{A}^-\mathbf{A} - \mathbf{I})\mathbf{W} \tag{10.26}$$

where $\mathbf{W}$ is arbitrary
Also

$$\mathbf{a}'\mathbf{a}\mathbf{A}^-\mathbf{a}' = \mathbf{a}'$$
$$\mathbf{a}\mathbf{A}^-\mathbf{a}'\mathbf{a} = \mathbf{a}$$

and if

$$\mathbf{P}_i = \mathbf{a}\mathbf{A}_i^-\mathbf{a}'$$

where $\mathbf{A}_i^-$ are generalized inverses of $\mathbf{A}$, then

$$\mathbf{P}_i = \mathbf{P}_j \qquad i \neq j$$
$$\mathbf{P}_i\mathbf{P}_i = \mathbf{P}_i$$

---

## Exercises 10.2

1. Show that Equation (10.17) is a solution of Equation (10.15), by multiplying (10.17) through by $\mathbf{A}$, and using the fact that $\mathbf{A}^-\mathbf{Y}$ is a solution.

2. Show that any solution of Equation (10.15) can be expressed in the form

(10.17) as follows. Let any solution be $\hat{\mathbf{X}}$, and choose

$$\mathbf{W} = (\mathbf{A}^- \mathbf{A} - \mathbf{I})\hat{\mathbf{X}}$$

Then show that Equation (10.17) reduces to $\hat{\mathbf{X}}$.

## 10.3 Least squares using generalized inverses

Suppose now we have a model (10.4) for which the design matrix $\mathbf{a}$ is not of full rank. That is, we are dealing with a model such as (10.2) for Example 10.1, leading to a matrix $\mathbf{a}'\mathbf{a}$ such as (10.6), which is singular.

For least squares we must minimize

$$\begin{aligned} S &= (\mathbf{y} - \mathbf{a}\theta)'(\mathbf{y} - \mathbf{a}\theta) \\ &= \mathbf{y}'\mathbf{y} - \mathbf{y}'\mathbf{a}\theta - \theta'\mathbf{a}'\mathbf{y} + \theta'\mathbf{a}'\mathbf{a}\theta \end{aligned} \qquad (10.27)$$

with respect to choice of $\theta$. Proceeding as in Chapter 5, this leads to the normal equations

$$\mathbf{a}'\mathbf{a}\hat{\theta} = \mathbf{a}'\mathbf{y}$$

for which a particular solution is

$$\hat{\theta} = \mathbf{A}^- \mathbf{a}'\mathbf{y} \qquad (10.28)$$

where $\mathbf{A}^-$ is a generalized inverse of $\mathbf{A} = \mathbf{a}'\mathbf{a}$. This particular solution depends on the generalized inverse chosen, and a general solution is given by

$$\hat{\theta} = \mathbf{A}^- \mathbf{a}'\mathbf{y} + (\mathbf{A}^- \mathbf{A} - \mathbf{I})\mathbf{W} \qquad (10.29)$$

where $\mathbf{W}$ is an arbitrary vector. We shall see below that only certain contrasts of $\theta$ can be estimated. However, many of the important properties of the sums of squares do not depend on the particular generalized inverse chosen.

If we put Equation (10.29) into Equation (10.27) we get

$$\begin{aligned} S_{\min} &= \mathbf{y}'\mathbf{y} - \mathbf{y}'\mathbf{a}\mathbf{A}^- \mathbf{a}'\mathbf{y} - \mathbf{y}'\mathbf{a}\mathbf{A}^{-\prime}\mathbf{a}'\mathbf{y} + \mathbf{y}'\mathbf{a}\mathbf{A}^{-\prime}\mathbf{A}\mathbf{A}^- \mathbf{a}'\mathbf{y} \\ &= \mathbf{y}'\mathbf{y} - \mathbf{y}'\mathbf{P}\mathbf{y} - \mathbf{y}'\mathbf{P}'\mathbf{y} + \mathbf{y}'\mathbf{P}'\mathbf{P}\mathbf{y} \\ &= \mathbf{y}'\mathbf{y} - \mathbf{y}'\mathbf{P}\mathbf{y} \end{aligned}$$

by using Equation (10.19), where $\mathbf{P}$ is defined by Equation (10.18). Therefore the values of $S_{\min}$ and of $S_{\mathrm{par}}$, the sum of squares due to the fitted model do not depend on the particular generalized inverse chosen. Further, we have

$$\mathbf{P}\mathbf{y} = \mathbf{a}\mathbf{A}^- \mathbf{a}'\mathbf{y} = \mathbf{a}\hat{\theta}$$

so that $S_{\mathrm{par}}$ and $S_{\min}$ can be written in the forms

$$\begin{aligned} S_{\mathrm{par}} &= \hat{\theta}'\mathbf{a}'\mathbf{y} \\ S_{\min} &= \mathbf{y}'\mathbf{y} - \hat{\theta}'\mathbf{a}'\mathbf{y} \end{aligned} \qquad (10.30)$$

The vector of residuals is given by

$$\mathbf{r} = \mathbf{y} - \mathbf{a}\hat{\boldsymbol{\theta}} = \mathbf{y} - \mathbf{a}\mathbf{A}^{-}\mathbf{a}'\mathbf{y}$$
$$= (\mathbf{I} - \mathbf{P})\mathbf{y} \tag{10.31}$$

which is independent of the particular inverse chosen, and we have

$$S_{\min} = \mathbf{r}'\mathbf{r}$$

We now need another set of theorems, parallel to those in Chapter 5, but applicable to models of the sort that we are dealing with in this chapter, which are of less than full rank, and where a generalized inverse is needed. The vital theorems for the analysis of variance are Theorems 5.4 to 5.8 and the only modification to these is that, in the notation of Chapter 5,

$$p = \text{rank }(a)$$

With this modification we still have, for example,

$$C(\mathbf{R}, \boldsymbol{\theta}) = \mathbf{0} \tag{10.32}$$

where we define $\mathbf{R}$ by replacing $\mathbf{y}$ by $\mathbf{Y}$ in Equation (10.31). We also have

$$E(S_{\min}) = (n - p)\sigma^2 \tag{10.33}$$

where $n$ is the number of observations, and

$$E(S_{\text{par}}) = p\sigma^2 + \hat{\boldsymbol{\theta}}'\mathbf{a}'\mathbf{a}\boldsymbol{\theta} \tag{10.34}$$

It follows from Equation (10.32) that $S_{\min}$ and $S_{\text{par}}$ are uncorrelated, since $S_{\min}$ is a function of $\mathbf{R}$ and $S_{\text{par}}$ is a function of $\hat{\boldsymbol{\theta}}$. We shall be assuming normality of the underlying random variables for most of the remainder of the text, and on adding this assumption, $S_{\min}$ and $S_{\text{par}}$ are seen to be statistically independent, by Result 6 of Appendix A. It can also be shown that the distribution of $S_{\min}/\sigma^2$ is $\chi^2$ on $(n - p)$ degrees of freedom, and that if $\boldsymbol{\theta} = \mathbf{0}$, $S_{\text{par}}/\sigma^2$ is (independently) distributed as $\chi^2$ on $p$ degrees of freedom. Confidence intervals for $\sigma^2$ can be obtained by using Theorem 5.7.

---

**Exercises 10.3**

1.* Show that $E(\mathbf{R}) = \mathbf{0}$

and            $V(\mathbf{R}) = (\mathbf{I} - \mathbf{P})\sigma^2$.

2.* Derive the results (10.32) to (10.34).

---

**10.4 One-way classification analysis of variance**

We shall now apply the theory of Section 10.3 to the one-way classification analysis of variance such as Example 10.1. Let there be $t$ treatments, each replicated $r$ times in a design similar to Example 10.1, and denote the observations

$$y_{ij} \qquad i = 1, 2, \ldots, t; \qquad j = 1, 2, \ldots, r$$

The methods we shall use here are much more elaborate than is necessary for data with such a simple structure, but the methods are of general application. We consider a sequence of models,

Model 1: $\qquad\qquad\qquad Y_{ij} = \mu + \varepsilon_{ij}$ $\qquad\qquad\qquad$ (10.35)

Model 2: $\qquad\qquad\qquad Y_{ij} = \mu + \gamma_i + \varepsilon_{ij}$ $\qquad\qquad\qquad$ (10.36)

where the $\varepsilon_{ij}$ are distributed independently $N(0, \sigma^2)$. Model 1 fits only a general mean whereas Model 2 fits a separate mean for each treatment. If we fit these two models, and difference the sums of squares due to the fitted model, we obtain a sum of squares due to the treatment effects.

The analysis of Model 1 is a straightforward application of the theory of Chapter 5. There is only one unknown parameter, $\mu$, and if we put the model in the form (10.4), the design matrix $\mathbf{a}$ is a column of ones,

$$\mathbf{a} = \mathbf{1}$$

Then we find that the estimated mean is

$$\hat{\mu} = \sum_{i=1}^{t} \sum_{j=1}^{r} y_{ij}/rt = \bar{y}_{..}$$

and the sum of squares due to the estimated parameters is

$$S_{\text{par}}^{(1)} = \left( \sum_{i=1}^{t} \sum_{j=1}^{r} y_{ij} \right)^2 \bigg/ rt \qquad\qquad (10.37)$$

and also

$$S_{\text{min}}^{(1)} = \sum_{i=1}^{t} \sum_{j=1}^{r} y_{ij}^2 - \frac{1}{rt} \left( \sum_{i=1}^{t} \sum_{j=1}^{r} y_{ij} \right)^2 \qquad\qquad (10.38)$$

By Theorem 5.7, the distribution of $S_{\text{min}}^{(1)}/\sigma^2$ is $\chi^2$ on $(rt - 1)$ degrees of freedom, since there are $rt$ observations, and only one parameter to estimate, but this assumes that Model 1 rather than Model 2 applies.

For Model 2 we write the model in form (10.4) with

$$\boldsymbol{\theta}' = (\mu, \gamma_1, \gamma_2, \ldots, \gamma_t) \qquad\qquad (10.39)$$
$$\mathbf{Y}' = (Y_{1,1}, \ldots, Y_{1,r}, \ldots, Y_{t,1}, \ldots, Y_{t,r})$$

and where the design matrix $\mathbf{a}$ is a matrix similar to (10.5) of dimensions

$tr \times (t + 1)$, as shown in Equation (10.40).

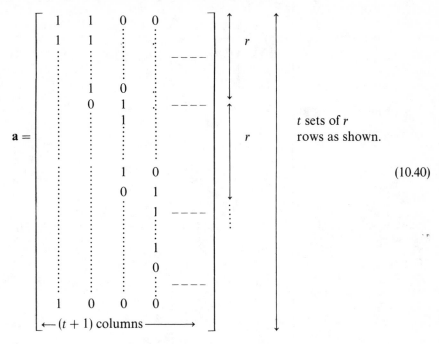

In this matrix the sum of columns 2 to $(t + 1)$ is equal to column one, and the matrix has rank $t$. Then we have

$$\mathbf{a'a} = \left( \begin{array}{c|c} tr & r\ \mathbf{1'} \\ \hline r\ \mathbf{1} & r\ \mathbf{I} \end{array} \right) \qquad \begin{array}{c} \text{No. of rows} \\ \updownarrow\ 1 \\ \updownarrow\ t \end{array} \tag{10.41}$$

No. of columns    1    $t$

and this matrix is singular, and also has rank $t$.
A convenient generalized inverse for this is

$$\mathbf{A}^{-} = \left[ \begin{array}{cc} \dfrac{1}{tr} & \mathbf{0} \\ \mathbf{0} & \dfrac{1}{r}\left(\mathbf{I} - \dfrac{1}{t}\mathbf{J}\right) \end{array} \right] \tag{10.42}$$

where $\mathbf{J}$ is a matrix of appropriate size having all entries unity.
Now the particular solution is Equation (10.28), and

$$\mathbf{a'y} = \begin{bmatrix} \sum_i \sum_j y_{ij} \\ \sum_j y_{1j} \\ \vdots \\ \sum_j y_{tj} \end{bmatrix}$$

so that Equation (10.28) gives

$$\hat{\boldsymbol{\theta}} = \begin{bmatrix} \hat{\mu} \\ \hat{\gamma}_1 \\ \vdots \\ \hat{\gamma}_t \end{bmatrix} = \begin{bmatrix} \bar{y}_{..} \\ \bar{y}_{1.} - \bar{y}_{..} \\ \vdots \\ \bar{y}_{t.} - \bar{y}_{..} \end{bmatrix} \tag{10.43}$$

At this point we must return to the general solution given by Equation (10.29). Now we have

$$\mathbf{A^- A} = \begin{bmatrix} \dfrac{1}{tr} & \mathbf{0} \\ \mathbf{0} & \dfrac{1}{r}\left(\mathbf{I} - \dfrac{1}{t}\mathbf{J}\right) \end{bmatrix} \begin{bmatrix} tr & r\mathbf{1}' \\ r\mathbf{1} & r\mathbf{I} \end{bmatrix}$$

$$= \begin{bmatrix} 1 & \dfrac{1}{t}\mathbf{1}' \\ \mathbf{0} & \left(\mathbf{I} - \dfrac{1}{t}\mathbf{J}\right) \end{bmatrix}$$

Therefore from Equation (10.29) the general solution is

$$\hat{\boldsymbol{\theta}} = \mathbf{A^- a'y} + (\mathbf{A^- A} - \mathbf{I})\mathbf{W}$$

$$= \begin{bmatrix} \bar{y}_{..} \\ \bar{y}_{1.} - \bar{y}_{..} \\ \vdots \\ \bar{y}_{t.} - \bar{y}_{..} \end{bmatrix} + \begin{bmatrix} 0 & \dfrac{1}{t}\mathbf{1}' \\ \mathbf{0} & -\dfrac{1}{t}\mathbf{J} \end{bmatrix}$$

$$= \begin{bmatrix} \bar{y}_{..} + \delta \\ \bar{y}_{1.} - \bar{y}_{..} - \delta \\ \vdots \\ \bar{y}_{t.} - \bar{y}_{..} - \delta \end{bmatrix} \tag{10.44}$$

where $\delta = \sum\limits_{2}^{t+1} W_i/t$ is an arbitrary constant. This result shows that with the model (10.36) we cannot estimate $\mu$, and we cannot estimate the absolute values of the $\gamma_i$. We can only estimate contrasts of the parameters which are independent of the parameter $\delta$, such as any comparison of the treatment means. That is, we can estimate any contrast $S$, such that

$$S = \sum c_i \gamma_i$$

where $$\sum c_i = 0 \tag{10.45}$$

or a similar linear combination of the parameters involving $\mu$ so that the arbitrary $\delta$ cancels out. This point is brought out in an example below by emphasizing methods of discussing treatment differences. Thus, when estimating the value for, say treatment one, we have

$$E(Y_1) = \hat{\mu} + \hat{\gamma}_1 = (\bar{y}_{..} + \delta) + (\bar{y}_{1.} - \bar{y}_{..} - \delta) = \bar{y}_{1.}.$$

and the terms in $\delta$ cancel.

The variance–convariance matrix of the estimators is obtained in a straightforward way. If we use generalized inverses which satisfy Equation (10.25), then we have

$$\begin{aligned} V(\hat{\theta}) &= V(\mathbf{A}^-\mathbf{a}'\mathbf{Y}) \\ &= \mathbf{A}^-\mathbf{a}'V(\mathbf{Y})\mathbf{a}\mathbf{A}^{-'} \\ &= \mathbf{A}^-\mathbf{A}\mathbf{A}^- \\ &= \mathbf{A}^-\sigma^2 \end{aligned} \tag{10.46}$$

This formula can only be used for estimable contrasts, since for all others, terms in $\delta$ enter. That is, we can use it for contrasts satisfying Equation (10.45). In effect, it means that we treat the parameter estimates as having variances and covariances

$$\begin{aligned} V(\hat{\mu}) &= \sigma^2/tr \\ V(\hat{\gamma}_i) &= (t-1)\sigma^2/tr \\ C(\hat{\gamma}_i, \hat{\gamma}_j) &= -\sigma^2/tr \qquad i \neq j \\ C(\hat{\mu}, \hat{\gamma}_i) &= 0 \end{aligned}$$

obtained from Equations (10.42) and (10.46).

From (10.30) and (10.43) we have

$$S^{(2)}_{\text{par}} = \frac{1}{tr}\left(\sum_{i=1}^{t}\sum_{j=1}^{r} y_{ij}\right)^2 + r\sum_{i=1}^{t}(\bar{y}_{i.} - \bar{y}_{..})^2 \tag{10.47}$$

and

$$S^{(2)}_{\text{min}} = \sum_{i=1}^{t}\sum_{j=1}^{r} y_{ij}^2 - \frac{1}{tr}\left(\sum_{i=1}^{t}\sum_{j=1}^{r} y_{ij}\right)^2 - r\sum_{i=1}^{t}(\bar{y}_{i.} - \bar{y}_{..})^2 \tag{10.48}$$

If now we consider $S_{min}$ and $S_{par}$ to be functions of random variables $Y_{ij}$ rather than of observations $y_{ij}$, their probability distributions follow from Theorems 5.7 and 5.8 as modified in Section 10.3. The distribution of $S_{min}/\sigma^2$ is $\chi^2$ with degrees of freedom given by

$$\text{d.f.} = (\text{no. of observations}) - (\text{rank } (10.40))$$
$$= rt - t = t(r - 1)$$

The matrix (10.40) is similar in structure to (10.5) but is of size $rt \times (t + 1)$, and it is easily seen that its rank is $t$. It reflects the fact that just $t$ parameters are required to represent the expectations of $t$ treatments, whereas the model (10.36) has used $(t + 1)$ parameters. Similarly, it follows that if $\boldsymbol{\theta} = \mathbf{0}$, the distribution of $S_{par}/\sigma^2$ is $\chi^2$ on $t$ degrees of freedom.

We can now obtain a sum of squares for testing the hypothesis that all $\gamma_i$ are zero, by applying the methods of Section 6.1. We calculate the difference between the sums of squares due to Models 1 and 2 of this section,

$$S_T = S_{par}^{(2)} - S_{par}^{(1)} \tag{10.49}$$

$$= r \sum_{i=1}^{t} (\bar{y}_{i.} - \bar{y}_{..})^2 \tag{10.50}$$

(We could get the same result by subtracting $S_{min}^{(2)}$ from $S_{min}^{(1)}$.)

Now we deal with the probability distributions of these sums of squares, so that the arguments should be treated as random variables $Y_{ij}$. We have just established that the first term on the right-hand side of Equation (10.49) has $t$ degrees of freedom. Earlier, we established that the second term on the right-hand side of Equation (10.49), which is Equation (10.37), has one degree of freedom. Therefore Equation (10.50) has $(t - 1)$ degrees of freedom. Physically, this represents the fact that, having fitted the parameter $\mu$, only $(t - 1)$ parameters are needed to specify $t$ independent treatment means. The number of degrees of freedom for $S_T$ can also be justified by subtracting the relevant numbers for $S_{min}^{(2)}$ from $S_{min}^{(1)}$.

The analysis-of-variance table now follows, and is shown in Table 10.2. The 'Total' sum of squares is $S_{min}^{(1)}$, and the 'Residual sum of squares' is $S_{min}^{(2)}$; by subtracting these we get Equation (10.50).

The numbers of degrees of freedom in Table 10.2 also follow from the reasoning given above. Those for the 'Total' line apply to $S_{min}^{(1)}$, those for the 'Residual' line are for $S_{min}^{(2)}$, and the 'Due to treatments' line degrees of freedom are obtained by subtraction. The numbers of degrees of freedom for 'Residual' can also be calculated directly. There are $t$ groups of $r$ observations, and a mean has been fitted to each group. Therefore, for estimation of the error variance, there are $(r - 1)$ independent contrasts in each of $t$ groups, making $t(r - 1)$ in all.

The expectations of the mean squares in Table 10.2 can be written down

Table 10.2 *One-way classification analysis of variance*

| Source | CSS | d.f. | E(MS) |
|---|---|---|---|
| Due to treatments | $r\sum_{1}^{t}(\bar{y}_{i.}-\bar{y}_{..})^2$ | $(t-1)$ | $\sigma^2 + \dfrac{r}{t-1}\sum(\alpha_i-\bar{\alpha}_.)^2$ |
| Residual | $\sum\sum(y_{ij}-\bar{y}_{i.})^2$ | $t(r-1)$ | $\sigma^2$ |
| Total | $\sum\sum(y_{ij}-\bar{y}_{..})^2$ | $tr-1$ | |

using the theory of Chapters 5 and 6. However, in this case it is easy to derive them directly. For example, if we insert the model (10.36) into the residuals we get

$$Y_{ij} - \bar{Y}_{i.} = \varepsilon_{ij} - \bar{\varepsilon}_{i.}$$

and this clearly only represents error. By squaring, adding and taking expectations, we can derive the result given in Table 10.2. The expectation of the mean square due to treatments follows in a similar way.

It now follows that a significance test of the hypothesis that all $\gamma_i$ are equal is given by referring the ratio

$$\frac{\text{Mean square due to treatments}}{\text{Residual mean square}} \qquad (10.51)$$

to $F$-tables on $\{(t-1), t(r-1)\}$ degrees of freedom. There is not usually very great interest in this significance test, as it is often known *a priori* that the $\gamma_i$ cannot be all equal. What is more important is to see where the difference between the treatment effects lie; one method for doing this is given in the next section.

The analysis-of-variance table (Table 10.2) is especially useful for two reasons. Firstly, it gives a summary of the data, showing the amount of variation attributable to different factors. Secondly, it provides an estimate of $\sigma^2$, and it is essential to have this in order to start discussing the possible significance of differences between treatment effects. A confidence interval for $\sigma^2$ can be obtained by using Theorem 5.7 on the residual mean square in Table 10.2.

---

## Exercises 10.4

1. Prove that the corrected sums of squares of Table 10.2 add up by writing

$$y_{ij} - \bar{y}_{..} = (y_{ij} - \bar{y}_{i.}) + (\bar{y}_{i.} - \bar{y}_{..})$$

and squaring and adding.

2. Obtain the expected mean square due to treatments, as shown in

Table 10.2: (a) by applying Theorems 5.6 and 6.1; and (b) by inserting the model (10.2) into the corrected sum of squares and taking expectations.

3. Obtain the expected residual mean square, by following the argument under Table 10.2.

## 10.5 A discussion of Example 10.1

There are three phases to the analysis of data such as that for Example 10.1 with models similar to (10.2). The first phase is the calculation of the treatment means, and of the analysis-of-variance table down to the $F$-test for significance of the treatment effects. In the second phase we need to use some kind of follow-up procedure to see where the differences between treatment effects are. In the third phase we need to examine the residuals for evidence of possible departures from assumptions made in the analysis. In the present discussion of Example 10.1 we shall carry out the first phase, and give one of the simplest follow-up procedures used in the second phase. We shall leave the third phase as an exercise( see Exercise 10.5.3).

### (i)  First phase – Calculations
The total corrected sum of squares is best calculated by noting that

$$\sum \sum (y_{ij} - \bar{y}_{..})^2 = \sum_{i=1}^{t} \sum_{j=1}^{r} y_{ij}^2 - \frac{(\sum \sum y_{ij})^2}{tr} \tag{10.52}$$

and the terms on the right are called the *uncorrected sum of squares* (USS) and the *correction factor* respectively. The corrected sum of squares due to treatments is

$$r \sum_{i=1}^{t} (\bar{y}_{i.} - \bar{y}_{..})^2 = \frac{\sum_{i=1}^{t} \left( \sum_{j=1}^{r} y_{ij} \right)^2}{r} - \frac{(\sum \sum y_{ij})^2}{tr} \tag{10.53}$$

and the first term on the right is the uncorrected sum of squares due to treatments. Having calculated the total and treatment sums of squares, the residual sum of squares can be calculated by subtraction. The main calcula-

Table 10.3

| Sum of squares | Total | Treatment |
|---|---|---|
| USS | 17 581.25 | 17 559.03 |
| Correction | 17 524.64 | 17 524.64 |
| CSS | 56.61 | 34.39 |

Table 10.4 *Analysis of variance for Example* 10.1

| Source | CSS | d.f. | MS |
|---|---|---|---|
| Due to treatments | 34.39 | 3 | 11.46 |
| Residual | 22.22 | 8 | 2.77 |
| Total | 56.61 | 11 | |

$F = 11.46/2.77 = 4.14$ on $(3, 8)$ d.f.

tions are set out in Tables 10.3 and 10.4. The terms can be identified with Equations (10.52) and (10.53).

The analysis of variance shows that a major part of the amount of variation in the data can be attributed to the treatment effects. In fact, the $F$-test of the hypothesis that all $\gamma_i$ are equal is only just significant at the 5 per cent level, since the 5 per cent point of $F(3,8)$ is 4.07. Since in this experiment treatment effects were expected *a priori*, this is reasonable evidence that treatment effects exist. It now follows for us to proceed with the second phase of the analysis, to see where these treatment effects are.

*(ii) Second phase – Least significant difference (LSD) method*

Here we discuss one of the simplest of the follow-up procedures used in the analysis of variance. We must examine the treatment means to see where the significant differences between them lie, and this can be achieved by comparing them in pairs. The variance of the difference between two means, each of $r$ observations, is $2\sigma^2/r$, so that if we applied the $t$-test, two means would differ at the 5 per cent level of significance if they are separated by more than

$$t_v(5\%)s\sqrt{(2/r)} \tag{10.54}$$

where $s^2$ is the residual mean square, $v$ is the number of degrees of freedom,

$$v = t(r - 1)$$

and $t_v(5\%)$ is the two-sided 5 per cent point of the $t$-distribution on $v$ degrees of freedom. The quantity (10.54) is called the *least significant difference* (LSD), and certainly anything less than this cannot possibly be regarded as significant. The $t$-test is only applicable for a comparison of two means, and there are $t(t - 1)/2$ pairs of means to be compared in the analysis of variance we have discussed. Some of these comparisons are likely to be greater than the quantity (10.54) by chance alone, especially if $t$ is moderate or large. The least significant difference therefore provides a lower bound to significant differences.

For Example 10.1 the standard error of the difference between two means is

$$\sqrt{\left(\frac{2 \times 2.77}{3}\right)} = 1.36 \qquad \text{on 8 d.f.}$$

The LSD at 5 per cent is 2.31 × 1.36 = 3.14, and Fig. 10.1 shows this LSD against the treatment results. A conclusion of the following kind can therefore be made:

The analysis of variance gives reasonable evidence of a difference between the treatment means, the $F$-test being just significant at 5 per cent. A comparison of the treatment means by the LSD method shows that treat-

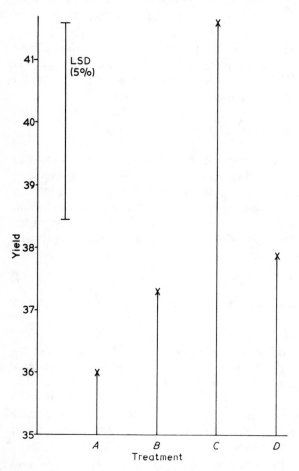

Fig. 10.1 The results of Example 10.1.

ment $C$ gives the highest yield, and there is no significant difference between the other three treatments. Treatment $A$ gave the lowest results.

Although there are many multiple comparison procedures available, the conclusions arrived at are usually very similar.

**Exercises 10.5**
1. Find the LSD at 1 per cent for Example 10.1.

2. Find the LSD for comparing $\bar{y}_C$ with $(\bar{y}_A + \bar{y}_B + \bar{y}_D)/3$ in Example 10.1. Comment on the validity of this procedure.

3. Suggest and carry out some analysis of residuals for the third phase of the analysis.

**10.6 Two-way classification**
In this section we apply the theory given at the beginning of Section 10.3 to data classified in two directions. Consider the following example.

*Example* 10.2  A sample of Portland cement was mixed and reduced to small samples for testing. The cement mortar was mixed with water and worked for a standard time; this process is known as 'gauging' and three men, called 'gaugers', took part in the experiment. The mortar was cast into cubes using moulds and stored at constant temperature for seven days. The cubes were then tested for compressive strength on a testing machine; this machine needed preliminary adjustment and three men, called 'breakers' did the work. The results given in Table 10.5 are pressures in pounds per square inch, and 5000 was subtracted from all results. [Modified form of an experiment reported by Davies (1967).]     □ □ □

It appears from the data shown in Table 10.5 that there could be an effect

Table 10.5 *Data for cement experiment (Example 10.2, 5000 subtracted)*

|  | Breaker 1 | Breaker 2 | Breaker 3 | Total | Mean |
|---|---|---|---|---|---|
| Gauger 1 | 340 | − 10 | − 185 | 145 | 48.3 |
| Gauger 2 | 45. | 345 | − 485 | − 95 | − 31.7 |
| Gauger 3 | 675 | 415 | − 82.5 | 1007.5 | 335.8 |
| Total | 1060 | 750 | − 752.5 | 1057.5 | 117.5 |
| Mean | 353.3 | 250 | − 250.8 |  |  |

on the observations due to both 'gaugers' and 'breakers'. There is a considerable amount of scatter in the results and we shall comment on this again later. We shall use this as an example of an experiment in which there are two meaningful classifications to take account of in the analysis.

Let us suppose that in general we have a row classification with $b$ rows, and a column classification with $t$ columns. Then by analogy with the one-way case we write a model as follows:

$$Y_{ij} = \mu + \beta_i + \gamma_j + \varepsilon_{ij} \qquad \begin{cases} i = 1, 2, \dots, b \\ j = 1, 2, \dots, t \end{cases} \qquad (10.55)$$

where $\varepsilon_{ij}$ represents error, $E(\varepsilon_{ij}) = 0$, $V(\varepsilon_{ij}) = \sigma^2$.

Again we have too many parameters; there are $(b + t + 1)$, and only $(b + t - 1)$ are needed to represent the row and column means, since both row and column means must add up to the same total. As in the one-way case, we could apply the side conditions

$$\sum_{j=1}^{t} \gamma_j = \sum_{i=1}^{b} \beta_i = 0 \qquad (10.56)$$

and proceed by the theory of Chapter 5. However, there is no need to do this; the approach using generalized inverses of Section 10.3 can deal with all models of this type, and what can be estimated will follow from the theory. We therefore reject the side conditions.

The model (10.55) can be rewritten in the form

$$E(\mathbf{Y}) = \mathbf{a}\boldsymbol{\theta} \qquad V(\mathbf{Y}) = \mathbf{I}\sigma^2$$

where

$$\boldsymbol{\theta}' = \{\mu, \beta_1, \dots, \beta_b, \gamma_1, \dots, \gamma_t\} \qquad (10.57)$$

and

$$\mathbf{Y}' = \{Y_{1,1}, \dots, Y_{1,t}, \dots Y_{b,1}, \dots, Y_{b,t}\}$$

Then the design matrix $\mathbf{a}$ is given by

$$\mathbf{a} = \begin{bmatrix} 1 & 1 & 0\dots0 & 1 & 0\dots0 \\ 1 & 1 & 0\dots0 & 0 & 1\dots0 \\ \vdots & \vdots & \vdots \quad \vdots & \vdots & \vdots \quad \vdots \\ 1 & 1 & 0\dots0 & 0 & 0\dots1 \\ 1 & 0 & 1\dots0 & 1 & 0\dots0 \\ 1 & 0 & 1\dots0 & 0 & 1\dots0 \\ \vdots & \vdots & \vdots \quad \vdots & \vdots & \vdots \quad \vdots \\ 1 & 0 & 0\dots1 & 0 & 0\dots1 \end{bmatrix}$$

with $t$ rows (first block) and $t$ rows (second block), $b$ sets of $t$ rows.

and

$$
\mathbf{a'a} = \begin{array}{c}
\overbrace{\phantom{xxxxxxx}}^{b} \quad \overbrace{\phantom{xxxxxxx}}^{t} \\
\left[\begin{array}{ccccc|ccc}
tb & t & t \;\ldots\, t & b & b \ldots b \\
t & t & 0 \;\ldots\, 0 & 1 & 1 \ldots 1 \\
\vdots & \vdots & \ddots \; \vdots & \vdots & \vdots \\
t & 0 & \ddots\, t & 1 & 1 \ldots 1 \\
\hline
b & 1 & \ldots 1 & b & 0 \ldots 0 \\
\vdots & \vdots & \vdots & \vdots & \ddots\; \vdots \\
b & 1 & \ldots 1 & 0 & 0 \ldots b
\end{array}\right]
\begin{array}{l} \Big\} b \\[20pt] \Big\} t \end{array}
\end{array}
\qquad (10.58)
$$

This can be written neatly in partitioned form,

$$
\mathbf{a'a} = \mathbf{A} = \begin{bmatrix}
tb & t\mathbf{1}' & b\mathbf{1}' \\
t\mathbf{1} & t\mathbf{I} & \mathbf{J} \\
b\mathbf{1} & \mathbf{J} & b\mathbf{I}
\end{bmatrix}
$$

where $\mathbf{J}$ is a matrix of appropriate size with all entries unity. Also we have,

$$
(\mathbf{a'y})' = \left(\sum_i \sum_j y_{ij}, \sum_j y_{1j}, \cdots \sum_j y_{bj}, \sum_i y_{i1}, \ldots, \sum_i y_{it}\right) \qquad (10.59)
$$

A generalized inverse is

$$
\mathbf{A}^- = \begin{bmatrix}
\dfrac{1}{tb} & \mathbf{0} & \mathbf{0} \\[10pt]
\mathbf{0} & \dfrac{1}{t}\left(\mathbf{I} - \dfrac{1}{b}\mathbf{J}\right) & \mathbf{0} \\[10pt]
\mathbf{0} & \mathbf{0} & \dfrac{1}{b}\left(\mathbf{I} - \dfrac{1}{t}\mathbf{J}\right)
\end{bmatrix} \qquad (10.60)
$$

A particular solution is $\mathbf{A}^-\mathbf{a'y}$, which is

$$
\hat{\boldsymbol{\theta}}' = \{\bar{y}_{..}, (\bar{y}_{1.} - \bar{y}_{..}), \ldots, (\bar{y}_{b.} - \bar{y}_{..}), (\bar{y}_{.1} - \bar{y}_{..}), \ldots, (\bar{y}_{.t} - \bar{y}_{..})\} \qquad (10.61)
$$

The variance–covariance matrix is $\mathbf{A}^-\sigma^2$, so that, for example,

$$
V(\hat{\gamma}_1) = V(\bar{Y}_{.1} - \bar{Y}_{..}) = \frac{(t-1)}{tb}\sigma^2
$$

and the row and column parameters are estimated independently.

In order to get the general solution we need $A^-A$, which is

$$A^-A = \begin{bmatrix} 1 & \dfrac{1}{b}\mathbf{1}' & \dfrac{1}{t}\mathbf{1}' \\[2ex] 0 & I - \dfrac{1}{b}J & 0 \\[2ex] 0 & 0 & I - \dfrac{1}{t}J \end{bmatrix}$$

so that a general solution is, from Equation (10.29),

$$\hat{\theta} + \begin{bmatrix} 0 & \dfrac{1}{b}\mathbf{1}' & \dfrac{1}{t}\mathbf{1}' \\[2ex] 0 & -\dfrac{1}{b}J & 0 \\[2ex] 0 & 0 & -\dfrac{1}{t}J \end{bmatrix} W \qquad (10.62)$$

where $\hat{\theta}$ is given by Equation (10.61). Now $W$ has $(t + b + 1)$ components. The general solution is therefore

$$\{(\bar{y}_{..} + \delta_1 + \delta_2), (\bar{y}_{1.} - \bar{y}_{..} - \delta_1), \dots, (\bar{y}_{b.} - \bar{y}_{..} - \delta_1),$$
$$(\bar{y}_{.1} - \bar{y}_{..} - \delta_2), \dots, (\bar{y}_{.t} - \bar{y}_{..} - \delta_2)\} \qquad (10.63)$$

where
$$\delta_1 = \sum_{2}^{b+1} W_i/b, \qquad \delta_2 = \sum_{b+2}^{b+t+1} W_i/t$$

This result is parallel to the result obtained in the one-way case, and it implies that we can only estimate contrasts

$$\sum_{1}^{b} c_i \beta_i \text{ such that } \sum_{1}^{b} c_i = 0$$

or
$$\sum_{1}^{t} d_i \gamma_i \text{ such that } \sum_{1}^{t} d_i = 0$$

That is, we can make comparisons among row or column effects separately.

We can now write down the sum of squares due to the fitted model, and $S_{\min}$, by using Equations (10.59) and (10.61) in Equation (10.30):

$$S_{par} = \bar{y}_{..}(\sum\sum y_{ij}) + \sum\sum y_{ij}(\bar{y}_{i.} - \bar{y}_{..}) + \sum\sum y_{ij}(\bar{y}_{.j} - \bar{y}_{..})$$
$$= tb\bar{y}_{..}^2 + t\sum_i(\bar{y}_{i.} - \bar{y}_{..})^2 + b\sum_j(\bar{y}_{.j} - \bar{y}_{..})^2 \qquad (10.64)$$
$$S_{\min} = \sum\sum y_{ij}^2 - S_{par}$$

In order to identify the terms of Equation (10.64) we consider a sequence of models, as in Section 10.4:

Model 1:  $Y_{ij} = \mu + \varepsilon_{ij}$

Model 2:  $Y_{ij} = \mu + \beta_i + \varepsilon_{ij}$

Model 3:  $Y_{ij} = \mu + \gamma_j + \varepsilon_{ij}$

Model 4:  $Y_{ij} = \mu + \beta_i + \gamma_j + \varepsilon_{ij}$

The sum of squares due to fitting Model 4 is given by Equation (10.64). The sums of squares due to fitting the other models can be written down by analogy with Equation (10.37) and (10.47):

$$S_{\text{par}}^{(1)} = tb\bar{y}_{..}^2$$

$$S_{\text{par}}^{(2)} = tb\bar{y}_{..}^2 + t\sum_i (\bar{y}_{i.} - \bar{y}_{..})^2$$

$$S_{\text{par}}^{(3)} = tb\bar{y}_{..}^2 + b\sum_j (\bar{y}_{.j} - \bar{y}_{..})^2$$

The sum of squares due to fitting the $\gamma_j$ terms is now

$$S_{\text{par}}^{(4)} - S_{\text{par}}^{(2)} = b\sum_j (\bar{y}_{.j} - \bar{y}_{..})^2$$

$$= S_{\text{par}}^{(3)} - S_{\text{par}}^{(1)}$$

That is, the sum of squares due to the column parameters is the same whether or not row parameters are included in the model; the same thing obviously holds with the roles of rows and columns reversed.

The orthogonal case of a least-squares analysis was described briefly at the end of Chapter 5, and it was applied in Chapter 7 to polynomial regression. In that model, all of the parameters are estimated independently. The usual use of the word 'orthogonal' is to the situation in which the parameters can be divided into subsets, such that there is independence between the subsets, but not within them. This enables us to list a separate sum of squares in the 'Anova' table for each subset.

The analysis above shows that given the parameter $\mu$ is fitted, the row and column parameters are estimated orthogonally, in the sense just described. The estimators of, say, the row parameters are correlated with each other, but the estimators of the row parameters are statistically independent of the estimators of the column parameters. The independence of the estimators of row and column parameters is also obvious from Equation (10.60), but the demonstration in terms of differencing sums of squares is more convincing. We therefore have the analysis-of-variance table as follows.

The analysis-of-variance table is shown in Table 10.6. The 'Total' line gives the sum of squares and number of degrees of freedom for $S_{\text{min}}^{(1)}$, which is for the model when only the parameter $\mu$ is fitted. The 'Row' and 'Column' lines give the extra sum of squares accounted for by adding row and column

Table 10.6 *Two-way analysis of variance*

| Source | CSS | d.f. | E(MS) |
|---|---|---|---|
| Rows | $t\sum(\bar{y}_{i.} - \bar{y}_{..})^2$ | $(b-1)$ | $\sigma^2 + \dfrac{t}{(b-1)}\sum(\beta_i - \bar{\beta}_.)$ |
| Columns | $b\sum(\bar{y}_{.j} - \bar{y}_{..})^2$ | $(t-1)$ | $\sigma^2 + \dfrac{b}{(t-1)}\sum(\gamma_i - \bar{\gamma}_.)^2$ |
| Residual | $S_{\min}$ | $(t-1)(b-1)$ | $\sigma^2$ |
| Total | $\displaystyle\sum_i\sum_j(y_{ij} - \bar{y}_{..})^2$ | $tb-1$ | |

parameters respectively, and the degrees of freedom for these. Thus, for example, given $\mu$ has been fitted, only $(b-1)$ parameters are required to represent $b$ independent expectations for rows. The 'Residual' line is the one corresponding to fitting Model 4.

It is easily verified that

$$S_{\min} = \sum_i\sum_j(y_{ij} - \bar{y}_{i.} - \bar{y}_{.j} + \bar{y}_{..})^2$$

but this term is usually calculated by subtraction. The expectations of the mean squares are readily verified by directly substituting the model (10.55) into the corrected sums of squares, and taking expectations.

## 10.7 A discussion of Example 10.2

The practical aspects of the two-way analysis of variance follow the same pattern as that laid down in Section 10.5 for the one-way analysis. The calculations are carried out in the same way. The following forms are useful for computing:

*Total sum of squares*

$$\sum\sum(y_{ij} - \bar{y}_{..})^2 = \sum\sum y_{ij}^2 - \text{correction}$$

where correction $= (\sum\sum y_{ij})^2/tb$

*Row sum of squares*

$$t\sum_{i=1}^{b}(\bar{y}_{i.} - \bar{y}_{..})^2 = \frac{1}{t}\sum_{i=1}^{b}\left(\sum_{j=1}^{t} y_{ij}\right)^2 - \text{correction}$$

*Column sum of squares*

$$b\sum_{j=1}^{t}(\bar{y}_{.j} - \bar{y}_{..})^2 = \frac{1}{b}\sum_{j=1}^{t}\left(\sum_{i=1}^{b} y_{ij}\right)^2 - \text{correction}$$

In these formulae notice that

$$\sum_{j=1}^{b} y_{ij} = \text{row} \quad i \quad \text{total} \quad \text{and} \quad \sum_{i=1}^{b} y_{ij} = \text{column} \quad j \quad \text{total}$$

Before setting out on any analysis it is important to examine the data critically. Table 10.5 shows that 8 of the 9 observations have a '0' or '5' in the units position. Clearly, there is something very odd about this set of data; patterns like this arise when the last digit has been rounded up. In this example, the pattern results largely from the way we have modified the original experiment, which was to average four observations in each cell. We shall proceed with the analysis and the calculations are as shown in Tables 10.7 and 10.8.

$F$-tests can now be carried out separately for the effect of 'gaugers' and 'breakers', using the mean square for residual as the denominator. Since the 5 per cent of $F(2,4)$ is 6.94, we conclude that the effect of gaugers is not significant, and that the data are consistent with there being no effect due to gaugers. The sum of squares due to 'breakers' is larger, and is just significant at 5 per cent, showing some indication of real differences between the breakers.

The next step in the analysis is to examine the row or column effects by a method such as the LSD method described in Section 10.5, to examine where the significant differences lie. There is no point in doing this for 'gaugers', since the 'gaugers' sum of squares is not significant.

Table 10.7

| Sum of squares | Total | Gauges | Breakers |
|---|---|---|---|
| USS | 1 140 856.2 | 348 368.7 | 750 785.4 |
| Correction | 124 256.2 | 124 256.2 | 124 256.2 |
| CSS | 1016 600.0 | 224 112.5 | 626 529.2 |

Table 10.8 *Analysis of variance for Example* 10.2

| Source | CSS | d.f. | MS | F |
|---|---|---|---|---|
| Gaugers | 224 113 | 2 | 112 056 | 2.70 |
| Breakers | 626 529 | 2 | 313 264 | 7.55 |
| Residuals | 165 958 | 4 | 41 490 | |
| Total | 1 016 600 | 8 | | |

The LSD at 5 per cent is, from Equation (10.54),

$$t_v(5\%)s\sqrt{(2/n)} = t_4(5\%)\sqrt{(41\,490)}\sqrt{(2/3)}$$
$$= 2.78 \times 203.7 \times 0.816 = 462.1$$

If we now look at the values of the means given in Table 10.5, we see that 'breaker 3' gives a lower response than the other two breakers. However, the mean square error is very large, and the LSD of 462.1 obtained above is also rather large. This means that the experiment was not very precise, and important effects could exist which have not been picked up. Indeed, there is some evidence that 'gauger 3' might be giving a consistently higher response than the other two gaugers, but no firm conclusion about this could be made from this set of data.

Results of this type would lead to the suggestion that another experiment should be run, with replicated observations within cells, in order to be able to make more clear conclusions. (The original experiment was in fact run with four observations in each cell.)

**Exercises 10.7**

1. Carry out an analysis of residuals on Example 10.2. Are these residuals strongly correlated?

2. Write up a conclusion for Example 10.2, complete with any necessary graphs, suitable to hand back to the experimenters.

3. An experimenter has available 24 Barred Columbian chicks all 35 days. He wishes to compare the percentage increase of lymphocytes in the circulating blood after applying grafts from other chickens of known breeds. The chicks were divided into 8 groups of 3 so that the chicks in each group were alike as far as possible as regards weight and size. Grafts of approximately 2 cm diameter were sutured in place on the outer median thigh of each chick. In each group the first chick, chosen randomly, was given a graft from a pure-bred Barred Columbian donor, the second a graft from a cross-bred Rhode Island Red and Barred Columbian donor, and the third a graft

Table 10.9 *Percentage increase of lymphocytes*

| Donor | Group number | | | | | | | |
|---|---|---|---|---|---|---|---|---|
| | 1 | 2 | 3 | 4 | 5 | 6 | 7 | 8 |
| Barred Columbian | 7.0 | 5.5 | 4.7 | 6.1 | 5.3 | 8.6 | 6.7 | 4.9 |
| R.I.R./B.C. | 7.8 | 5.2 | 5.9 | 7.4 | 6.8 | 9.3 | 6.4 | 5.6 |
| Rhode Island Red | 11.8 | 8.6 | 9.3 | 10.9 | 9.7 | 11.9 | 9.9 | 10.3 |

Table 10.10

| Experiment | Group | | | |
| | Control | X-ray | Beta | Total |
|---|---|---|---|---|
| 1 | a  12 | 3 | 4 | 19 |
|   | b  10 | 15 | 12 | 37 |
| 2 | a  14 | 5 | 6 | 25 |
|   | b   3 | 10 | 9 | 22 |
| 3 | a   9 | 5 | 2 | 16 |
|   | b  11 | 17 | 17 | 45 |
| 4 | a  17 | 5 | 7 | 29 |
|   | b   2 | 14 | 13 | 29 |

(a denotes the number of cells reaching mid-mitosis, b the number not reaching mid-mitosis.)

from a pure-bred Rhode Island Red. Blood smears were taken immediately before the grafting operation and again 5 days afterwards. The percentage increase of lymphocytes is recorded in Table 10.9.

Analyse these data to see whether the percentage increase of lymphocytes in the blood of Barred Columbian chicks is the same whatever type of donor is used. [London B.Sc. General, 1958.]

4. Table 10.10 contains the results of four radiation experiments in a series to investigate the effects of X-rays and beta-rays on mitotic rates in grasshopper neuroblasts. Embryos from the same egg were divided into three groups, one serving as a control and the other two being exposed to physically equivalent doses of X- and beta-radiation. After irradiation, approximately equal numbers of cells in each group were examined for a specified time, and the number passing through mid-mitosis noted.

Use a suitable transformation in order to examine what conclusions can be drawn about the effect of irradiation on the proportion of cells reaching mid-mitosis. Is there any evidence of a difference between the effect of the two types of radiation? What special difficulties do you have in this analysis? Suggest other methods of approach. [Modified question from Cambridge Diploma, 1959.]

## 10.8 General method for analysis of variance

The method of using generalized inverses gives a useful link with the least-squares theory of Chapter 5, but it is unnecessarily complicated for the models to be discussed. We must recognize that where we set up a model such as (10.2), we cannot estimate all of the parameters, but only certain functions of them. Bearing this in mind, the least-squares equations can be written down from first principles, and we shall find that all functions of the parameters that we need are estimable. The separation of the sum of squares can also be justified from first principles.

Let our model be

$$Y_{ij} = \mu + \alpha_i + \varepsilon_{ij} \qquad i = 1, 2, \ldots, t; \quad j = 1, 2, \ldots, r \qquad (10.65)$$

where $\varepsilon_{ij}$ is $N(0, \sigma^2)$.

The sum of squares to be minimized is

$$S = \sum_{i=1}^{t} \sum_{j=1}^{r} (y_{ij} - \mu - \alpha_i)^2$$

from which we have

$$\frac{\partial S}{\partial \mu} = -2 \sum_i \sum_j (y_{ij} - \mu - \alpha_i)$$

$$\frac{\partial S}{\partial \alpha_i} = -2 \sum_j (y_{ij} - \mu - \alpha_i) \qquad i = 1, 2, \ldots, t$$

This leads to the normal equations

$$\bar{y}_{..} = \hat{\mu} + \frac{1}{t} \sum \hat{\alpha}_i \qquad\qquad (10.66)$$

$$\bar{y}_{i.} = \hat{\mu} + \hat{\alpha}_i$$

Therefore, for example, the residuals are defined by

$$y_{ij} - \hat{\mu} - \hat{\alpha}_i = y_{ij} - \bar{y}_{i.}.$$

Now the corrected sum of squares can readily be separated, since

$$\sum_i \sum_j (y_{ij} - \bar{y}_{..})^2 = \sum_i \sum_j \{(y_{ij} - \bar{y}_{i.}) + (\bar{y}_{i.} - \bar{y}_{..})\}^2$$

$$= \sum_i \sum_j (y_{ij} - \bar{y}_{i.})^2 + r \sum_i (\bar{y}_{i.} - \bar{y}_{..})^2 \qquad (10.67)$$

This yields the CSS terms of Table 10.2. The first term on the right-hand side of Equation (10.67) is the sum of squares of the residuals, and the second term is the sum of squares of the fitted values. The expectations of the mean squares can easily be obtained directly. For example, by inserting the model we have

$$\bar{y}_{i.} - \bar{y}_{..} = \mu + \alpha_i - (\mu + \bar{\alpha}) + \bar{\varepsilon}_{i.} - \bar{\varepsilon}_{..}$$

$$= (\alpha_i - \bar{\alpha}) + (\bar{\varepsilon}_{i.} - \bar{\varepsilon}_{..}) \qquad\qquad (10.68)$$

so that

$$\sum_i \sum_j (\bar{y}_{i.} - \bar{y}_{..})^2 = r \sum_i (\alpha_i - \bar{\alpha})^2 + r \sum_i (\bar{\varepsilon}_{i.} - \bar{\varepsilon}_{..})^2 + 2r \sum_i (\alpha_i - \bar{\alpha})(\bar{\varepsilon}_{i.} - \bar{\varepsilon}_{..})$$

If now we take expectations over $\varepsilon$ we obtain

$$E(\sum_i \sum_j (\bar{y}_{i.} - \bar{y}_{..})^2) = (t-1)\sigma^2 + r\sum_i (\alpha_i - \bar{\alpha}_.)^2 \qquad (10.69)$$

which is the result given in Table 10.2.

This method of justifying the analysis of variance will be used in subsequent sections rather than the full generalized inverse approach.

### Sufficient statistics*

There is another way of obtaining least-squares estimates which is equivalent to the method just described. The normal equations are, from Section 10.3,

$$\mathbf{a'a}\hat{\boldsymbol{\theta}} = \mathbf{a'y}$$

and the quantities on the right-hand side are the sufficient statistics (see Exercise 5.1.4). In any problem, the sufficient statistics are usually obvious, being quantities such as the treatment totals. Now there is a general result which states that if we can find a linear function of the sufficient statistics which is an unbiased estimator of the function we require, then this is the least-squares estimator. In some problems, therefore, it is easier to write down the sufficient statistics, and then arrange a linear combination to have an expectation equal to the function we wish to estimate. This applies particularly to more advanced types of experimental design.

---

### Exercises 10.8

1. Use the technique given in this section to obtain the normal equations and the analysis of variance for the one-way unequally replicated case. Let the model be

$$Y_{ij} = \mu + \alpha_i + \varepsilon_{ij} \qquad i = 1, 2, \ldots, t; \quad j = 1, 2, \ldots, n_i$$

and obtain the normal equations equivalent to (10.66). Show that the minimized sum of squares is

$$\sum_i \sum_j (y_{ij} - \bar{y}_{i.})^2$$

and hence that the analysis-of-variance table is as shown in Table 10.11.

Table 10.11

| Source | CSS | d.f. |
|---|---|---|
| Due to treatments | $\sum n_i (\bar{y}_{i.} - \bar{y}_{..})^2$ | $(t-1)$ |
| Residual | $\sum \sum (y_{ij} - \bar{y}_{i.})^2$ | $\sum (n_i - 1)$ |
| Total | $\sum \sum (y_{ij} - \bar{y}_{..})^2$ | $\sum n_i - 1$ |

What are the $E(MS)$ for this table?

**Further reading**

For generalized inverses, see Ben-Israel and Greville (1974), Rao (1971) or Searle (1971). For alternative approaches to the analysis of variance see Davies (1967) for a practical emphasis, and Searle (1971) for a more theoretical account.

For further numerical illustration see the references given in Appendix B.

# CHAPTER ELEVEN

# Designs with regressions in the treatment effects

## 11.1 One-way analysis

In Example 10.1 the four treatments being compared were four qualitatively different treatments, but in some experiments the treatments are just different quantities of a given fertilizer, different temperatures for operating an industrial process, etc. With such experiments, the analysis proceeds by combining the ideas of regression with the methods given in Chapter 10 for 'The analysis of variance'. The following example was reported by Williams (1959).

*Example* 11.1   A laboratory experiment was conducted into the effect of different pressures during the sheet pressing phase of the manufacture of paper. One of the properties measured was called the 'tear factor', and roughly speaking this was the pressure required to tear the paper (of specified thickness) in a standardized experiment. The pressures used were recorded on a scale $x = 2 \log_2$ (pressure/71), and data obtained are given in Table 11.1. This set of data can be viewed as either a one-way analysis of variance with five 'treatments', or else as a regression analysis in which the observations are replicated at each value of the explanatory variable $x$. In fact it is just another example of application of the theory for models linear in the parameters. □ □ □

Figure 11.1 shows a certain amount of scatter in the results, but this is

Table 11.1

| Pressure scale | Tear factor | | | | Total |
|---|---|---|---|---|---|
| −2 | 112 | 119 | 117 | 113 | 461 |
| −1 | 108 | 99 | 112 | 118 | 437 |
| 0 | 120 | 106 | 102 | 109 | 437 |
| 1 | 110 | 101 | 99 | 104 | 414 |
| 2 | 100 | 102 | 96 | 101 | 399 |

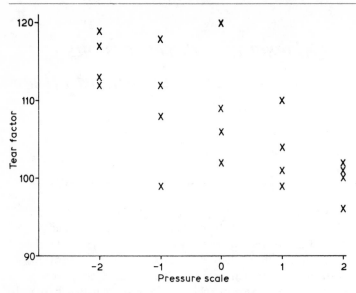

Fig. 11.1 Tear factor values plotted against pressure.

quite usual in experiments of this type. There is clear evidence of a regression relationship between 'tear factor' and pressure. The fact that the observations are replicated enables us to check quickly to see if there is any evidence of heterogeneity of the error variance. If we calculate the range and variance at each value of $x$ we obtain the following values:

| $x$ | $-2$ | $-1$ | 0 | 1 | 2 |
|---|---|---|---|---|---|
| Range | 7 | 19 | 18 | 11 | 6 |
| Variance within groups | 10.92 | 63.58 | 59.58 | 23.00 | 6.92 |

Methods of testing for the presence of heterogeneity of variances are discussed in Chapter 13, but for this problem a very simple procedure will suffice. For comparing just two variances we use the $F$-test, and the 5 per cent value for $F(3, 3)$ is 9.28. It is clear from this that the 5 per cent value for the maximum ratio of two out of four variances would be somewhat larger than 9.28. The maximum ratio of variances in Example 11.1 is 9.19, so that there is very little evidence of heterogeneity of the variance. We shall proceed with the analysis on the assumption of homogeneous variance. (If the details were available to us we would certainly want to check to see if there were reasons for some of the features of Fig. 11.1. The questions we would ask would include asking whether there was any reason for the presence of an outlier in the $-1$ and 0 pressure groups, and whether there were reasons for large differences in variance between the groups. In the event, the differences in the variances are not significant, and somewhat erratic results from this type of experiment were to be expected.)

A reasonable model for the data is therefore to write

$$Y_{ij} = \mu + \gamma_i + \varepsilon_{ij}, \qquad i = 1, 2, \ldots 5; \quad j = 1, 2, \ldots 4 \qquad (11.1)$$

where $E(\varepsilon) = 0$, $V(\varepsilon) = \sigma^2$, and where the $\gamma_i$ could be written as a (possibly polynomial) regression on $x$.

If for the moment we take the $\gamma_i$ to be simply constants for each group of four observations, we have the one-way analysis of variance discussed in Section 10.4 and the analysis-of-variance table is as shown in Table 11.2 (cf. Table 10.2).

We now notice that a regression analysis can be performed, regressing the treatment means $\bar{y}_{i.}$ on the pressure scale $x_i$. If we do the calculations as if the treatment means $\bar{y}_{i.}$ are single observations, then the total sum of squares is

$$\sum (\bar{y}_{i.} - \bar{y}_{..})^2$$

and the variance of each treatment mean is $\sigma^2/4$, where $\sigma^2$ is defined in Equation (11.1). The analysis of variance for regression must therefore be multiplied by four (the number of observations in each group) in order to present it on the same basis as Table 11.2. By doing this the total sum of squares for the regression analysis is the same as the sum of squares due to treatments in Table 11.2, and the expectations of the means squares all read $\sigma^2$ plus some quantity. If we try linear regression only, and write $\gamma_i = \beta x_i$,

Table 11.2 *Analysis of variance for Example* 11.1

| Source | CSS | d.f. | E(MS) |
|---|---|---|---|
| Due to 'treatments' | $4\sum (\bar{y}_{i.} - \bar{y}_{..})^2$ | 4 | $\sigma^2 + \sum (\alpha_i - \bar{\alpha})^2$ |
| Residual | $\sum_i \sum_j (y_{ij} - \bar{y}_{i.})^2$ | 15 | $\sigma^2$ |
| Total | $\sum_i \sum_j (y_{ij} - \not{y})^2$ | 19 | |

Table 11.3 *Analysis of variance for regression of group means* (×4)

| Source | CSS | d.f. | E(MS) |
|---|---|---|---|
| Due to regression | $4\{CS(\bar{y}_{i.}, x_i)\}^2/CS(x, x)$ | 1 | $\sigma^2 + 4\beta^2 CS(x, x)$ |
| Deviations | by subtraction | 3 | $\sigma^2 + \text{(non-linear)}$ |
| Total (due to treatments) | $4\sum (\bar{y}_{i.} - \bar{y}_{..})^2$ | 4 | $\sigma^2 + \sum \gamma_i^2$ |

then, by comparison with Table 2.3, this regression analysis is as shown in Table 11.3.

Another way of justifying the sums of squares due to regression is as follows. If we carry out a regression of observations $y_{ij}$ on $x_i$, $j = 1, 2, \ldots, n$, then the corrected sum of products is

$$\sum_i \sum_j (y_{ij} - \bar{y}_{..})(x_i - \bar{x}_.) = n \sum_i (\bar{y}_{i.} - \bar{y}_{..})(x_i - \bar{x}_.) = nCS(\bar{y}_{i.}, x_i)$$

Since the data are replicated $n$ times at each $x$-value, the corrected sum of squares for $x$ is $nCS(x, x)$, so that the sum of squares due to regression is as given in Table 11.3. Clearly, the 'Deviations' mean square will be greater than $\sigma^2$ if the regression is not linear. If we put Tables 11.2 and 11.3 together, we have the combined analysis-of-variance table for $t$ groups of $n$ observations each, shown in Table 11.4.

These tables show that if we have regressions with replicated observations, then we have a separate sum of squares, which is the sum of squares from within groups, which will lead to an unbiased estimate of $\sigma^2$ whether or not the regression is linear. The 'Deviations' mean square will only be an unbiased estimator of $\sigma^2$ if there are no non-linear terms in the regression, and otherwise it will tend to be inflated.

It is now clear that a test of the significance of the regression is obtained by refering the ratio $A/D$ to tables of $F(1, t(n-1))$. Similarly, a test of the significance of non-linear terms in the regression is obtained by refering the ratio $B/D$ to tables of $F((t-2), t(n-1))$. If the mean square $B$ is at all large, it would be advisable to include at least a quadratic term in the regression; in Example 11.1 this would be easy as the $x$-values are equally spaced, and orthogonal polynomials can be used. Two of the advantages of replication in regression are that a more reliable estimate of $\sigma^2$ is obtained, and there is a better test of non-linearity than in the unreplicated situation.

The calculations for Example 11.1 lead to Table 11.5, and we see that the mean square for deviations from regression is not significant, but the signi-

Table 11.4 *Analysis of variance for regression with replication*

| Source | CSS | d.f. | Mean square |
|---|---|---|---|
| Due to regression | $n\{CS(\bar{y}_{i.}, x_i)\}^2 / CS(x, x)$ | 1 | $A$ |
| Deviations | by subtraction | $(t-2)$ | $B$ |
| Due to treatments | $n \sum_i (\bar{y}_{i.} - \bar{y}_{..})^2$ | $(t-1)$ | $C$ |
| Residual | $\sum_i \sum_j (y_{ij} - \bar{y}_{i.})^2$ | $t(n-1)$ | $D$ |
| Total | $\sum_i \sum_j (y_{ij} - \bar{y}_{..})^2$ | $(tn-1)$ | |

Table 11.5 *Analysis of variance for Example* 11.1

| Source | CSS | d.f. | Mean square |
|---|---|---|---|
| Due to regression | 540.225 | 1 | 540.2 |
| Deviations | 28.575 | 3 | 9.5 |
| Due to 'treatments' | 568.800 | 4 | — |
| Within groups | 492.000 | 15 | 32.8 |
| Total | 1060.800 | 19 | |

ficance of the regression is tested by $F = 540.225/32.800 = 16.47$, on $(1, 15)$ d.f., which is very highly significant.

Clearly, the principle involved in carrying out this analysis can be extended. We began by fitting the model (11.1), and arrived at Table 11.5 by using a linear regression model for the five treatments on the pressure scale $x$ as the explanatory variable. We could extend this by fitting quadratic, cubic and quartic terms in $x$, so that Equation (11.1) becomes

$$Y_{ij} = \mu + \beta_1 x + \beta_2 x^2 + \beta_3 x^3 + \beta_4 x^4 + \varepsilon_{ij} \qquad (11.2)$$

However, we notice that the pressure scale $x$ is equally spaced, so that orthogonal polynomials can easily be used, and we can write

$$Y_{ij} = \mu + \beta_1 f_1(x) + \beta_2 f_2(x) + \beta_3 f_3(x) + \beta_4 f_4(x) + \varepsilon_{ij} \qquad (11.3)$$

where $f_i(x)$ is an orthogonal polynomial of degree $i$ in $x$, defined in Chapter 7. If we proceed with model (11.3), we can enter Chapter 5 theory directly, and the resulting analysis-of-variance table is similar to Table 11.5, but with the sums of squares due to the linear, quadratic, cubic and quartic terms listed separately, with one degree of freedom each; we leave it as an exercise to work through this approach, and instead we proceed on a more intuitive basis.

If we wish to write down the analysis for the model (11.3), the formulae for the sums of squares due to each of the polynomial terms will be similar to those given in Chapter 7, but slightly different due to the fact that observations are replicated at each value of the explanatory variable. The contrasts of the group means to be used for the polynomial terms are clear from Chapter 7 and Table C.11 of Appendix C; what is not so clear is how to calculate the corrected sums of squares. The method of linear contrasts is very helpful at this point; it enables us to separate off single degrees of freedom from the sum of squares due to treatments. This approach therefore starts with Table 11.2.

*Method of linear contrasts*

Let there be $t$ groups of $n$ observations each, and all observations have a variance of $\sigma^2$. Denote the totals $T_i$, $i = 1, 2, \ldots t$, and suppose that we are interested in a linear contrast

$$W_1 = \sum_{i=1}^{t} f_i T_i \tag{11.4}$$

where

$$\sum_{i=1}^{t} f_i = 0$$

Then if all groups have the same expectation we have

$$
\begin{aligned}
E(W_1) &= E(T)\sum f_i = 0 \\
E(W_1^2) &= V(W_1) \\
&= V(\sum f_i T_i) = \sum f_i^2 V(T_i) \\
&= n\sigma^2 \sum f_i^2
\end{aligned}
\tag{11.5}
$$

since $V(T_i) = n\sigma^2$

Therefore under the null hypothesis that all group means are equal,

$$W_i^2 / \{n \sum f_i^2\} \tag{11.6}$$

has expectation $\sigma^2$, and is the appropriate sum of squares to enter in an analysis-of-variance table.

If now we are interested in another linear contrast

$$W_2 = \sum_{i=1}^{t} g_i T_i$$

this will be uncorrelated with $W_1$ if the covariance of $W_1$ and $W_2$ is zero. From Appendix A, Exercise 2, we have

$$
\begin{aligned}
C(W_1, W_2) &= C(\sum f_i T_i, \sum g_i T_i) \\
&= \sum f_i g_i V(T_i)
\end{aligned}
$$

since the totals are independent.

Therefore the covariance is

$$C(W_1, W_2) = n\sigma^2 \sum f_i g_i$$

and the contrasts are uncorrelated if

$$\sum_{i=1}^{t} f_i g_i = 0 \tag{11.7}$$

The sums of squares for the contrasts can only be listed separately in an analysis-of-variance table for orthogonal sets of contrasts, so that all contrasts must be mutually uncorrelated.

*Application to Example* 11.1

As remarked above, the contrasts of the group totals to be used for linear,

Table 11.6 *Subdivision of CSS between groups*

| Polynomial | group total | | | | | $\dfrac{(\sum f_i T_i)^2}{n \sum f_i^2}$ |
|---|---|---|---|---|---|---|
| | 461 | 437 | 437 | 414 | 399 | |
| Linear | $-2$ | $-1$ | 0 | 1 | 2 | 540.225 |
| Quadratic | 2 | $-1$ | $-2$ | $-1$ | 2 | 0.446 |
| Cubic | $-1$ | 2 | 0 | $-2$ | 1 | 6.400 |
| Quartic | 1 | $-4$ | 6 | $-4$ | 1 | 21.729 |
| *Total of non-linear terms* | | | | | | 28.575 |

quadratic, etc., contributions are clear from Table C.11 of Appendix C. The relevant coefficients are listed in Table 11.6, and the group totals are repeated from Table 11.2. The final column gives values of the expression (11.6), and these are the corrected sums of squares due to each of the polynomial terms in the model, each on one degree of freedom. These sums of squares can be entered into Table 11.5, and tested individually against the mean square within groups. In this case the quadratic component, and possibly the cubic component, are suspiciously low, or alternatively, within-groups CSS has been inflated in some way. In theory, each CSS in Table 11.6 has an expectation equal to or greater than the expectation of the CSS within groups, so that an $F$-ratio as small as $0.466/32.8$ is cause for investigation.

### Exercises 11.1

1. The set of data listed in Table 11.7 is based upon experiments by Bain and Batty (1956), concerning the destruction of adrenaline by liver tissue. Three replicate determinations were made at each of five times. The adrenaline concentration is given in units $\mu g/ml$.

Analyse the data, bearing in mind that a logarithmic transformation may be desirable.

Table 11.7

| Time (min) | 6 | 18 | 30 | 42 | 54 |
|---|---|---|---|---|---|
| | 30.0 | 8.9 | 4.1 | 1.8 | 0.8 |
| | 28.6 | 8.0 | 4.6 | 2.6 | 0.6 |
| | 28.5 | 10.8 | 4.7 | 2.2 | 1.0 |

2. The data given below were reported by Jeffers *et al.* (1976), and they were derived from an experimental programme to determine the rates of decomposition of leaf litter in field conditions, but protected from animals. Accurately weighed samples of leaf litter were allowed to decompose in tubes which were exposed in field conditions. Two tubes were sampled randomly at certain intervals, and the data given are the percentage weight losses and the number of days decomposition on which the samples were taken. Analyse the data and report on your conclusions. Fit a polynomial regression model to the data. Can you think of a more suitable model?

| Days | 14 | 28 | 50 | 78 | 148 | 181 | 232 | 274 |
|------|-----|-----|-----|-----|------|------|------|------|
| % Wt. | 3.61 | 3.36 | 7.78 | 5.12 | 14.65 | 16.31 | 17.96 | 28.05 |
| loss | 3.55 | 3.80 | 5.07 | 7.85 | 10.21 | 14.28 | 16.55 | 25.14 |

| Days | 323 | 365 | 421 | 484 | 540 | 596 | 659 | 729 |
|------|------|------|------|------|------|------|------|------|
| % Wt. | 27.29 | 28.06 | 26.43 | 31.28 | 38.03 | 33.79 | 37.03 | 36.74 |
| loss | 28.83 | 35.06 | 35.67 | 30.45 | 36.35 | 33.75 | 38.15 | 38.66 |

3. Work directly from the model (11.3) through the theory of Chapter 5 to obtain the appropriate analysis-of-variance table. In particular, derive the algebraic form of the sums of squares due to each of the polynomial terms.

Table 11.8

| Group | Treatment variable | Observations | Mean |
|-------|--------------------|--------------| -----|
| 1 | $x_1$ | $y_{1,1} \cdots y_{1,n_1}$ | $\bar{y}_1$ |
| $\vdots$ | $\vdots$ | $\vdots \quad \vdots$ | $\vdots$ |
| $t$ | $x_t$ | $y_{t,1} \cdots y_{t,n_t}$ | $\bar{y}_t$ |

Table 11.9 *Analysis of variance*

| Source | CSS |
|--------|-----|
| Due to regression | $\{\sum n_i(\bar{y}_{i.} - \bar{y}_{..})(x_i - \bar{x})\}^2 / \{\sum n_i(x_i - \bar{x})^2\}$ |
| Deviations | by subtraction |
| Between groups | $\sum n_i(\bar{y}_{i.} - \bar{y}_{..})^2$ |
| Within groups | $\sum \sum (y_{ij} - \bar{y}_{i.})^2$ |
| Total | $CS(y, y)$ |

4. Examine the case where the groups are of unequal size, and show that the analysis-of-variance table is as set out below in Table 11.9 where the terms are defined in Table 11.8. List the degrees of freedom appropriate to each term.

## 11.2 Parallel regressions

We shall return to Example 11.1 in the next section, but in this section we discuss a set of data in which there are two regressions, and a question arises as to whether or not they are parallel. The method we use is an application of the extra sum-of-squares principle discussed in Section 6.1. However, here it is convenient to use a slight modification.

Suppose we are carrying out a multiple regression analysis, and at some point in the analysis the two 'Anova' tables given in Table 11.10 are considered (for the same set of data). In Table A, four regressor variables are used, but in Table B, $x_3$ and $x_4$ are ignored. The adjusted sum of squares due to $x_3$ and $x_4$ can be obtained by subtracting the sum of squares due to regression for Table B from that for Table A. However, it can also equivalently be calculated by subtracting the 'Deviations' sums of squares for Table A from that for Table B. The two methods of calculation are equivalent, and we can use whichever is more convenient. Indeed, for certain theoretical purposes it is better to use the 'Deviations' sum of squares method, and in the example given below we shall use this method throughout.

*Example* 11.2    The set of data recorded in Table 11.11 was observed in an experiment to compare two different methods, A and B, of carrying out calculations of $\chi^2$ for a contingency table. The data given are times in seconds.

Table 11.10 *Skeleton analysis-of-variance tables*

| Table A | | Table B | |
|---|---|---|---|
| Source | d.f. | Source | d.f. |
| Regr. on $x_1, x_2, x_3, x_4$ | 4 | Regr. on $x_1, x_2$ | 2 |
| Deviations | $(n - 5)$ | Deviations | $(n - 3)$ |
| Total | $(n - 1)$ | Total | $(n - 1)$ |

Table 11.11

| Replication | 1 | 2 | 3 | 4 | 5 | 6 |
|---|---|---|---|---|---|---|
| Method A | 7.7 | 7.9 | 7.5 | 7.5 | 6.6 | 6.3 |
| Method B | 5.6 | 5.0 | 5.0 | 4.7 | 4.4 | 3.6 |
| Scale for repl. | $-5$ | $-3$ | $-1$ | 1 | 3 | 5 |

For each replication, the decision as to whether method A or B was to be carried out first was randomized. [Artificial data, based on a real experiment.]                                                                    □ □ □

Two facts are very clear; the experimenter improves his speed with replications, but nevertheless method B is much faster than method A. It would appear from Fig. 11.2 that we could try fitting a model in which there is a linear regression of time on replication number, and the regression lines for the two methods may be parallel. Since the $x$-values are equally spaced, it is obvious that regression calculations will be simplified by using the scale shown; let this scale be denoted $x$, and times as $y$. The change in the scale of $x$ has a direct effect on the estimated slope, but no effect on the sum of squares due to regression.

Three of the models we could fit are set out in Fig. 11.3 together with skeleton Anova tables. In this figure model 1 is just two simple linear regression analyses carried out on the two sets of results separately. In model 3, 'Methods' is being treated simply as a duplicate measurement, and the anova table follows from the discussion of this case in Section 11.1. In model 2 there are three parameters, $\alpha_A$ and $\alpha_B$, and $\beta$, which describe the regression lines, and if we take out the mean, 2 d.f. are left for regression, and 9 d.f. for deviations. The anova table for model 2 can be expanded (see below). To complete all three models we add the assumption $V(Y) = \sigma^2$.

At this point, various sums of squares can be calculated and given a

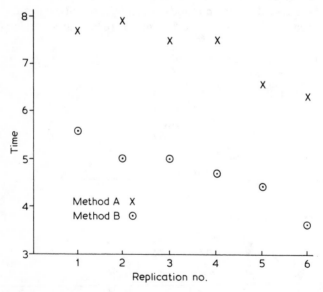

Fig 11.2 Graph of $\chi^2$ times data.

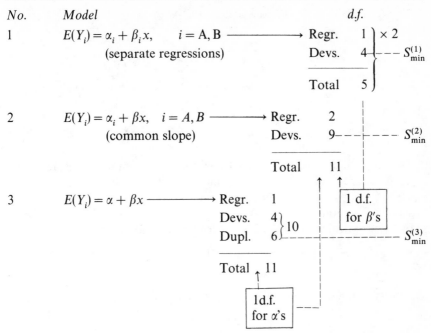

Fig.   11.3 Models and anova for $\chi^2$ times data

direct interpretation. The 'extra sum of squares' principle is used and it will be found helpful to keep Fig. 11.3 in mind. In model 1, two separate lines are fitted, with two constant terms $\alpha_A$ and $\alpha_B$, and two slopes, $\beta_A$ and $\beta_B$. The minimized sum of squares, $S^{(1)}_{min}$, is therefore obtained by adding together the Deviation sums of squares from the two separate regression 'Anova' tables, and so $S^{(1)}_{min}$ has 8 degrees of freedom. In model 2 we fit two $\alpha$'s, but only one slope $\beta$, and so the minimized sum of squares, usually written as 'Deviations' sum of squares in the 'Anova' table, has 9 degrees of freedom. Therefore the term

$$S^{(2)}_{min} - S^{(1)}_{min}$$

is a sum of squares on 1 degree of freedom, due to the difference between $\beta_A$ and $\beta_B$.

In model 3 only one regression line is fitted, so that $S^{(3)}_{min}$ has 10 degrees of freedom. In fact, $S^{(3)}_{min}$ can be separated into 6 degrees of freedom due to duplications, and 4 degrees of freedom due to deviations from the model. The term

$$S^{(3)}_{min} - S^{(2)}_{min}$$

is therefore a sum of squares on 1 degree of freedom due to the difference between the $\alpha$'s, assuming parallel regressions.

Table 11.12 *Anova for model 1 treatment of Example* 11.2

| Source | d.f. | Method A | | Method B | |
|---|---|---|---|---|---|
| | | CSS | MS | CSS | MS |
| Due to regression | 1 | 1.6973 | 1.6973 | 2.0916 | 2.0916 |
| Deviations | 4 | 0.3777 | 0.0944 | 0.1967 | 0.0492 |
| Total | 5 | 2.0750 | | 2.2883 | |

The anova tables for the model 1 analysis are shown in Table 11.12. The regressions are obviously highly significant, and an $F$-test of the two mean squares for 'Deviations' is not significant, so that we can proceed on the basis of a common $\sigma^2$.

In order to deal with model 2 we rewrite it in the following form:

$$E(Y_{ij}) = \mu + \delta z_i + \beta x, \qquad i = A, B; \qquad j = 1, \dots, 6$$
$$z_A = 1, z_B = -1$$

Therefore in the notation of Chapter 5 we have

$$E(\mathbf{Y}) = \mathbf{a}\boldsymbol{\theta}$$

where

$$\boldsymbol{\theta}' = (\mu, \delta, \beta)$$

and

$$\mathbf{a} = \begin{pmatrix} 1 & 1 & -5 \\ \vdots & \vdots & \vdots \\ 1 & 1 & 5 \\ 1 & -1 & -5 \\ \vdots & \vdots & \vdots \\ 1 & -1 & 5 \end{pmatrix} \qquad \mathbf{a}'\mathbf{a} = \begin{pmatrix} 12 & 0 & 0 \\ 0 & 12 & 0 \\ 0 & 0 & 140 \end{pmatrix}$$

From this last matrix we see that the contrast due to methods ($\delta$) is orthogonal to that due to regression, and it would also be orthogonal to the quadratic, cubic, etc., components if these were introduced. This eliminates the need for the model 3 analysis, and the final anova table for model 2 is shown in Table 11.13.

We see from this table there is no evidence of deviations from linearity of the regression, but that 'Methods' is very highly significant. In order to obtain a CSS due to the difference between the $\beta$'s we can now compare Tables 11.12 and 11.13 to get

$$(1.6973 + 2.0916) - 3.7786 = 0.0103 \qquad \text{(on 1 d.f.)}$$

This is less than any of the mean squares for error, so that we can conclude that there is no evidence that the lines differ in slope.

Table 11.13 *Anova for model 2 treatment of Example* 11.2

| Source | CSS | d.f. | Mean square |
|---|---|---|---|
| Due to regression | 3.7786 | 1 | 3.7786 |
| Deviations | 0.3181 | 4 | 0.0795 |
| Replication | 4.0967 | 5 | |
| Methods | 19.2534 | 1 | 19.2534 |
| Residual | 0.2666 | 5 | 0.0533 |
| Total | 23.6167 | 11 | |

It should be emphasized that Example 11.2 has been introduced for illustrative purposes. The reader should follow through the theory for the model 2 analysis, and check the calculations of Tables 11.12 and 11.13.

### Exercises 11.2

1. Derive the estimate of $\beta$ for model 2, and its variance.

2. Check the calculations of Tables 11.12 and 11.13.

3. At the beginning of this section two methods of calculation were explained for using the 'extra sum of squares' principle. Show how to follow through the above analysis by subtracting the sums of squares due to the fitted model. For which part of the analysis is the other method of calculation easier, and why?

### 11.3 The two-way analysis

The experiment mentioned in Example 11.1 was carried out on four separate days, with one observation at each pressure level on each day (so that the columns of the data set represent days). We can therefore separate 3 d.f. for days, leaving 12 d.f. for 'Residual'.

In order to obtain the CSS for days we proceed as follows:

| Day | 1 | 2 | 3 | 4 | Total |
|---|---|---|---|---|---|
| Day Total | 550 | 527 | 526 | 545 | 2148 |

$$\text{CSS} = (550^2 + 527^2 + 526^2 + 545^2)/5 - 2148^2/20 = 90.80$$

The analysis-of-variance table is then as set out in Table 11.14.

Table 11.14 *Analysis of variance for Example* 11.1

| Source | CSS | d.f. | Mean square |
|---|---|---|---|
| Linear | 540.225 | 1 | 540.225 |
| Quadratic | 0.466 | 1 | 0.466 |
| Cubic | 6.400 | 1 | 6.400 |
| Quartic | 21.729 | 1 | 21.729 |
| Between pressures | 568.800 | 4 | — |
| Between days | 90.800 | 3 | 30.267 |
| Residual | 401.200 | 12 | 33.433 |
| Total | 1060.800 | 19 | |

This table has not altered our conclusions. The between-days mean square is not significant, and the mean square we use to estimate the error variance is almost unchanged. The term which has been called 'between groups' is renamed 'between pressures', because of the two-way analysis.

At this point it is useful to introduce the term *interaction*, although a full discussion of this term will be delayed until Chapter 15. The model we have used for a two-way analysis is (10.55), and this can be written

$$\begin{bmatrix} \text{Expectation of} \\ \text{observation at} \\ i\text{th row and } j\text{th} \\ \text{column} \end{bmatrix} = \text{Constant} + \begin{bmatrix} \text{Component} \\ \text{due to} \\ \text{row } i \end{bmatrix} + \begin{bmatrix} \text{Component} \\ \text{due to} \\ \text{column } j \end{bmatrix}$$

Under this model, the effects on the mean of the observation $(i, j)$ due to

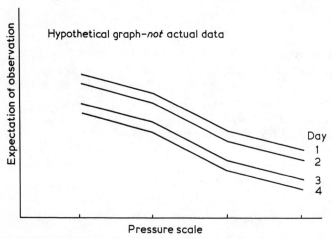

Fig. 11.4 Illustration of additive model.

row $i$ and column $j$ are purely additive. For our present example, the model would produce the parallel-lines picture shown in Fig. 11.4. If the data are such that the model cannot be represented in this additive form and the effect of, say, column $j$ depends on which row is used, then we say that there is an *interaction*, between the row variable and the column variable.

For example, in Table 11.14 we have assumed that the linear regression on pressures is the same on all four days. If this is not the case, we say that there is an interaction between the linear effect due to pressures, and 'days'. The calculations involved in obtaining a corrected sum of squares for this particular interaction component are straightforward. We fit separate linear regressions to each day and add the sums of squares; from this total we subtract the sum of squares due to the parallel regression on days. This will give a sum of squares on 3 degrees of freedom, which is due to the comparison between the four (day) linear regressions. The easiest way to set out the calculations and to see the meaning of the terms is to proceed as follows. The data is reproduced in Table 11.15 for convenience, together with the coefficients for the linear contrast on pressures. We first calculate the linear contrast for day 1, and this is

$$( - 2 \times 112) - (1 \times 108) + (1 \times 110) + (2 \times 100) = - 22$$

We can now refer back to the method of linear contrasts in Section 11.1 to see how to convert this into a corrected sum of squares for our anova table. Here we have only one observation in each group, so that $n = 1$, and from the expression (11.6) the sum of squares is

$$( - 22)^2/10 = 48.4$$

Proceeding in this way we see that separate regressions for each day account for a corrected sum of squares of

$$48.4 + 102.4 + 302.5 + 144.4 = 597.7 \tag{11.8}$$

and since four parameters have been fitted, there are 4 degrees of freedom.

We now wish to fit a common regression slope to all four days, and calculate the corrected sum of squares accounted for in this way. The common

Table 11.15 *Calculation of linear polynomials for each day*

| Day | 1 | 2 | 3 | 4 | Linear polynomial ($f_i$) | |
|---|---|---|---|---|---|---|
| | 112 | 119 | 117 | 113 | $-2$ | |
| | 108 | 99 | 112 | 118 | $-1$ | |
| | 120 | 106 | 102 | 109 | 0 | $\sum f_j^2 = 10$ |
| | 110 | 101 | 99 | 104 | 1 | |
| | 100 | 102 | 96 | 101 | 2 | |
| $\sum f_j y_{ij}$ | $-22$ | $-32$ | $-55$ | $-38$ | Grand total $= - 147$ *(linear only)* | |

slope is

$$(\sum f_i T_i)/\{4\sum f_i^2\}$$

where the $f_i$ are the linear polynomial coefficients listed in Table 11.15, and the sum of squares is

$$\frac{(\sum f_i T_i)^2}{4\sum f_i^2} = \frac{(-147)^2}{4 \times 10} = 540.225 \qquad (11.9)$$

This is, of course, the value given in Table 11.6. It has 1 degree of freedom since one parameter has been fitted.

It is now clear that by subtracting Equation (11.9) from Equation (11.8) we obtain CSS due to the differences on the day regressions. This leads to

$$597.7 - 540.225 = 57.475 \qquad \text{on } 4 - 1 = 3 \text{ d.f.}$$

The other interaction components are obtained in a similar way, and quite separately from each other. The fact that these calculations are quite separate is due to our use of orthogonal polynomials. The method cannot be used for a quite general set of linear contrasts. The results are as shown in Table 11.16. This table reveals that the small overall quadratic component of pressure in Table 11.14 is very much an average, and there are rather wild fluctuations of the component with days. The linear component, however, is relatively stable.

The chief difficulty with separating the interaction this far in Example 11.1 is that there is no valid estimate of $\sigma^2$, and hence no tests of significance that we can carry out on Table 11.16. What we usually do in effect is to assume that all these interaction effects represent error, and use the mean square on 12 d.f. (Table 11.14) to estimate $\sigma^2$.

An analysis such as that given in Table 11.16 is useful in the following circumstances:

(i) When there is a separate estimate of $\sigma^2$ available against which to test these interaction components. This usually means replicating observa-

Table 11.16 *Separation of residual sum of squares in Table* 11.14

| Source | CSS | d.f. | Mean square |
|--------|-----|------|-------------|
| Days × pressures (linear) | 57.48 | 3 | 19.16 |
| (quadratic) | 165.34 | 3 | 55.11 |
| (cubic) | 91.40 | 3 | 30.47 |
| (quartic) | 86.99 | 3 | 29.00 |
| Days × pressures | 401.21 | 12 | 33.43 |

tions within cells; the analysis of experiments with replicated observations is dealt with in Chapter 15.

(ii) If it is thought that one component, say the linear interaction component, may be significant on *a priori* grounds, so that it would be unsafe to pool it with the other interaction components to use as error.

## Exercise 11.3

1. Suppose that in Example 11.1 the experiment was carried out on four successive days, then the between-days CSS can be split up into linear, quadratic and cubic components.

   (*a*) Carry out the calculations.
   (*b*) Consider how to split up, say, the days × pressures (linear) component into three components:
   > days (linear) × pressures (linear)
   > days (quadratic) × pressures (linear)
   > days (cubic) × pressures (linear)
   > each on 1 d.f.

[*Hint for part* (*b*) : In each case you will need a linear contrast).]

# CHAPTER TWELVE

# An analysis of data on trees

## 12.1 The data

The preceding chapters have already made several applications of the extra sum-of-squares principle discussed in Section 6.1. This chapter contains an analysis of some data published by Pearce (1965), and it serves to give some further illustrations of the use of this principle.

The data were collected in an investigation into ways of measuring the weight of a plum tree without damaging it. One possibility was to pull up a group of mature trees, weigh each one, and see how far a relationship could be established with the trunk circumferences. It was decided to use all measurements in logarithmic transformation because experience had shown that log (tree weight) plotted against log (trunk circumference) usually gave a straight-line relationship. Of course, logarithms of size show up growth rates and it is not surprising that a tree with a fast-growing trunk should have a proportionately fast-increasing weight. Trunk dimensions were measured in centimetres and the tree weights measured in pounds.

The three variates in Table 12.1 are:

$w$, log (circumference at the base of the trunk)
$x$, log (circumference at the top of the trunk)
$y$, log (weight of tree above ground)

*Aims and procedure of the analysis*

In analysing these data, the chief objective is clearly to obtain an equation which can predict the weight of a tree above ground, and to give a measure of error to be used with the equation. In doing this we shall have to demonstrate that one equation is satisfactory to use for all four rootstocks. A number of subsidiary questions are also of interest, such as to test whether the explanatory variables alone account for the difference between rootstocks.

We shall start by using multiple regression on the data for each rootstock independently, using model 1:

Table 12.1 *Data for the tree-growth experiment.*
*Size data for plum trees grown on four kinds of roots*

| Own roots | | | Common plum | | |
|---|---|---|---|---|---|
| w | x | y | w | x | y |
| 1.690 | 1.663 | 2.318 | 1.525 | 1.515 | 1.940 |
| 1.583 | 1.568 | 2.100 | 1.517 | 1.501 | 1.940 |
| 1.693 | 1.643 | 2.225 | 1.582 | 1.542 | 2.045 |
| 1.648 | 1.609 | 2.140 | 1.537 | 1.529 | 1.919 |
| 1.628 | 1.599 | 2.107 | 1.504 | 1.487 | 1.869 |
| 1.600 | 1.603 | 2.049 | 1.589 | 1.562 | 2.049 |
| 1.677 | 1.640 | 2.228 | 1.567 | 1.548 | 1.954 |
| 1.588 | 1.535 | 2.029 | 1.480 | 1.449 | 1.820 |
| 1.645 | 1.606 | 2.140 | 1.498 | 1.490 | 1.892 |
| 14.752 | 14.466 | 19.336 | 13.799 | 13.623 | 17.428 |

| Common mussel | | | Myrobalan B | | |
|---|---|---|---|---|---|
| w | x | y | w | x | y |
| 1.604 | 1.540 | 2.104 | 1.747 | 1.712 | 2.382 |
| 1.650 | 1.633 | 2.233 | 1.769 | 1.754 | 2.413 |
| 1.591 | 1.581 | 2.090 | 1.763 | 1.747 | 2.446 |
| 1.598 | 1.579 | 2.079 | 1.766 | 1.738 | 2.450 |
| 1.672 | 1.629 | 2.255 | 1.739 | 1.719 | 2.387 |
| 1.585 | 1.555 | 2.064 | 1.749 | 1.742 | 2.396 |
| 1.645 | 1.610 | 2.185 | 1.738 | 1.710 | 2.310 |
| 1.628 | 1.612 | 2.068 | 1.731 | 1.689 | 2.332 |
| 1.653 | 1.642 | 2.201 | 1.707 | 1.650 | 2.204 |
| 14.626 | 14.381 | 19.279 | 15.709 | 15.461 | 21.320 |

Model 1:

$$Y_{ij} = \alpha_i + \beta_{1i}(w_{ij} - \bar{w}_{i.}) + \beta_{2i}(x_{ij} - \bar{x}_{i.}) + \varepsilon_{ij}, \quad \begin{matrix} i = 1, 2, 3, 4 \\ j = 1, 2, \dots, 9 \end{matrix} \quad (12.1)$$

where $\varepsilon_{ij}$ is independently $N(0, \sigma^2)$ and where $i$ denotes the rootstock, and where $\bar{w}_i$ and $\bar{x}_i$ are the means for the root stock $i$. This model has twelve parameters. We then follow this by further analyses, which fit less than twelve parameters, and which can be used to generate certain sums of squares which are of interest.

*Graphs of the data*
Figure 12.1 shows a graph of $w$ plotted against $x$ for the four rootstocks. As one would expect, there is a very strong relationship between the two variables and this feature is likely to give some trouble in interpreting the results of the multiple regression analysis.

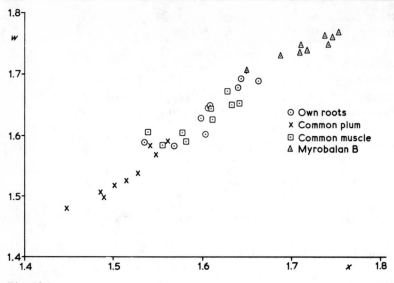

Fig. 12.1

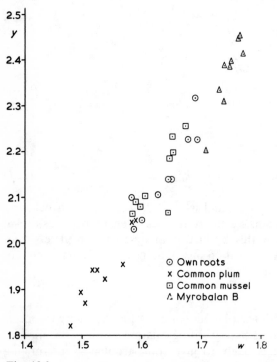

Fig. 12.2

In Fig. 12.2, $y$ is plotted against $w$ for the four rootstocks. There appears to be an approximately linear relationship between the variables, and the same line seems to account for the relationship for all four rootstocks. We shall test this more formally below.

There are two observations which straggle below the line a little; the eighth observation of common mussel, and the last observation of myrobalan B. Both of these points are picked out by the analysis of residuals below.

## 12.2 Regression analyses

*The individual regressions*
Table 12.2 shows some results on the individual regressions. In all cases the regressions are highly significant. Tests of significance on coefficients show only $\beta_1$ for common mussel to be significant, and in all cases $\beta_2$ contributes less to the regression than $\beta_1$. Since some of the $\beta_2$ values are negative as well, we might suspect that variable $x$ could be eliminated.

Table 12.2 *The individual regressions*

| Source | CSS | d.f. | MS | | Standard errors |
|---|---|---|---|---|---|
| (a) *Own roots* | | | | | |
| Regression | 0.059 068 | 2 | 0.029 534 | $\beta_1 = 1.2473$ | 0.837 |
| Deviations | 0.010 234 | 6 | 0.001 706 | $\beta_2 = 0.8865$ | 0.907 |
| Total | 0.069 302 | 8 | | $F = 17.3$ on (2, 6) d.f. | |
| (b) *Common plum* | | | | | |
| Regression | 0.039 737 | 2 | 0.019 869 | $\beta_1 = 1.9053$ | 0.967 |
| Deviations | 0.005 437 | 6 | 0.000 906 | $\beta_2 = -0.0885$ | 1.047 |
| Total | 0.045 174 | 8 | | $F = 21.9$ on (2, 6) d.f. | |
| (c) *Common mussel* | | | | | |
| Regression | 0.037 661 | 2 | 0.018 830 | $\beta_1 = 2.4414$ | 0.848 |
| Deviations | 0.008 396 | 6 | 0.001 399 | $\beta_2 = -0.2734$ | 0.744 |
| Total | 0.046 057 | 8 | | $F = 13.5$ on (2, 6) d.f. | |
| (d) *Myrobalan B* | | | | | |
| Regression | 0.042 390 | 2 | 0.021 195 | $\beta_1 = 2.4570$ | 1.771 |
| Deviations | 0.005 332 | 6 | 0.008 89 | $\beta_2 = 0.7692$ | 1.056 |
| Total | 0.047 723 | 8 | | $F = 23.8$ on (2, 6) d.f. | |
| (e) *Combined analysis* | | | | | |
| Regression | 0.178 856 | 8 | 0.022 357 | | |
| Deviations | 0.029 399 | 24 | 0.001 225 | | |
| Total | 0.208 255 | 32 | | | |

The final entry of Table 12.2 is simply obtained by adding the four separate tables, and we use this in the next step in the analysis. Before making this combined analysis, we should note that the deviations mean squares do not differ significantly.

*The difference between regression coefficients*
Suppose now we fit model 2:

$$\text{Model 2:} \qquad Y_{ij} = \alpha_i + \beta_1 (w_{ij} - \bar{w}_{i.}) + \beta_2 (x_{ij} - \bar{x}_{i.}) + \varepsilon_{ij} \qquad (12.2)$$

which involves only six parameters; then we obtain Table 12.3.

The regression on $w$ in this table has four degrees of freedom since model (12.2) has four block constants $\alpha_i$. The contribution due to $x$ is not significant. If the above analysis is extended to calculate the sum of squares due to $w$ adjusting for $x$, we find that this also is non-significant, though it is slightly greater than the sum of squares due to $x$ adjusting for $w$. Clearly the explanatory variables are very strongly correlated, but $w$ accounts for a slightly greater sum of squares than $x$.

By subtracting the deviations sum of squares in Table 12.2(e) from that in Table 12.3, we obtain a sum of squares on six degrees of freedom representing the difference between models 1 and 2, that is, due to the differences between the two sets of four $\beta$'s.

| Table 12.3 Devs. | Table 12.2(e) Devs. | CSS | d.f. | MS |
|---|---|---|---|---|
| 0.036 973 | 0.029 399 | 0.007 574 | 6 | 0.001 262 |

$$(12.3)$$

If we use the mean square for deviations in Table 12.2(e) as our estimate of the error variance, this mean square is perfectly consistent with there being common values of $\beta_1$ and $\beta_2$ for the different rootstocks.

The sum of squares (12.3) can be separated further by fitting the models:

$$\text{Model 3:} \qquad Y_{ij} = \alpha_i + \beta_1 (w_{ij} - \bar{w}_{i.}) + \beta_{2i} (x_{ij} - \bar{x}_{i.}) + \varepsilon_{ij} \qquad (12.4)$$

Table 12.3 *Anova for model 2*

| Source | CSS | d.f. | MS | |
|---|---|---|---|---|
| Regression on $w$ and planting methods | 1.009 925 | 4 | 0.252 481 | |
| Due to $x$, adj. for $w$ | 0.003 567 | 1 | 0.003 567 | $F = \quad 2.89$ |
| Regression on $w$ & $x$ | 1.013 492 | 5 | 0.202 698 | $F = 164.5$ |
| Deviations | 0.036 973 | 30 | 0.001 232 | |
| Total | 1.050 465 | 35 | | |

Table 12.4 *Anova for models 3 and 4*

| Source | CSS | d.f. | MS |
|---|---|---|---|
| Model 3 | | | |
| Regression | 1.019 394 | 8 | 0.127 424 |
| Deviations | 0.031 071 | 27 | 0.001 151 |
| Total | 1.050 465 | 35 | |
| Model 4 | | | |
| Regression | 1.019 226 | 8 | 0.127 403 |
| Deviations | 0.031 239 | 27 | 0.001 157 |
| Total | 1.050 465 | 35 | |

Table 12.5 *Differences between regression coefficients*

| Source | CSS | d.f. | MS |
|---|---|---|---|
| Differences between $\beta_{1i}$ | 0.001 672 | 3 | 0.000 557 |
| Differences between $\beta_{2i}$ | 0.001 840 | 3 | 0.000 613 |

and

Model 4: $\qquad Y_{ij} = \alpha_i + \beta_{1i}(w_{ij} - \bar{w}_{i.}) + \beta_2(x_{ij} - \bar{x}_{i.}) + \varepsilon_{ij}$ (12.5)

each of which has nine parameters; and we obtain Table 12.4.

If we subtract the deviations sum of squares from Table 12.2(e) in each case we obtain the sums of squares as shown in Table 12.5.

These sums of squares are quite small, and add up to a lot less than (12.3), and they are both considerably less than our error mean square (though not dramatically so). Clearly, whatever difference there is between rootstock regressions is accounted for if one of the explanatory variables is given different slopes for the rootstocks.

## 12.3 The analysis of covariance

First we carry out a one-way analysis of variance on the response variable $y$, ignoring the explanatory variables $w$ and $x$. The results are shown in Table 12.6, and we see that the rootstocks are very highly significant, as we would expect.

Now we obtain an adjusted sum of squares due to rootstocks, given the explanatory variables are used in the model. That is, we obtain the sum of squares due to the model:

Model 5: $\qquad Y_{ij} = \alpha + \beta_1(w_{ij} - \bar{w}_{i.}) + \beta_2(x_{ij} - \bar{x}_{i.}) + \varepsilon_{ij}$ (12.6)

Table 12.6 *Analysis of variance*

| Source | CSS | d.f. | MS | |
|---|---|---|---|---|
| Between rootstocks | 0.842 2 | 3 | 0.2807 | $F = 43.2$ |
| Within rootstocks | 0.208 3 | 32 | 0.0065 | |
| Total | 1.050 465 | 35 | | |

Table 12.7 *Anova for model 5*

| Source | CSS | d.f. | MS | | Standard error |
|---|---|---|---|---|---|
| Regression on w & x | 1.010 315 | 2 | 0.505 158 | $\beta_1 = 1.430$ | 0.377 |
| Deviations | 0.040 150 | 33 | 0.001 217 | $\beta_2 = 0.638$ | 0.386 |
| Total | 1.050 465 | 35 | | | |

Table 12.8 *Adjusted sum of squares due to rootstocks*

| Source | CSS | d.f. | MS |
|---|---|---|---|
| Due to fitting (12.2) | 1.013 492 | 5 | |
| Due to fitting (12.6) | 1.010 315 | 2 | |
| Due to rootstocks (adjusted for w and x) | 0.003 177 | 3 | 0.001 059 |

and subtract this from the sum of squares due to the model (12.2). The regression analysis for model (12.6) is given in Table 12.7; this table simply treats the data as one set of 36 observations.

Now we subtract the sum of squares due to the model (12.6) from that due to the model (12.2) given in Table 12.3, to obtain Table 12.8.

This shows that the highly significant difference between rootstocks which was shown in Table 12.6 has completely disappeared when the explanatory variables are brought into the model. In this problem this is an important result, as it shows how well the explanatory variables account for differences in log (weight); there is no need to bring rootstocks into the equation.

The procedure which leads to the adjusted sum of squares due to rootstocks in Table 12.8 is known as the *analysis of covariance*.

## 12.4 Residuals

Table 12.9 shows the residuals from the common regression. The standard deviation is 0.035, so that the two observations referred to earlier are just

Table 12.9  *Residuals from model* (12.6)

| Own roots | Common plum | Common mussel | Myrobalan B |
|---|---|---|---|
| 0.057 | 0.009 | 0.044 | 0.008 |
| 0.053 | 0.030 | 0.048 | − 0.019 |
| − 0.027 | 0.016 | 0.023 | 0.027 |
| − 0.026 | − 0.038 | 0.003 | 0.033 |
| − 0.024 | − 0.014 | 0.042 | 0.020 |
| − 0.045 | − 0.003 | 0.022 | 0.000 |
| 0.000 | − 0.058 | 0.022 | − 0.049 |
| − 0.004 | − 0.004 | − 0.071 | − 0.004 |
| − 0.020 | 0.016 | 0.006 | − 0.073 |

over two standard deviations out, and in a sample of 36 observations this is not too surprising. Apart from this, there is some bunching of the residuals which could be investigated. A normal plot shows no apparent deviation from normality.

## Exercises 12.4

1. Subtract the 'Deviations' sums of squares in Table 12.2(e) from the 'Deviations' sum of squares in Table 12.7. What is the interpretation of the resulting sum of squares?

2. Make a normal plot of the residuals, and carry out any other tests on the residuals that you think necessary.

3. Write out a statement of conclusions for this analysis suitable to submit to the experimenter. Include a recommended equation and a formula for its variance, as well as verbal statements of the main results.

4. Identify the sums of squares due to the models, $S_{par}^{(i)}$, and the minimized sums of squares $S_{min}^{(i)}$, throughout this chapter.

5. How do you test the hypothesis that in model 2, $\beta_1 = \beta_2 = 0$?

# CHAPTER THIRTEEN

# The analysis of variance: Subsidiary analyses

## 13.1 Multiple comparisons: Introduction

In Chapter 10 we developed the theory of the analysis of variance, and gave some methods for following up the calculation of the analysis of variance (anova) table. This present chapter is devoted to a further discussion of follow-up techniques, and certain other related points.

In Section 10.5 the data of Example 10.1 were analysed. The four treatment means were

$$A: 36.00 \quad B: 37.30 \quad C: 41.60 \quad D: 37.88$$

and the $F$-test showed reasonable evidence for the existence of real differences between those means. The problem of deciding precisely where these real differences are is the problem of multiple comparisons, and in Section 10.5(ii) the LSD (least significant difference) method was used as one of the simplest multiple comparison techniques.

The LSD method can be used more widely than many other multiple comparison procedures. All we have to do to compare two means is to multiply the standard error of the difference of the means by the appropriate percentage point of the $t$-distribution. Therefore to compare two means based on $n_1$ and $n_2$ observations and for a 5 per cent level test, we use as an LSD

$$t_v(5\%)s \sqrt{\left(\frac{1}{n_1} + \frac{1}{n_2}\right)} \tag{13.1}$$

where $s^2$ is an estimate of the variance $\sigma^2$, and $v$ is the number of degrees of freedom.

However, if in a given experiment we have, say, 7 means to compare, then there are 21 separate comparisons of pairs of means. None of these comparisons is independently made, because a common estimate of $\sigma^2$ is used, and some comparisons involve common means. Nevertheless, with 21 comparisons all made at the 5 per cent level we shall tend to get too many false positives. It is normal practice only to use the LSD method if the overall $F$-test is significant but this gives very little safeguard. Sir Ronald Fisher suggested using a significance level of $(5/m)$ per cent, where $m$ is the number of comparisons, but this tends to be too conservative. This is the core of the multiple comparisons problem, and over the years many different techniques have been devised. We shall limit our discussion here to a few of the most popular methods.

Before going any further, several points should be made. In some experiments, such as Example 11.1, the different treatments can be connected by a regression model which can be estimated, and the multiple comparisons problem doesn't arise. In many other experiments, where the treatments are qualitatively distinct, the comparisons to be made are clear before the experiment is run. For example, we may be interested in comparing several new treatments with a standard simply to see whether or not the new treatments are worth further investigation. In such an experiment the contrasts to be considered can be taken separately, and there is no requirement for *simultaneously* making many comparisons.

We shall not attempt to list the many different situations arising, but the multiple comparisons problem only really arises in one limited case. The treatments must be qualitatively distinct, and not structured in some way. The aim of the experiment is one of *fundamental exploration*. That is, we are examining an unstructured set of means in order to see what patterns we can find. We wish to make statements of the sort: 'The treatments separate into three groups, $(A, F)$, $(B, G, E)$, and $(C, D)$.' This involves the simultaneous correctness of a large number of different comparisons.

The next major point is that most multiple comparisons literature is occupied with discussions of *error rates*, and two are particularly important:

*Comparisonwise error rate* = (number of erroneous inferences)/
(number of inferences)

*Experimentwise error rate* = (number of experiments with one or more
erroneous inferences)/(number of experiments)

There is even talk of 'budgeting' and 'spending' error rates. However, if an error rate is spent at different rates on different comparisons, the overall error rate is irrelevant to the interpretation of one of the comparisons. All of this is very much like an attempt to make the Neyman–Pearson theory of testing hypotheses fit into somewhere where it will not fit – practical data snooping. For most situations it is convenient to adopt the following view. By all means let us make use of 'multiple comparisons techniques' in order to look for patterns in an unstructured set of means, but let us not try to be too precise about the statements that result. If taken in this light, some of the techniques are probably quite useful and we shall outline a few.

## 13.2 Multiple comparisons: Various techniques

*Linear contrasts*

The method of linear contrasts was described in Section 11.1, and illustrated there using orthogonal polynomials. The method has very wide applications. Suppose a chemical experiment was run with four treatments involving two different catalysts, $A$ and $B$, and two operating temperatures, 200 °C and 250 °C, then we can pick out three orthogonal contrasts as in Table 13.1.

Contrast 1 represents the effect of the catalysts, contrast 2 the effect due to temperatures. Contrast 3 represents the difference between the temperature effect for the two catalysts, and this would usually be called the interaction effect between the effects of temperatures and catalysts. The important point here is that there are *a priori* grounds for selecting and looking at the contrasts just described. Sums of squares can be obtained for each and their significance can be tested against an error sum of squares in the usual way. (Indeed, if the contrasts we are interested in are not an orthogonal set, this should not prevent us calculating them but they cannot be listed neatly in an 'anova' table, and they will not be independent.) This method should not be used for contrasts suggested by the data.

*Scheffé's S method*

Suppose that there are $t$ treatments, with $n_i$ observations on the $i$th treatment, and denote the observations $y_{ij}$, for $i = 1, 2, \dots, t$; $j = 1, 2, \dots, n_i$, and the relevant random variables $Y_{ij}$. The contrasts we are interested in are of the form

$$W_k = \sum_{i=1}^{t} f_{ik} E(\bar{Y}_{i.})$$ (13.2)

where

$$\sum_{i=1}^{t} f_{ik} = 0$$

Table 13.1 *Some linear contrasts*

|  | Contrast | | |
|---|---|---|---|
| *Treatment combinations* | 1 | 2 | 3 |
| Catalyst  *A*,  250 °C | 1 | 1 | 1 |
| Catalyst  *A*,  200 °C | 1 | −1 | −1 |
| Catalyst  *B*,  250 °C | −1 | 1 | −1 |
| Catalyst  *B*,  200 °C | −1 | −1 | 1 |

Then we use the estimator

$$\hat{W}_k = \sum_{i=1}^{t} f_{ik} \, \bar{Y}_{i.} \tag{13.3}$$

and its variance is

$$V(\hat{W}_k) = \sigma^2 \sum_{i=1}^{t} (f_{ik}^2/n_i) \tag{13.4}$$

so that if $s^2$ is an estimate of $\sigma^2$ on $v$ degrees of freedom, an estimate of $V(\hat{W}_k)$ is

$$\hat{\sigma}_{W_k}^2 = s^2 \sum_{i=1}^{t} (f_{ik}^2/n_i) \tag{13.5}$$

Then Scheffé (1959) showed that the probability is $(1 - \alpha)$ that all contrasts $W_k$ *simultaneously* satisfy

$$\hat{W}_k - \hat{\sigma}_{W_k} S \le W_k \le \hat{W}_k + \hat{\sigma}_{W_k} S \tag{13.6}$$

where

$$S^2 = (t - 1) F_\alpha(t - 1, v) \tag{13.7}$$

For $t = 2$, Equation (13.7) reduces to $t_v^2(\alpha)$, and the interval (13.6) is the same as that obtained by the LSD approach. However, for comparing $t > 2$ means in pairs, the interval (13.6) is much wider than that obtained by the LSD method. If the interval (13.6) does not include zero, then we say that the contrast is significant.

The fact that inequality (13.6) holds simultaneously for all possible contrasts is a great advantage. The price paid for this is that the intervals are rather wide. If we are comparing means of four observations each (3 d.f.) we have the following values for $\alpha = 5\%$:

| $t$ | 2 | 4 | 6 | 8 | 10 |
|---|---|---|---|---|---|
| $v$ | 6 | 12 | 18 | 24 | 30 |
| $S$ | 2.267 | 3.127 | 3.647 | 4.064 | 4.409 |

Therefore for comparing ten means, contrasts of two means will have an interval nearly double that given by the LSD method.

One other important property of the $S$-method is that it is precisely equivalent to the analysis-of-variance $F$-test in the following sense. The usual $F$-test will be significant at a level $\alpha$ if and only if there is at least one contrast $W$ which is significantly different from zero at the same significance level. (Clearly, on occasions, cases will arise when the significant contrasts are of no practical importance.)

Scheffé suggested that if the length of the $S$-method intervals causes difficulty, then $\alpha$ should be increased, e.g., to 10 per cent. The argument is that it is better to use guaranteed 90 per cent intervals than to use 95 per cent

LSD-method intervals, and not know how much in excess of 5 per cent the true error rate is.

The S-method is very versatile. In fact, it can be applied to correlated means, by making the obvious modifications to Equations (13.4) and (13.5).

*Example* 13.1 We shall illustrate our methods by using some data given by Duncan (1955). This was a randomized block experiment involving six replicates of seven varieties of barley. The treatment means (in bushels per acre) were as follows:

| A | F | G | D | C | B | E |
|---|---|---|---|---|---|---|
| 49.6 | 58.1 | 61.0 | 61.5 | 67.6 | 71.2 | 71.3 |

The analysis of variance table is summarized in Table 13.2.

The F-ratio for testing the varieties effect is 4.61 on (6,30) d.f., and this is very highly significant, showing strong evidence of a real difference between the treatment means. The standard error of a treatment mean is

$$\sqrt{\left(\frac{79.64}{6}\right)} = 3.643$$

We can now use the above techniques to try to find out where the differences are.

The LSD for a 5 per cent significance level is

$$t_{30}(5\%)\sqrt{(79.64)}\sqrt{(2/6)} = 10.51$$

If we apply this LSD to the means given above, the results can be presented by

Table 13.2

| Source | d.f. | Mean square |
|---|---|---|
| Between varieties | 6 | 366.97 |
| Between blocks | 5 | 141.95 |
| Residual | 30 | 79.64 |

Table 13.3 *Results of applying some multiple comparisons procedures to Duncan's data*

| Procedure | A 49.6 | F 58.1 | G 61.0 | D 61.5 | C 67.6 | B 71.2 | E 71.3 |
|---|---|---|---|---|---|---|---|
| (i)   LSD | | | | | | | |
| (ii)  S-method | | | | | | | |
| (iii) T-method | | | | | | | |
| (iv)  Normal plotting | | | | | | | |

underlining means (means underlined are not significantly different). Table 13.3 (i) shows the results, and we see that, for example, $(G, D, C, B, E)$ are all significantly different from $A$, whereas only $(B, E)$ are significantly different from $F$, etc.

If we now apply the $S$-method we obtain from Equations (13.5) and (13.7), the factor for comparing pairs of means is (at 5 per cent)

$$\hat{\sigma}_w S = \sqrt{(79.64 \times 2/6)} \sqrt{(6 \times 2.42)} = 19.63$$

This leads to the results (ii) of Table 13.1.                    ☐ ☐ ☐

*Tukey's T-method*
This method also gives a guaranteed interval for all possible contrasts, but it requires that the means be of the same number $(n)$ of observations, and be statistically independent. It is based on the distribution of the Studentized range $q(t, v)$, rather than on the $F$-distribution, where $t$ is the number of variables (means in this case) being considered, and $v$ is the number of degrees of freedom of the error estimate. The $\alpha$ percentage point will be written $q(\alpha, t, v)$.

The result is that the probability is $(1 - \alpha)$ that all contrasts satisfy

$$\hat{W}_j - sq(\alpha, t, v)\, T \le W_j \le \hat{W}_j + sq(\alpha, t, v)T \qquad (13.8)$$

where

$$T = \sum_{i=1}^{t} |f_{ij}|/2\sqrt{n} \qquad (13.9)$$

For contrasts of pairs of means, this reduces to

$$\hat{W}_j - sq(\alpha, t, v)/\sqrt{n} < W_j < \hat{W}_j + sq(\alpha, t, v)/\sqrt{n} \qquad (13.10)$$

Any pair of means differing by more than $sq(\alpha, t, v)/\sqrt{n}$ is said to be significantly different at the $100\alpha$ per cent level. However, if used in this way, the $T$-method may on occasion give results incompatible with the analysis-of-variance $F$-test. Tables of $q(\alpha, t, v)$ are given in Appendix C.

*Example* 13.2  For comparisons of pairs of means, the factor for the $T$-method is

$$q(\alpha, t, v)s/\sqrt{n} = 4.46 \times 3.643 = 16.25$$

The results are shown in Table 13.3(iii).                    ☐ ☐ ☐

*Normal plotting*
A very good method of examining a set of means for patterns is to make a normal plot as described in Section 1.4(a). We simply plot the ordered means

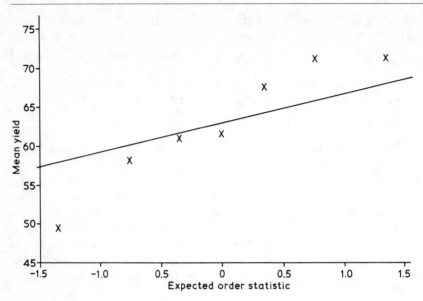

Fig. 13.1 Normal plot for one group of seven.

against the expected normal order statistics,

$$E(Z_{(i)}) = u_{i,m}$$

for the $i$th order statistic in a sample of $m$ means. Some values of $u_{i,m}$ are given in Appendix C, and Fig. 13.1 shows the resulting graph for Duncan's data (Example 13.1).

Now if the means $\bar{x}_i$ are all means of $n$ observations, and there are no real differences due to the varieties, the means should all lie close to the line

$$y_i = \bar{x} + s\,u_{i,m}/\sqrt{n}$$

where $\bar{x}$ is the overall mean. If the variety means can be broken into two or three groups, with no significant differences within groups, then the means will lie close to two or three lines, each line having the same slope $s/\sqrt{n}$. The breaks between the lines will be shown by slopes significantly different from $s/\sqrt{n}$.

For Duncan's data the overall mean is 62.90 and $s/\sqrt{n} = 3.64$, and the line

$$y = 62.9 + 3.64\,u_{i,m}$$

is drawn in. The slopes are much greater than 3.64 between means $A$ and $F$ (1 and 2), and means $D$ and $C$ (4 and 5). The slope between means $i$ and $(i+1)$ is

$$q_{i,i+1} = (\bar{x}_{(i+1)} - \bar{x}_{(i)})/(u_{i+1,7} - u_{i,7}) \qquad (13.11)$$

where $\bar{x}_i$ is the $i$th ordered mean. Now it turns out that these slopes have approximately constant variance and relatively small covariances.

We see from Fig. 13.1 that the two large slopes are $q_{1,2}$ and $q_{4,5}$, and these are 14.3 and 17.3 respectively, and these are both much larger than the values 3.64 estimated from the anova table.

$$q_{1,2} = \frac{58.1 \times 49.6}{1.352 - 0.757} = 14.3 \qquad q_{4,5} = \frac{67.6 - 61.5}{0.353 - 0.000} = 17.6$$

If we consider the larger of these, $q_{4,5}$, then

$$V(q_{4,5}) = \frac{1}{(0.353)^2} \{V(x_{(4)}) + V(x_{(5)}) - 2C(x_{(4)}, x_{(5)})\} \frac{\sigma^2}{n}$$

Variances and covariances of order statistics up to $n = 20$ are tabulated by Sarhan and Greenberg (1956). Using their values, the standard error of $q_{4,5}$ is estimated at 3.248, so that there is very strong evidence that $q_{4,5}$ differs from the slope estimated from the anova table (3.64). We therefore split the means into two groups at this point, and Fig. 13.2 shows the means replotted using groups of size 4 and 3. The largest three means now all lie on a line of slope approximately 3.64, and there is no basis for further splitting of these. For the lower group, $q_{1,2}$ now becomes 11.6, with a standard error of 2.99. This slope is still 2.66 standard errors from 3.64, so that there is some reasonable evidence for separating off treatment $A$ as having the lowest mean yield. The final result of this is presented in Table 13.3(iv).

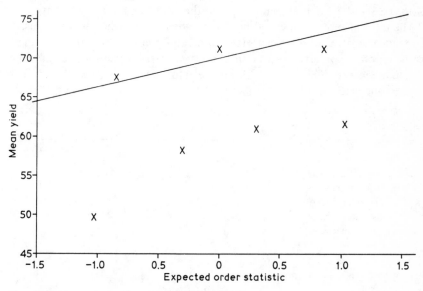

Fig. 13.2 Normal plot for groups of four and three.

Some comments should be made on this procedure. It requires tables of variances and covariances or order statistics, which are not readily accessible. It is probably rather sensitive to non-normality, but there are no investigations on this to date. Finally, this method is necessarily rather subjective and imprecise. On the other hand, it is a good visual method, which enables us to make tentative conclusions as to what patterns might be present in the data.

*Discussion*
There are many other multiple comparison techniques which have been devised, but those mentioned above are the most popular. The book by Miller (1966) contains an extensive discussion of the topic, with a detailed description of many procedures. The review paper by O'Neill and Wetherill (1971) contains a large number of references, classified under different headings. For another review see Chew (1976), and for a recent comparison of some procedures see Ury (1976). An excellent description of probability-plotting methods is given by Gerson (1975).

The properties of the procedures we have mentioned can be summarized in the Table 13.4, in which $\sqrt{} = $ Yes, $\times = $ No, and a dash means Not applicable.

However, this table does not tell the whole story. For comparisons of pairs of means, the $T$-method gives shorter confidence intervals, but for more complicated contrasts the $S$-method generally gives shorter intervals (see Exercise 13.2.1). The $S$-method is also known to be fairly insensitive to violations of the assumptions of normality or of homogeneity of variance. Both the $S$ and $T$ methods tend to give rather wide confidence intervals.

The normal-plotting method has many advantages, when it can be used. The fact that it has no precise error rate can be considered an advantage, as the emphasis on error-rates in multiple comparison methods seems to be rather misplaced. This feature of 'normal plotting' tends to lead to rather vague conclusions, but it does facilitate the sort of data-snooping that we

Table 13.4

|  | LSD | S | T | Normal plotting |
|---|---|---|---|---|
| Handles correlated means | $\sqrt{}$ | $\sqrt{}$ | $\times$ | $\times$ |
| "        unequal replications | $\sqrt{}$ | $\sqrt{}$ | $\times$ | $\times$ |
| "        general contrasts | – | $\sqrt{}$ | $\sqrt{}$ | $\times$ |
| Error rate: experimentwise | – | $\sqrt{}$ | $\sqrt{}$ | – |
| "    "    comparisonwise | $\sqrt{}$ | – | – | – |
| "    "    not precise | – | – | – | $\sqrt{}$ |
| Simultaneous confidence intervals | $\times$ | $\sqrt{}$ | $\sqrt{}$ | $\times$ |
| Good visual display | $\times$ | $\times$ | $\times$ | $\sqrt{}$ |

would usually like to do. 'Half-normal plotting' can be considered as an extension of the normal-plotting technique to handling some more general contrasts (see Gerson, 1975).

Finally, as a caution against being too certain about the conclusions resulting from these procedures, the reader should take another look at Fig. 13.1.

---

### Exercises 13.2

1. Denote the contrast

$$\frac{1}{3}(\theta_1 + \theta_3 + \theta_4) - \frac{1}{4}(\theta_2 + \theta_5 + \theta_6 + \theta_7)$$

as a $(3, 4)$ contrast. Calculate the relative length of $S$-method and $T$-method intervals for a series of contrasts including $(1, 1)$, $(1, 2)$, $(1, 3)$, $(1, 4)$, $(2, 2)$, $(2, 3)$, $(2, 4)$. Assume an infinite number of degrees of freedom for error. [For results see Scheffé, 1959, p. 77.]

2. Are your values in Exercise 1 above altered substantially when the number of degrees of freedom is small?

---

### 13.3 Departures from underlying assumptions

It is important to ask what effect departures from underlying assumptions have on analysis-of-variance procedures. Clearly, the algebra of calculating the various corrected sums of squares always holds, but once we calculate a significance test, the technique we are using depends on assumptions which in any given case may be unlikely to hold. Some of the relevant results on this question of sensitivity to assumptions are already given or implied in Chapter 1, but three points for discussion are the effects of non-normality, lack of independence, and lack of homogeneity of the variance. The material of this section depends heavily on Scheffé (1959, Chapter 10), Box (1954) and Box and Andersen (1955).

*(a) Non-normality*
Suppose that we have a one-way analysis-of-variance model such as that given in Equation (10.2), in which the error $\varepsilon$ is not normally distributed but has coefficients of skewness and kurtosis of $\gamma_1$ and $\gamma_2$; then Equations (1.11) and (1.12) suggest that inferences about means are unlikely to be affected much and several sources confirm this. Pearson (1931) reported some empirical sampling trials of the one-way analysis-of-variance model with values of $\gamma_1$ in the range $(0, 1)$ and values of $\gamma_2$ in the range $(-1.2, 4.1)$, and his results

Table 13.5 *Probability that the F-test exceeds the nominal 5% point (5 rows, 5 columns, serial correlation within columns.)*

| $\rho$ | $-0.4$ | $-0.2$ | 0 | 0.2 | 0.4 |
|---|---|---|---|---|---|
| Row test | 0.059* | 0.053* | 0.050 | 0.054* | 0.064* |
| Column test | 0.0003 | 0.010 | 0.050 | 0.13 | 0.25 |

*Approximate result.

demonstrated that the standard $F$-distribution was still a good fit to the distribution of the $F$-statistic. Also, Box and Andersen (1955) used a novel approach to obtain approximations to the $F$-distribution. Their results showed that for five groups of five observations each, the actual probability that $F$ exceeded the nominal 5 per cent value lay in the range (0.048, 0.052) for $\gamma_1$ and $\gamma_2$ in the range (0, 1). Thus the $F$-test for differences between means is robust to departures from normality, and this is a feature also shared by Scheffé's $S$-method of multiple comparisons. Methods for testing for normality were given in Chapter 1, and the procedures given there could be used on residuals provided that they are not too strongly correlated.

### (b) Lack of independence
In Chapter 1 the effect of lack of independence was studied by considering serial correlation, and even small amounts of serial correlation were shown to have a powerful effect on the distribution of $t$. It is intuitively clear that this result extends to the $F$-test in an analysis of variance. This is confirmed by some results tabled below, which were derived by Box (1954). If we have a two-way analysis of variance, with five rows and five columns, then suppose that the successive elements in each column are serially correlated, but that columns are independent. Table 13.5 shows that the $F$-test for rows is affected very little, but the $F$-test for columns is drastically affected. If therefore the experiment is by nature, or has been carried out in a manner such that correlations are possible, then the residuals should be checked carefully for this.

### (c) Lack of homogeneity of the variance
Consider a one-way analysis of variance in which there are $t$ groups of $n_i$ observations each. Denote the observations $y_{i,j}$, for $i = 1, 2, \ldots, t$ and $j = 1, 2, \ldots, n_i$, and suppose that the $i$th group is drawn from a normal population with expectation $\mu_i$ and variance $\sigma_i^2$. The usual $F$-test between groups given in Section 10.4, which assumes homogeneity of variance, is based on the ratio

$$F = \sum n_i (\bar{y}_{i.} - \bar{y}_{..})^2 / \{(t-1) \, \text{MS(Res)}\} \qquad (13.12)$$

where

$$\mathrm{MS(Res)} = \sum_i (n_i - 1)s_i^2 \bigg/ \bigg\{ \sum_i (n_i - 1) \bigg\} \tag{13.13}$$

and $s_i^2$ is the sample variance of the $i$th group. Now if we suppose the sample size to be large, we can replace the $s_i^2$ by $\sigma_i^2$, and if we write

$$\bar{\sigma}^2 = \sum_i (n_i - 1)\sigma_i^2 \bigg/ \bigg\{ \sum_i (n_i - 1) \bigg\} \simeq \sum_i n_i \sigma_i^2 \bigg/ \sum_i n_i$$

then Equation (13.12) becomes

$$F = \sum_i n_i (\bar{y}_{i.} - \bar{y}_{..})^2 / \{(t - 1)\bar{\sigma}^2\} \tag{13.14}$$

Now under the null hypothesis, the $\mu_i$ are all equal, and the numerator of Equation (13.14) is the sum of squares of normally distributed quantities with the same expectation. After some algebra it can be shown that, for large $n$, the expected value of $F$ is

$$E(F) = \{t(\bar{\sigma}^2/\tilde{\sigma}^2) - 1\}/(t - 1) \tag{13.15}$$

where

$$\tilde{\sigma}^2 = \sum_i \sigma_i^2 / t$$

and

$$V(F) = \{2t(1 + V_u)(\bar{\sigma}^2/\tilde{\sigma}^2) - 4V_w - 2\}/(t - 1)^2 \tag{13.16}$$

where $V_u$ and $V_w$ are the unweighted and weighted coefficients of variation of the $\sigma_i^2$,

$$V_u = \sum_i (\sigma_i^2 - \tilde{\sigma}^2)/(t\tilde{\sigma}^4) \tag{13.17}$$

$$V_w = \sum_i n_i (\sigma_i^2 - \tilde{\sigma}^2)/(\tilde{\sigma}^4 \sum_i n_i) \tag{13.18}$$

When all the variances are equal, $E(F)$ is unity and $V(F)$ becomes $2/(t - 1)$. Now if all groups are of the same size, $\bar{\sigma}^2 = \tilde{\sigma}^2$, and $E(F)$ still has expectation unity, and $V(F)$ becomes

$$V(R) = \{2/(t - 1)\}\{1 + V_u(t - 2)/(t - 1)\}$$
$$= \{2/(t - 1)\}Q, \quad \text{say} \tag{13.19}$$

Since $V_u$ is always positive if all $\sigma_i^2$ are not equal, this variance is too large. From this we see that in large samples, even with equal group sizes, the actual probability of exceeding the nominal $F$ percentage point exceeds the true value. This is in sharp contrast to the results shown in Table 1.6 for tests between two groups, where there is no effect for equal group sizes. Table 13.6 is extracted from Box (1954), and illustrates this result. This table shows how the probability increases with the value of $Q$ of Equation (13.19).

Table 13.6 *Probability that* F *exceeds the nominal* 5% *value*

| No. of groups | Group size | Ratio of variances | Q (Eq. (13.19)) | Prob. |
|---|---|---|---|---|
| 3 | 5 | 1 : 2 : 3 | 1.12 | 0.056 |
| 3 | 5 | 1 : 1 : 3 | 1.24 | 0.059 |
| 5 | 5 | 1 : 1 : 1 : 1 : 3 | 1.31 | 0.074 |
| 7 | 3 | 1 : 1 : 1 : 1 : 1 : 1 : 3 | 2.24 | 0.12 |

Table 13.7 *Probability that* F *exceeds the nominal* 5% *value*

| No. of groups | Group sizes | Ratio of variances | Prob. |
|---|---|---|---|
| 3 | 7, 5, 3 | 1 : 2 : 3 | 0.092 |
| 3 | 3, 5, 7 | 1 : 2 : 3 | 0.040 |
| 3 | 9, 5, 1 | 1 : 1 : 3 | 0.17 |
| 3 | 1, 5, 9 | 1 : 1 : 3 | 0.013 |

If the group sizes are not equal, neither $E(F)$ nor $V(F)$ are equal to the value obtained for equal $\sigma_i^2$, and it is clear that the effect can be very serious. Table 13.7 shows some values, also extracted from Box (1954).

Clearly there is a need to check for homogeneity of the variance, and if there is replication within cells, the procedures given in the next section can be used. If there is no replication, and the design is a complicated one, the presence of heterogeneity of variances may not be easy to detect, but a detailed study of residuals should be revealing.

*Discussion*
Two other points should be noted here. Firstly, it sometimes happens that problems of non-normality and/or of heterogeneity of variances can be avoided by transforming the observations, and for this we refer back to Chapter 8. Secondly, there can also be problems over the additivity assumed in the linear model. While this also can sometimes be avoided by using transformations, some discussion on interactions in Chapter 15 is also relevant.

Those who wish to study further the effect of deviations from assumptions of analysis-of-variance procedures should read Chapter 10 of Scheffé (1959), and consult the references mentioned there.

Finally, it should be noted that, if in an analysis of variance we carry out tests of normality, for homogeneity of variance, this amounts to a sequence of tests, and affects the size of the test in the resulting procedure. However, so long as we are aware of this, there is little in practice to be done about it.

### 13.4 Tests for heteroscedasticity

On occasions it is necessary to test for homogeneity of variance, as in Chapter 11, Example 11.1, and some methods of doing this are discussed below. Unfortunately, it is not a neat story, and it is necessary to use some judgement in the selection of the technique to use. Let us suppose that we have $t$ groups or cells, each containing replicate observations, so that the $i$th cell yields a sample variance $s_i^2$ on $v_i$ degrees of freedom, and let the true variances be $\sigma_i^2, i = 1, 2, \ldots, t$.

#### (a) Bartlett's test

Bartlett (1937) derived a test for homogeneity of variance which has been shown to be so sensitive to normality that it can only be used safely if we are sure that normality holds. His test is based calculating the statistic

$$B = (v \log s^2 - \sum v_i \log s_i^2)/G \tag{13.20}$$

where

$$v = \sum v_i$$
$$s^2 = \sum v_i s_i^2 / v$$
$$G = 1 + \{\sum v_i^{-1} - v^{-1}\}/\{3(t-1)\}$$

and referring $B$ to $\chi_{t-1}^2$ tables.

#### (b) Cochran's test

Cochran (1941) suggested using the statistic

$$C = \max_i s_i^2 / \{\sum s_i^2\} \tag{13.21}$$

and percentage points of this statistic are tabulated in Pearson and Hartley (1970). This test is very sensitive to departures from normality.

#### (c) Hartley's test

If all $v_i$ are equal, a test proposed by Hartley (1950) can be used. The statistic is

$$H = \{\max_i s_i^2\}/\{\min_i s_t^2\} \tag{13.22}$$

Percentage points are tabulated in Pearson and Hartley (1970) and this test is also very sensitive to departures from normality.

#### Discussion

Hartley's test is the simplest to use, and it was the one (in effect) used in Chapter 11. However, it can only be used with equal $v_i$, and it is rather inefficient for large $t$. Cochran's test is very sensitive to the case where all variances except one are equal. Bartlett's test is the most powerful against general alternatives.

It should be noted that gamma plots can be used to detect heteroscedasticity; an account of the technique is given in Gerson (1975). Also, if the $v_i$ are equal, a normal plot of log $s_i^2$ may be helpful.

## 13.5 Residuals and outliers

The topic of residuals has been mentioned a number of times, and here we remark that suitable tables or plots of residuals can detect non-normality, heteroscedasticity and outliers. There are many computer packages available for assisting in analyses of variance, and it is usually very simple to have the residuals tabulated ready for plotting, or even to have the plotting done by the computer.

Sometimes an analysis of residuals reveals 'outliers', and in dealing with these we refer back to the cautionary remarks of Section 1.3. If we are regularly dealing with data in which a percentage of outliers is expected, it is possible to devise automated methods for dealing with them on a computer. One suggestion was put forward by Tukey (1962, Sections 12 to 20, pp. 14–32), and we refer the reader to the paper, which contains detailed numerical illustration.

A useful test which can be used for the two-way classification with one observation per cell, discussed in Section 10.5, has been given by Stefansky (1972a, 1972b), called the maximum normed residual (MNR) test. The procedure is to calculate

$$w = \max |r_{ij}| / \sqrt{(\sum\sum r_{ij}^2)} \tag{13.23}$$

where $r_{ij}$ are the residuals

$$r_{ij} = y_{ij} - \bar{y}_{i.} - \bar{y}_{.j} + \bar{y}_{..} \tag{13.24}$$

Large values of $w$ are regarded as significant, and some tables giving significant levels of the statistic are given in Appendix C. The significance levels are calculated on the assumptions that the response variables are normally distributed, and that all residuals have a common variance.

*Example* 13.3 The residuals for Example 10.2 are as shown in Table 13.8.

Table 13.8

|  | *Breaker* | | |
| *Gauger* | 1 | 2 | 3 |
| 1 | 55.9 | − 190.8 | 135.0 |
| 2 | − 159.1 | 244.2 | − 85.0 |
| 3 | 103.4 | − 53.3 | − 50.0 |

Table 13.9 *Pattern of residuals produced by an outlier*

| Observations | | | Total | Residuals | | |
|---|---|---|---|---|---|---|
| 0 | 0 | 0 | 0 | 1 | $-2$ | 1 |
| 0 | 9 | 0 | 9 | $-2$ | 4 | $-2$ |
| 0 | 0 | 0 | 0 | 1 | $-2$ | 1 |
| Total    0 | 9 | 0 | 9 | | | |

The statistic given in Equation (13.23) is

$$w = 244.2/\sqrt{(165\,958)} = 0.599$$

Reference to Appendix C shows that the 5 per cent point for $r = c = 3$ is 0.648, so that the largest residual is not quite significant.    □ □ □

Even if a residual is found to be significant by Stefansky's MNR test, there remains a problem of what to do with it. If the experiment is one in which outliers are fairly common, we might well delete or modify it, or present the analysis with and without it. If possible, detailed investigations should be made of the reasons for an outlying observation. As was remarked earlier, outliers are sometimes the most important observations.

Another point to be emphasized here is the pattern produced by an outlying observation. Suppose that in Example 13.3 the centre observation was an outlier, then by use of Equation (13.24) this would have some effect on all of the residuals. The pattern of disturbances produced is easily seen by putting all observations zero, except a 9 (to avoid fractions) at the centre (see Table 13.9).

It is interesting to see that the pattern of positive and negative residuals shown in Table 13.9 is almost reproduced in Example 13.3. This gives further support to our suspicions that in Example 10.2, notwithstanding the result of the Stefansky test, something unusual about the response to Gauger 2, Breaker 2 is at least part of the problem in the analysis.

For further reading on outliers, see Grubbs (1969), Daniel (1976), Anscombe (1960) and Tukey (1962).

### 13.6  Some points of experimental design: General points

The aim of this section is to draw attention to some essential aspects of the planning of experiments, if we are to hope for useful conclusions from the analyses. Our comments can be brief, as there are many excellent texts on the subject, of which perhaps the best for our present purpose is Cox (1958). However, the brevity of this section should not be taken in any way as an indication that the points mentioned are not important. Whole experiments have been ruined at the 'planning' stage, before any observations were taken, and Cox's book gives examples of this. When doing any analysis, there are certain points about the experimental procedure that it is vital to check on, and it is also frequently relevant to make comments about the suitability of the design, or to make suggestions for improvements. Clearly, it is not possible to give any extensive coverage of these points, but instead a few important headings are given, and the reader is referred to Cox's book.

Two technical terms which are important are the terms *treatment* and *experiment unit*. We have already used the term 'treatment' several times; the term orginates from the agricultural context in which the subject of experimental design arose, and is used for different fertilizers, different dressing with the same fertilizer, different strains of a crop, etc. The term 'experimental unit' is the smallest division of the experimental material such that any two units in the actual experiment may receive a different treatment (Cox). For example, in an experiment designed to compare the effect of three different feeds on pigs, the pigs were fed in pens containing ten animals each. The experimental unit here is the pen, not the pig, since any two pens could differ in the feed given, but two pigs from the same pen must have the same feed. An essential point about an experiment is that the allocation of treatments to experimental units is under the experimenter's control, e.g., the allocation of feeds to pens.

Some requirements of a good experiment can be set out under the following headings:

*Absence of systematic error*    If, for example, pens of pigs given one feed were grouped together at the draughty end of the pig house, it would be impossible to separate out the feed effect from the effect of draughts. Thus this requirement usually implies that some form of randomization must be used in allocating treatments.

*Precision*    The experiment should give the desired precision on the estimation of contrasts of treatment effects, etc. Further, there should be a valid estimate of the precision achieved from the data obtained. The first point here raises a number of issues such as the use of designs to reduce error, and the type and number of observations made. The second point again usually implies a need for randomization and also for replication of the treatments.

*Range of validity*    It is important to notice that the conclusions made will be limited by the particular experimental units used and the conditions under which the experiment is performed. This affects the choice of experimental material used and the interpretation of any results. There are classic examples of attempts to generalize results too widely, and it may sometimes be necessary to note this point when stating conclusions.

*Choice of treatments*  The treatments used should differ uniquely in specified ways, so that differences between treatment effects can have unique interpretations. Sometimes this leads to the need for a 'control' treatment, like the experimental ones in all respects except the feature we wish to study.

## 13.7  Some points of experimental design: Randomized blocks

We shall discuss this very common form of experimental design enough to emphasize some of the points made above, and we shall do this in terms of an example. Suppose we wish to study the effect of four different fertilizers on the fibre strength of a cotton crop, and one large field is available for the experiment. If the field were simply divided into four strips, with one fertilizer used in each, differences of soil, lighting and drainage would bias the results, and it would not be possible to make unambiguous conclusions. Two possibilities are as follows.

*Completely randomized design*

The field is divided into twelve plots, and the four treatments are each allocated randomly to three of the plots. The analysis would follow that of Example 10.1 and would be as in Table 13.10. One problem with this design is that the residual term is inflated by possibly systematic differences of fertility, drainage, etc., across the field. The next design attempts to overcome this.

*Randomized block design*

The twelve plots are divided into three 'blocks' of four plots each, so that the four plots in each block are as homogeneous as possible, and as much as possible of the difference in drainage, soil conditions, etc., lay *between* blocks. Then in each block, the four treatments are randomly allocated to the four

Table 13.10 *Skeleton analysis of variance completely randomized design.*

| Souce | d.f. |
|---|---|
| Due to treatments | 3 |
| Residual | 8 |
| Total | 11 |

Table 13.11  *A randomized block design*

| | Block 1 | A | C | B | D |
|---|---|---|---|---|---|
| Fertility | 2 | D | B | A | C |
| difference | 3 | C | A | D | B |

plots. An illustration of how this design might look is shown in Table 13.11. This should be compared with Example 10.1.

When analysing a randomized block design we have to take account of both block and treatment differences, and for the design shown in Table 13.11 a suitable model is

$$Y_{ij} = \mu + \beta_i + \gamma_j + \varepsilon_{ij}$$

where the $\gamma_j$ are constants associated with treatments, and the $\beta_i$ are constants associated with blocks. This is the model for the two-way analysis-of-variance design discussed in Section 10.6, and the skeleton analysis-of-variance table is given in Table 13.12.

This table shows that in comparison with Table 13.10, the degrees of freedom for 'Residual' have been reduced by two, which are 'Due to blocks'. If the blocking has been successful, there will be large differences between block averages, and the blocks sum of squares will be large. This in turn means that the sum of squares for 'Residual' will be greatly reduced, leading to increased precision.

However, 'blocking' is not always successful. Unusual weather, for example, could easily produce quite different patterns in an agricultural experiment. Sometimes, therefore, the sum of squares due to blocks is quite small, and then all we have achieved by the 'randomized blocks' technique is to reduce the number of degrees of freedom for error. Careful design of the blocks is important, and the aim should be to confound as much as possible of the uncontrolled variation with blocks.

When analysing the results of an experiment such as that outlined in Table 13.11, the calculations proceed as for the two-way analysis down to the analysis-of-variance table. At this point it is only differences between treatments which are of interest. There is not much interest in testing the significance of the sum of squares due to blocks, except as a guide to the appropriate design for any repetition of the experiment. It is *not* valid to combine the sum of squares due to blocks with residual if it is not significant. The analysis depends on the way the design was layed out, and not on the results obtained.

Table 13.12 *Skeleton analysis of variance for randomized block design*

| Source | d.f. |
| --- | --- |
| Due to blocks | 2 |
| Due to treatments | 3 |
| Residual | 6 |
| Total | 11 |

Table 13.13  *Layout and results of weed-control experiment*

| Block 1 | A | D | B | E | C | F |
|---|---|---|---|---|---|---|
| | 438 | 17 | 538 | 18 | 77 | 115 |
| Block 2 | C | B | F | A | E | D |
| | 61 | 422 | 57 | 442 | 26 | 31 |
| Block 3 | E | C | D | F | B | A |
| | 77 | 157 | 87 | 100 | 377 | 319 |
| Block 4 | B | A | E | C | D | F |
| | 315 | 380 | 20 | 52 | 16 | 45 |

Table 13.14  *Transformation $\sqrt{(X + \frac{3}{8})}$, block and treatment totals for Example 13.4*

| Block | Treatment | | | | | | Total |
|---|---|---|---|---|---|---|---|
| | A | B | C | D | E | F | |
| 1 | 20.94 | 23.20 | 8.80 | 4.17 | 4.29 | 10.74 | 72.14 |
| 2 | 21.03 | 20.55 | 7.83 | 5.60 | 5.14 | 7.57 | 67.72 |
| 3 | 17.87 | 19.43 | 12.54 | 9.35 | 8.80 | 10.02 | 78.01 |
| 4 | 19.50 | 17.76 | 7.24 | 4.05 | 4.51 | 6.74 | 59.80 |
| Total | 79.34 | 80.94 | 36.41 | 23.17 | 22.74 | 35.07 | 277.67 |
| Mean | 19.83 | 20.24 | 9.10 | 5.79 | 5.69 | 8.76 | 11.57 |

*Example* 13.4    Bartlett (1936) reported the results, given in Table 13.13, of an experiment on the weed control of cereals. The observations recorded were the numbers of poppy plants per plot of $3\frac{3}{4}$ square feet. Treatment A was a control treatment, and there were five other treatments to be compared. A randomized block design was used, with four blocks of six plots per block.

It is obvious from looking at the results that some transformation is necessary to stabilize the variance. The ranges of the response variable for the different treatments vary considerably. Reference back to Section 8.3 suggests that the square root transformation should be used. The point of Bartlett's reference to the experiment, which was one of a series, was to illustrate the success of the square root transformation in stabilizing variance, and this can be seen by comparing the responses in Table 13.13 with Table 13.14.

We see from the figures that there is some slight evidence of a difference between blocks, as we might expect in an experiment of this type. The calculations for the analysis of variance are as shown in Table 13.15. The residual line can be filled in by subtraction to obtain the results in Table 13.16.

From the 'anova' table we see that treatment is so very highly significant that it is pointless computing the F-ratio. A test of the hypothesis that there

Table 13.15

| Sum of squares | Total | Treatments | Blocks |
|---|---|---|---|
| USS | 4195.893 | 4113.917 | 3241.963 |
| Correction | 3212.526 | 3212.526 | 3212.526 |
| CSS | 983.367 | 901.391 | 29.437 |

Table 13.16 *Anova of Example* 13.4 *data*

| Source | CSS | d.f. | MS |
|---|---|---|---|
| Between treatments | 901.39 | 5 | 180.3 |
| Between blocks | 29.44 | 3 | 9.81 |
| Residual | 52.54 | 15 | 3.50 |
| Total | 983.37 | 23 | |

are no block effects is made by referring the ratio

$$\frac{\text{Mean square due to blocks}}{\text{Mean square residual}} = \frac{9.81}{3.50} = 2.80$$

to $F$-tables on $(3, 15)$ d.f. The 10 per cent value is 2.49 and the 5 per cent value is 3.29, and the observed value falls in between these. Thus the 'blocking' was not very effective, although its mean square is much greater than the error.

The least significant difference between pairs of treatment means is, at the 5 per cent level,

$$2.13 \times \sqrt{(3.50/4)} = 1.99$$

From this it can be seen that the treatments can be split up into pairs

$$(A, B), \ (C, F), \ (D, E)$$

such that there are significant differences between, but not within pairs, and $D$ and $E$ are the most effective treatments.  □ □ □

Although the blocking in Example 13.4 was not highly effective, the example illustrates the method of the analysis. If the blocking system is carefully designed, use of blocks can lead to greatly increased precision over completely randomized designs.

*Discussion*

This brief section is no substitute for a course on the design of experiments, but the design and analysis of an experiment are inseparably linked. It is important in the analysis stage to look for errors in design, such as lack of randomization, or we may be likely to make an invalid analysis. Further aspects of the effect of design on analysis will be obvious from discussion in the next two chapters.

### Exercises 13.7

1. Examine the use of the Box–Cox technique on Example 13.4.

2. The following data, reported by Wooding (1969), are the results of an irritation experiment on animals. Animals formed blocks for the experiment, and the 'plots' were four sites on the backs of the animals. The treatments applied were a base material plus varying amounts of a test substance, as follows:

$A$: base only
$B$: base plus 0.01 per cent of the test substance
$C$: base plus 0.10 per cent of the test substance
$D$: base plus 1.00 per cent of the test substance

The results are scores for the irritation of the skin as compared to adjacent untreated areas, as measured by the degree of redness observed. A single judge took all observations and used a scale ranging from 0 to 12. Analyse the results and report on the conclusions.

| Treatment | Animal (Block) | | | | | | | |
|---|---|---|---|---|---|---|---|---|
| | 1 | 2 | 3 | 4 | 5 | 6 | 7 | 8 |
| $A$ | 1.00 | 1.50 | 2.50 | 3.00 | 1.50 | 1.50 | 1.50 | 2.00 |
| $B$ | 1.25 | 1.75 | 3.00 | 2.00 | 2.00 | 1.25 | 2.00 | 2.00 |
| $C$ | 1.50 | 2.50 | 4.00 | 3.00 | 2.50 | 2.00 | 2.50 | 2.50 |
| $D$ | 2.00 | 3.50 | 3.50 | 2.50 | 3.50 | 3.00 | 2.50 | 4.00 |

### Further reading on experimental design

See Cox (1958), Finney (1960), John and Quenouille (1977) and Cochran and Cox (1957).

# CHAPTER FOURTEEN

# Components of variance

## 14.1 Components of variance

So far, the examples of analysis-of-variance problems that we have considered all fall within the scope of what we call the 'fixed effects' analysis, or the 'model I' situation. In these problems the questions being asked are questions about contrasts of sets of means. For example, in Example 10.1 we wanted to know what differences there were between the mean yields of treatments $A$, $B$, $C$ and $D$, regarded as specific individuals. We now consider a different type of analysis-of-variance problem, suitable for application of what we call the 'random effects' or the 'model II' analysis. Consider the following example.

*Example* 14.1   When cotton yarn is produced it is wound on to large bobbins. One quality characteristic of considerable importance is the breaking load of the yarn since, for example, yarn with a low breaking load is likely to snap in weaving. The data given in Table 14.1 are extracted from the results of a larger experiment devised to investigate variations in breaking load between one bobbin and another, and between random lengths of yarn from the same bobbin.

Table 14.1  *Breaking load (in ounces) of cotton yarn*

|  | Bobbin | | | | | |
| --- | --- | --- | --- | --- | --- | --- |
|  | 1 | 2 | 3 | 4 | 5 | 6 |
|  | 14.8 | 16.0 | 15.1 | 14.8 | 12.4 | 16.2 |
|  | 16.8 | 15.9 | 16.1 | 14.1 | 15.6 | 15.9 |
|  | 13.6 | 17.9 | 18.5 | 14.9 | 15.8 | 15.3 |
|  | 15.3 | 17.4 | 16.3 | 12.7 | 14.2 | 15.8 |
| Totals | 60.5 | 67.2 | 66.0 | 56.5 | 58.0 | 63.2 |
| Means | 15.1 | 16.8 | 16.5 | 14.1 | 14.5 | 15.8 |

Example 14.1 presents a one-way classification analysis-of-variance problem, for which the model (10.2) might be appropriate, and for convenience we rewrite this

$$Y_{ij} = \mu + \alpha_i + \varepsilon_{ij} \qquad \begin{cases} i = 1, 2, \ldots, 6 \\ j = 1, 2, \ldots, 4 \end{cases} \qquad (14.1)$$

where we usually take $\varepsilon_{ij}$ to be independently distributed $N(0, \sigma^2)$. This model assumes that the variation of breaking load from one length to another within a bobbin is represented by sampling from a normal distribution with variance $\sigma^2$, and that $\sigma^2$ is constant from one bobbin to another. (Some comments on these assumptions are made below.) However, in Example 14.1 we are not normally interested in analysing differences in breaking load from one individual bobbin to another. It is known that the mean breaking loads for bobbins vary due to factors such as differences in the raw cotton and the conditions of production of the yarn. In fact the six bobbins were selected at random from a very large number of bobbins, and while for some purposes we might be interested in the bobbins as individuals, more commonly we are interested in making inferences about the population of bobbins from which the six used were drawn. Therefore the parameters $\alpha_1, \ldots, \alpha_6$ can be considered as a random sample of six from a population of $\alpha$'s, and we shall approximate this population by a normal distribution with expectation $\mu_\alpha$ and variance $\sigma_\alpha^2$.

The model we have just introduced has four parameters, $\mu, \mu_\alpha, \sigma^2$ and $\sigma_\alpha^2$, and two phases of random sampling are involved:

| Sample | Population |
|---|---|
| Six bobbins | Possible bobbins for test (supposed infinite) |
| Four lengths | Possible lengths for test (supposed infinite) |

In such a problem we are usually interested in making inferences about the populations sampled, and not about differences between particular bobbins.

If we take expectations of Equation (14.1) over both $\varepsilon$ and $\alpha$ and write these $E_\varepsilon$ and $E_\alpha$ respectively, we have

$$E_\alpha E_\varepsilon (Y_{ij}) = \mu + \mu_\alpha$$

for all $i, j$. Therefore it is impossible to estimate $\mu$ and $\mu_\alpha$ separately, and one of them, say $\mu_\alpha$, could be set to zero. We shall not do this here, but simply notice that the general mean is $(\mu + \mu_\alpha)$.

The variance of any function of the $Y_{ij}$ will depend on $\sigma^2$ and $\sigma_\alpha^2$, and these are called *components of variance*. One aim of the random-effects analysis is to estimate these components of variance, their ratio, or some

Table 14.2 *Analysis of variance for Example* 14.1

| Source | CSS | d. f. | MS | E(MS) |
|---|---|---|---|---|
| Between bobbins | 23.230 | 5 | 4.646 | $\sigma^2 + 4\sigma_\alpha^2$ |
| Within bobbins | 25.355 | 18 | 1.409 | $\sigma^2$ |
| Total | 48.585 | 23 | | |

other function of them. The calculation down to the analysis-of-variance table is identical to that in Chapter 10, and its structure is unchanged, but the procedure which follows is different, since the objectives are different.

By following the method of calculation given in Section 10.5 we obtain the analysis-of-variance table shown in Table 14.2. It also follows from Section 10.4 and Table 10.2 that with the model (14.1) we have

$$E_\varepsilon \text{ (Mean square within groups)} = \sigma^2$$

and

$$E_\varepsilon \text{ (Mean square between groups)} = \sigma^2 + \tfrac{4}{5} \sum_i (\alpha_i - \bar{\alpha})^2$$

The mean square within groups does not involve $\alpha$, but if we take the expectation of the mean square between groups over the distribution of $\alpha$ we obtain

$$E_\alpha E_\varepsilon \text{ (Mean square between groups)} = \sigma^2 + 4\sigma_\alpha^2$$

This establishes the results shown in Table 14.2.

The hypothesis $\sigma_\alpha^2 = 0$ is the hypothesis that the data are a random sample from a single normal population $N(\mu, \sigma^2)$. This hypothesis is tested by calculating the ratio

$$\frac{\text{Mean square between bobbins}}{\text{Mean square within bobbins}} = \frac{4.646}{1.409} = 3.3$$

and referring to $F$-tables. Since the 5 per cent point of $F(5, 18)$ is 2.77 and the $2\tfrac{1}{2}$ per cent point is 3.38, this result is significant, but not strongly so, and therefore there is some evidence that $\sigma_\alpha^2 > 0$.

The sort of analyses which follow the calculation of the analysis-of-variance table are discussed in the next section.

## 14.2 Components of variance: Follow-up analysis

In this section we discuss the sorts of analyses appropriate to a components-of-variance model which are carried out subsequent to the analysis-of-variance $F$-test. We shall see how radically it differs from a fixed-effects analysis at this stage.

In order to present the results in more general context, we assume that we have data similar to those in Example 14.1, but with $t$ groups and $r$ observations per group. The mean squares between and within groups are denoted $s_b^2$ and $s_w^2$ respectively, where

$$s_b^2 = r \sum_{i=1}^{t} (\bar{y}_{i.} - \bar{y}_{..})^2/(t-1) \tag{14.2}$$

and

$$s_w^2 = \sum_{i=1}^{t} \sum_{j=1}^{r} (y_{ij} - \bar{y}_{i.})^2/\{t(r-1)\} \tag{14.3}$$

These are the functions used to calculate the mean squares from data. However, in the discussion below we shall be studying the properties of estimators of the components of variance, and we shall regard the arguments of these mean squares as random variables $Y_{ij}$, etc. (see the convention about this in the preface).

*Estimation of the components of variance*
From Section 14.1 we see that

$$E_\alpha E_\varepsilon(s_b^2) = \sigma^2 + r\sigma_\alpha^2$$

and

$$E_\varepsilon(s_w^2) = \sigma^2$$

so that unbiased estimators of the components of variance are

$$\hat{\sigma}^2 = s_w^2$$

and

$$\hat{\sigma}_\alpha^2 = (s_b^2 - s_w^2)/r \tag{14.4}$$

For Example 14.1 the estimates are as follows:

$$\hat{\sigma}^2 = 1.409, \qquad \hat{\sigma}_\alpha^2 = (4.646 - 1.409)/4 = 0.809$$

Now although under this model $s_b^2$ has a larger expectation than $s_w^2$, it may not be much larger if $\sigma_\alpha^2$ is small, and the *observed* $s_b^2$ may sometimes be less than $s_w^2$. However, if $s_b^2 < s_w^2$, then Equation (14.4) gives a negative estimate for $\sigma_\alpha^2$! The best procedure is to record $\hat{\sigma}_\alpha^2 = 0$ on these occasions, and recognize that this introduces a bias into the estimator (14.4).

Sometimes it is the ratio of the components of variance, $\lambda = \sigma_\alpha^2/\sigma^2$, which is of interest, and this is estimated by

$$\hat{\lambda} = \hat{\sigma}_\alpha^2/\hat{\sigma}^2 = \left(\frac{s_b^2}{s_w^2} - 1\right)\bigg/r \tag{14.5}$$

*Variance of the estimators*
If the observations are independently distributed, and the distribution of the $\varepsilon$

is normal, then $(v_w s_w^2 / \sigma^2)$ has a $\chi^2$-distribution on

$$v_w = t(r - 1)$$

degrees of freedom. It follows from Exercise 5.4.2 that

$$V(\hat{\sigma}^2) = V(s_w^2) = 2\sigma^4 / v_w \qquad (14.6)$$

which we can estimate by inserting $\hat{\sigma}^2$.

One difference between the fixed- and random-effects models is that with the latter, the distribution of the mean square between groups is also easy to derive. The mean square $s_b^2$ involves $\bar{Y}_{i.}$ and $\bar{Y}_{..}$ but from Equation (14.1) we have

$$\bar{Y}_{i.} = \mu + \alpha_i + \bar{\varepsilon}_{i.}$$

so that $\bar{Y}_{i.}$ is normally distributed with expectation $(\mu + \mu_\alpha)$ and variance $(\sigma_\alpha^2 + \sigma^2 / r)$. Therefore by Result 7 of Appendix A the distribution of

$$\frac{v_b(s_b^2 / r)}{(\sigma_\alpha^2 + \sigma^2 / r)} \qquad \text{where} \quad v_b = t - 1$$

is $\chi^2$ on $v_b$ degrees of freedom. Therefore it follows by using Exercise 5.4.2 that

$$V(s_b^2) = 2(\sigma^2 + r \sigma_\alpha^2)^2 / v_b \qquad (14.7)$$

which we can estimate by inserting

$$\hat{\sigma}^2 + r\hat{\sigma}_b^2 = s_b^2$$

In Section 10.3 we established that $s_b^2$ and $s_w^2$ are statistically independent, so that we have

$$V(\hat{\sigma}_\alpha^2) = \{V(s_b^2) + V(s_w^2)\} / r^2$$

Therefore we obtain

$$V(\hat{\sigma}_\alpha^2) = \frac{2}{r^2} \left\{ \frac{(\sigma^2 + r\sigma_\alpha^2)^2}{v_b} + \frac{\sigma^4}{v_w} \right\} \qquad (14.8)$$

and an estimate of this variance is given by

$$\text{Est } V(\hat{\sigma}_\alpha^2) = \frac{2}{r^2} \left\{ \frac{s_b^4}{v_b} + \frac{s_w^4}{v_w} \right\} \qquad (14.9)$$

When using this variance it is important to note that the estimator $\hat{\sigma}_\alpha^2$ is made up of quantities which have $\chi^2$-distributions. Therefore $\hat{\sigma}_\alpha^2$ is likely to have a distribution which is strongly positively skewed, except for large values of $v_b$ and $v_w$.

An approximate variance of the estimator (14.5) is given in Exercise 14.2.3.

*Confidence intervals*
A $100(1 - \alpha)$ per cent confidence interval for $\sigma^2$ is provided by a straight-forward application of the method used in Section 1.2(a), but with different degrees of freedom; see also Theorem 5.7 and Section 10.3 and Exercise 5.5.1. The result is

$$\left( \frac{v_w s_w^2}{\chi_{v_w}^2 (1 - \alpha/2)}, \frac{v_w s_w^2}{\chi_{v_w}^2 (\alpha/2)} \right) \qquad (14.10)$$

for the notation see the relevant section of the preface.

The estimator (14.5) for $\lambda$ is proportional to the ratio of $s_b^2$ to $s_w^2$, and we have just shown that the mean squares $s_b^2$ and $s_w^2$ both have distributions proportional to $\chi^2$-distributions. It follows that the distribution of the ratio

$$\frac{s_b^2/(\sigma^2 + r\sigma_\alpha^2)}{s_w^2/\sigma^2} = \frac{s_b^2}{s_w^2 (1 + r\lambda)}$$

is exactly $F(v_b, v_w)$. Therefore we have

$$\Pr \left\{ F_{\alpha/2}(v_b, v_w) < \frac{s_b^2}{s_w^2 (1 + r\lambda)} < F_{1-\alpha/2}(v_b, v_w) \right\} = 1 - \alpha$$

and a $100 (1 - \alpha)$ per cent confidence interval for $\lambda$ is given by

$$\left( \frac{1}{r} \left\{ \frac{s_b^2}{s_w^2 F_{1-\alpha/2}(v_b, v_w)} - 1 \right\}, \frac{1}{r} \left\{ \frac{s_b^2}{s_w^2 F_{\alpha/2}(v_b, v_w)} - 1 \right\} \right) \qquad (14.11)$$

Again, for the notation, see the relevant section of the preface.

*Example 14.2* For Example 14.1, 90 per cent confidence intervals for $\lambda = \sigma_\alpha^2/\sigma^2$ are

$$\left( \frac{1}{4} \left\{ \frac{4.646}{(1.409)(2.77)} - 1 \right\}, \frac{1}{4} \left\{ \frac{(4.646)(4.58)}{1.409} - 1 \right\} \right)$$
$$= (0.048, 3.53)$$

The confidence interval (14.11) is not without problems. If we use an $\alpha$ such that

$$F_{1-\alpha/2}(v_b, v_w) > s_b^2/s_w^2$$

then the lower bound is negative! (In fact the upper bound can be negative as well, but this is rather rare.) Opinions differ on whether or not to replace a bound by zero if it is negative, on the grounds that truncating the interval can give a misleading impression of precision; see for example the discussion in Scheffé (1959, Section 7.2, pp. 229–31). A reasonable procedure is to truncate the interval at zero when a negative lower bound is obtained, but to note in writing up that this was done.

If a confidence interval for $\sigma_\alpha^2$ is required, an approximate result is available. The most convenient source for details of the derivation is Scheffé (1959, Section 7.2, pp. 231–5), but the result is as follows. Suppose that there are two mean squares, $s_1^2$ and $s_2^2$ such that

$$\frac{\nu_1 s_1^2}{(\sigma^2 + r\phi)} \quad \text{and} \quad \frac{\nu_2 s_2^2}{\sigma^2}$$

have $\chi^2$-distributions on $\nu_1$ and $\nu_2$ degrees of freedom respectively, and a confidence interval is required for $\phi$. If we write

$$F_1 = F_{1-\alpha/2}(\nu_1, \nu_2) \qquad F_2 = F_{1-\alpha/2}(\nu_1, \infty)$$
$$F_3 = F_{1-\alpha/2}(\nu_2, \nu_1) \qquad F_4 = F_{1-\alpha/2}(\infty, \nu_1)$$
$$F = s_1^2/s_2^2$$

then the $100(1-\alpha)$ per cent approximate confidence interval is given by

$$\left( \left\{ (F - F_1)(F + F_1 - F_2)/FF_2 \right\} \frac{s_2^2}{r}, \left\{ FF_4 - 1 + \frac{(F_3 - F_4)}{FF_3^2} \right\} \frac{s_2^2}{r} \right) \qquad (14.12)$$

All results on confidence intervals are sensitive to normality, especially in the distribution of the $\alpha$'s.

*Estimating the overall mean*

Frequently we find that the main objective of a 'components of variance' type of experiment is to estimate the overall mean. For example, an estimate of the mean breaking load for the yarn used in Example 14.1 was required for comparison with the mean breaking load for another type of yarn. Clearly, an estimate of the overall mean is $\bar{y}_{..}$, and the expectation is given by

$$E_\alpha E_\varepsilon(\bar{Y}_{..}) = \mu + \mu_\alpha$$

By substituting the model (14.1), we find that the variance is given by

$$V(\bar{Y}_{..}) = E_\alpha E_\varepsilon(\bar{Y}_{..} - \mu - \mu_\alpha)^2 = E_\varepsilon(\bar{\varepsilon}^2) + E_\alpha(\bar{\alpha}_. - \mu_\alpha)^2$$
$$= \sigma^2/rt + \sigma_\alpha^2/t = (\sigma^2 + r\sigma_\alpha^2)/rt \qquad (14.13)$$

Clearly, this can be estimated by calculating

$$\frac{\text{Between bobbins mean square}}{\text{Total no. of observations}}$$

However, we can ask whether a different design would have been better for estimating the overall mean. Suppose that it costs an amount $\gamma$ to sample a bobbin in terms of the amount it costs to sample a length of yarn from a given bobbin, then the total cost is

$$C = \gamma t + rt$$

and we can ask for the values of $r$ and $t$ which give a minimum variance for a set cost $C$. Given $C$, the variance (14.13) is

$$V = \frac{\sigma^2}{(C - \gamma t)} + \frac{\sigma_\alpha^2}{t}$$

so that

$$\frac{dV}{dt} = \frac{\gamma \sigma^2}{(C - \gamma t)^2} - \frac{\sigma_\alpha^2}{t^2}$$

This leads to optimum values

$$t = C\sqrt{\lambda}/\{[1 + \sqrt{(\gamma\lambda)}]\sqrt{\gamma}\} \qquad r = \sqrt{(\gamma/\lambda)} \qquad (14.14)$$

The actual values of $r$ and $t$ must be integral, and close to those in Equation (14.14).

The optimum values of $t$ and $r$ depend on $\lambda$, which is not usually known. Therefore in practice it would usually be necessary to carry out a small experiment in order to estimate $\lambda$, and then the major experiment would be designed with approximately optimal allocation.

*Some examples of use*
A good illustration of a problem in which this technique was used is given by Cameron (1951). The price paid for a shipment of wool depended on the average clean content, which is defined as the ratio of the clean weight to the raw weight. It is not practicable to determine this quantity for the entire shipment at the dockside, and sampling must be used. There are two components of variance. Firstly, the average clean content varies from bale to bale representing variation over farms in a large area and possibly over time as well. Secondly, the average clean content varies within each bale, representing variation from one sheep to another in the same flock, or from one part to another of the same sheep. The variation within a bale was measured by taking 'cores' of wool from each bale selected. The procedure used was to estimate components of variance between and within bales from a small sample. Then a larger sample was taken in order to estimate the average clean content more precisely. See Cameron (1951) for details.

The discussion of Cameron (1951) above gives one illustration of the objectives of a 'components of variance' situation. Another good illustration with a different aim in view is given by Desmond (1954). Here the problem concerned the quality control of voltage regulators for private motor cars. These regulators were required to operate within the range 15.8 to 16.4 volts, and the production system was as follows. At one stage in the production the regulators were set to work in the given range at one of a number of 'setting stations'. The regulators then passed to a 'testing station', and those not operating in the correct range were returned for resetting. The problem was that sometimes over 50 per cent of regulators were returned for resetting, creating severe bottlenecks. A 'product team' investigated the problem, and one thing they did was to conduct an experiment in which a sample of regulators from each setting station was passed through each testing station. As a result it was possible to estimate the amount of variation arising from several different parts of the process:

(a)  between regulators from the same setting station;
(b)  biases of testing stations;
(c)  measurement error;
(d)  bias due to setting stations.

of these items, (b) and (d) are fixed effects, and items (a) and (c) are components of variance. Only after all these sources of variation are estimated is it possible to start discussing the problem of the 'bottlenecks'. In the event it was found that measurement error was so large that certain regulators were rejected at one testing station as being below the specified range; rejected at another as being above the specified range; and accepted at another testing station as satisfactory! The definition and estimation of components of variance was a vital part of studying this problem; see the source paper for details.

*Non-normality and 'random effects' model analyses*
In analysing Example 14.1 we assumed that both $t$ and $\alpha$ were normally distributed, and it is important to know what the effect of non-normality would be on the procedures we have suggested. Some discussion relevant to these points has already been given in Sections 1.7 and 13.3. In particular, the effect of non-normality of the distribution of $\varepsilon$ on the validity of the $F$-test of the hypothesis $H_0 : \sigma_\alpha^2 = 0$ is very small, see Section 13.3. As regards to the confidence-interval procedures described in Section 14.2 the position is not so satisfactory. It is immediately apparent from Equation (1.14) that skewness of the distribution of $\alpha$ could seriously affect the validity of confidence-interval statements such as (14.10) and (14.12). In a discussion of this point, Scheffé (1959) concludes that positive kurtosis of the distribution of $\alpha$ reduces the effective confidence coefficient of any confidence statement, and negative kurtosis of the distribution of $\alpha$ increases it.

One of the problems with random-effects models is that there are some-times very few degrees of freedom available for estimation of $\sigma_\alpha^2$ (five in Example 14.1). This means that it is often impossible to make adequate checks on the distribution of $\alpha$, and it is not at all satisfactory to have results strongly dependent upon assumptions that we cannot check. There are two implications of all this:

(i)  When designing experiments according to a 'random effects' model, we should try to allow for a good number of degrees of freedom to be available for variance components.

(ii) We should take great care that there was not something special, for example, about any particular bobbin selected in Example 14.1, or about the trials conducted on any bobbin.

---

## Exercise 14.2

1. Follow the method of Example 14.2 and obtain a 99 per cent confidence interval for $\lambda$, for Example 14.1 data. Comment on the result.

2. Obtain an approximate 95 per cent confidence interval for $\sigma_\alpha^2$ in Example 14.1.

3. If $Y_1$ and $Y_2$ are independent, you are given that

$$V\{f(Y_1, Y_2)\} \simeq \left(\frac{df}{dy_1}\right)^2 V(Y_1) + \left(\frac{df}{dy_2}\right)^2 V(Y_2)$$

where the differential coefficients are evaluated at the point $(E(Y_1), E(Y_2))$. Use this result to show that an approximate variance of $\hat{\lambda}$ in Equation (14.5) is

$$V(\hat{\lambda}) = \left(\lambda + \frac{1}{r}\right)^2 \left\{\frac{2}{v_b} + \frac{2}{v_w}\right\}$$

4. Check the derivation of results (14.14).

5. Obtain approximate variances of the estimated optimum values of $t$ and $r$ obtained by putting $\hat{\lambda}$ into (14.14).

6. The data in Table 14.3 were recorded in the routine control of stock weight in a textile mill. Twelve spinning machines were selected randomly, and four lengths of yarn were tested from each machine. The lengths of yarn were all randomly selected from one week's production. The observa-

Table 14.3 *Test results less 50*

| Machine | | | | | | | | | | | |
|---|---|---|---|---|---|---|---|---|---|---|---|
| 5.5 | 1.8 | 6.0 | 2.0 | 0.3 | 9.9 | 5.5 | 8.8 | 0.4 | 5.4 | 8.8 | 8.6 |
| 0.2 | 4.2 | 7.1 | 6.1 | 8.8 | 8.2 | 0.2 | 6.1 | 2.1 | 6.3 | 7.3 | 9.2 |
| 4.8 | 0.1 | 5.2 | 4.1 | 4.0 | 6.3 | 1.6 | 9.2 | 1.6 | 6.1 | 6.8 | 5.7 |
| 2.1 | 4.1 | 1.1 | 3.0 | 3.8 | 7.0 | 4.7 | 7.1 | 3.5 | 8.1 | 3.9 | 7.1 |

tion recorded is the 'yarn number', which is the number of 840-yard lengths of yarn per pound weight. Estimate components of variance between and within machines, and report on your analysis of the data. If you were told that machines 1 to 4, 5 to 8 and 9 to 12 were sampled in three separate weeks, how would your analysis alter? [Data modified from an experiment reported by Enrick (1962).]

7. In a study of a technique for measuring a certain quantity in dental research, eight X-ray films are made of each patient at different times, but under nominally controlled conditions, and three independent readings of each film are made.

| Film | | | | | | | |
|---|---|---|---|---|---|---|---|
| 1 | 2 | 3 | 4 | 5 | 6 | 7 | 8 |
| 8.1 | 7.2 | 9.1 | 10.1 | 7.1 | 6.7 | 11.2 | 8.1 |
| 7.6 | 7.3 | 9.9 | 8.6 | 7.0 | 6.6 | 8.9 | 8.3 |
| 9.3 | 7.9 | 9.6 | 8.6 | 6.8 | 8.4 | 9.3 | 9.1 |

(a) Estimate the components of variance, and give a 90 per cent confidence interval for the ratio.
(b) If the film costs 7 times as much as a measurement, find the number of measurements which should be taken on each film which minimizes the variance of the overall average $\bar{X}$ for a given expenditure.
(c) For the same expenditure as the above experiment, suggest a better design.

## 14.3 Nested classifications
A good example of the use of random-effects model analysis of variance is given by its use in nested classifications. The following example is given by Davies (1967) and the description he gives is quoted below.

*Example* 14.3  'The example refers to deliveries of a chemical paste contained in casks where, in addition to sampling and testing errors, there are variations in quality between deliveries which require to be estimated. As a routine, three casks selected at random from each delivery were sampled, and the samples were kept for reference. It was desired to estimate the variability in the paste strength from cask to cask, and from one delivery to another. Ten of the delivery batches were chosen at random, and two analytical tests carried out on each of the 30 samples. In order to ensure that the tests were independent, all 60 strength determinations were carried out in random order.' The resulting data are given in Table 14.4.   ☐ ☐ ☐

As the data are laid out in Table 14.4 it looks rather like a two-way analysis of variance, but it is not because for example, cask 1 of batch 1 has nothing especially in common with cask 1 of batch 2. The ten batches were selected at random, and then three casks were sampled at random from a large number in each batch. This can all be laid out in a nested or hierarchical design as illustrated in Fig. 14.1.

The variation between determinations is almost entirely due to analytical error. For any given batch, the paste varies in strength from one cask to another due perhaps to variations resulting from the operation of the production process. However, variation in paste strength from one batch to another is affected by a quite separate set of factors, such as the use of different batches of raw material, different operating temperatures, etc. In this problem the batches were not taken successively from the process and there was no overall trend over the period observed. We therefore have a situation in which three separate populations have been sampled, as set out in Table 14.5.

Any model for this set of data must represent the three separate phases of sampling, and the primary aim of any analysis is usually to estimate the amount of variability in the three populations. If the populations sampled are all large or infinite we frequently assume that we have separate and

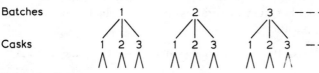

Fig. 14.1 Nested structure of Example 14.3.

Table 14.4 *Example 14.3 data. Percentage paste strength*

| Batch | Cask 1 | | Cask 2 | | Cask 3 | | Batch total |
|---|---|---|---|---|---|---|---|
| 1 | 62.8 | 62.6 | 60.1 | 62.3 | 62.7 | 63.1 | 373.6 |
| 2 | 60.0 | 61.4 | 57.5 | 56.9 | 61.1 | 58.9 | 355.8 |
| 3 | 58.7 | 57.5 | 63.9 | 63.1 | 65.4 | 63.7 | 372.3 |
| 4 | 57.1 | 56.4 | 56.9 | 58.6 | 64.7 | 64.5 | 358.2 |
| 5 | 55.1 | 55.1 | 54.7 | 54.2 | 58.8 | 57.5 | 335.4 |
| 6 | 63.4 | 64.9 | 59.3 | 58.1 | 60.5 | 60.0 | 366.2 |
| 7 | 62.5 | 62.6 | 61.0 | 58.7 | 56.9 | 57.7 | 359.4 |
| 8 | 59.2 | 59.4 | 65.2 | 66.0 | 64.8 | 64.1 | 378.7 |
| 9 | 54.8 | 54.8 | 64.0 | 64.0 | 57.7 | 56.8 | 352.1 |
| 10 | 58.3 | 59.3 | 59.2 | 59.2 | 58.9 | 56.6 | 351.5 |

Table 14.5

| Sample | Population |
|---|---|
| 10 Batches | Collection of batches available for sampling |
| 3 Casks | The casks in one delivery batch |
| 2 Determinations | Hypothetical population of observations resulting from repeat analytical determinations from samples from a given cask. |

independent normal populations, but sometimes it is necessary to use finite population models. This section deals with the infinite population model; the finite population model is discussed in Section 14.5. We shall set out the algebra in a general setting, and return to Example 14.3 later.

Suppose that a random sample of $p$ 'classes' is drawn from a population $P_c$ of classes. From each class, a random sample of $q$ subclasses is taken from a population $P_{sc}$ of subclasses. We then take $r$ multiple determinations, and the observations are denoted $y_{ijk}$, $i = 1, 2, \ldots, p$; $j = 1, 2, \ldots, q$; $k = 1, 2, \ldots, r$. The model we shall use is

$$Y_{ijk} = \mu + \alpha_i + \beta_{j(i)} + \varepsilon_{ijk} \tag{14.15}$$

where $\alpha_i$ is a random variable distributed $N(\mu_\alpha, \sigma_\alpha^2)$, $\beta_{j(i)}$ is a random variable distributed $N(\mu_\beta, \sigma_\beta^2)$ and $\varepsilon$ is a random variable distributed $N(0, \sigma^2)$. All random variables are assumed to be statistically independent, and the $\alpha, \beta$ and $\varepsilon$ terms represent sampling from $P_c$, $P_{sc}$ and determinations respectively. This model has made several assumptions which need checking when making an application:

 (i) that the variance of the error of determinations is the same in all sub-classes;
 (ii) that the variance of subclasses within classes is the same for all classes;
 (iii) that all the populations are normal.

Before discussing the analysis of variance appropriate to the model (14.15), it is important to realize the meaning of averages of some of the terms. The average $\bar{\alpha}$ represents the average effect on the observations due to sampling from classes. The average $\bar{\beta}_{.(i)}$ represents the average effect on the observations due to sampling subclasses within the $i$th class. However, there is *no meaning to* the term $\bar{\beta}_{j(.)}$. This would be the average effect on the observations due to the $j$th subclass drawn from each class and our statements above make it clear that subclasses are randomly drawn from each class. The $j$th samples from each class therefore have nothing is common.

Another point to emphasize about the model (14.15) is that it contains redundant parameters, but following the methods of Chapter 10 this causes no problem. The analysis-of-variance table is easily built up from similar tables which have been developed earlier, and a brief outline of the argument follows.

Suppose we represent the data simply as between and within subclasses, ignoring the classes effect in the analysis; then we have a one-way equally replicated analysis of variance (see Table 14.6).

If we now ignore the multiple determinations, and suppose that we had simply $pq$ observations $\bar{y}_{ij.}$, then we have a similar analysis of variance, leading to Table 14.7.

We now follow the argument in Section 11.1, and combine the two tables, remembering that the corrected sums of squares in Table 14.7 must be multiplied by $r$ to present them on the same basis as Table 14.6 (see

Table 14.6 *Analysis of variance between and within subclasses*

| Source | CSS | d. f. |
|---|---|---|
| Between subclasses | $r\sum_i\sum_j(\bar{y}_{ij.} - \bar{y}_{...})^2$ | $pq - 1$ |
| Between determination within subclasses | $\sum_i\sum_j\sum_k(y_{ijk} - \bar{y}_{ij.})^2$ | $pq(r - 1)$ |
| Total | $\sum_i\sum_j\sum_k(y_{ijk} - \bar{y}_{...})^2$ | $pqr - 1$ |

Table 14.7 *Analysis of variance ignoring multiple determinations*

| Source | CSS | d. f. |
|---|---|---|
| Between classes | $q\sum_i(y_{i..} - \bar{y}_{...})^2$ | $p - 1$ |
| Between subclasses within classes | $\sum_i\sum_j(\bar{y}_{ij.} - \bar{y}_{i..})^2$ | $p(q - 1)$ |
| Total | $\sum_i\sum_j(\bar{y}_{ij.} - \bar{y}_{...})^2$ | $pq - 1$ |

Table 14.8 *Analysis of variance for a nested classification*

| Source | CSS | d.f. | MS |
|---|---|---|---|
| Between classes | $rq\sum_i(\bar{y}_{i..} - \bar{y}_{...})^2$ | $p - 1$ | $A$ |
| Between sub classes within classes | $r\sum_i\sum_j(\bar{y}_{ij.} - \bar{y}_{i..})^2$ | $p(q - 1)$ | $B$ |
| Between determinations within subclasses | $\sum_i\sum_j\sum_k(y_{ijk} - \bar{y}_{ij.})^2$ | $pq(r - 1)$ | $C$ |
| Total | $\sum_i\sum_j\sum_k(y_{ijk} - \bar{y}_{...})^2$ | $pqr - 1$ | |

Section 11.1 for details). This leads to the analysis-of-variance table shown in Table 14.8.

The expected mean squares can be obtained readily by substituting the model (14.15) and taking expectations over the three distributions. First we take expectations over the distribution of the $\varepsilon$'s to obtain:

$$E_\varepsilon\{MS(C)\} = \sigma^2$$

$$E_\varepsilon\{MS(B)\} = \sigma^2 + \frac{r}{p(q-1)} \sum_{i=1}^{p} \sum_{j=1}^{q} (\beta_{j(i)} - \bar{\beta}_{\cdot(i)})^2 \qquad (14.16)$$

$$E_\varepsilon\{MS(A)\} = \sigma^2 + \frac{rq}{(p-1)} \sum_{i=1}^{p} (\bar{\beta}_{\cdot(i)} - \bar{\beta}_{\cdot(\cdot)})^2$$

$$+ \frac{rq}{(p-1)} \sum_{i=1}^{p} (\alpha_i - \bar{\alpha})^2 + \text{terms in } \alpha\beta \qquad (14.17)$$

When we take expectations over $\beta$, only one term gives any difficulty and this is the second term on the right-hand side of Equation (14.17). However, $\bar{\beta}_{\cdot(i)}$ is a mean of $q$ random variables, and has a variance of $\sigma_\beta^2/q$. Therefore we have

$$E_\beta\left\{\frac{1}{(p-1)} \sum_{i=1}^{p} (\bar{\beta}_{\cdot(i)} - \bar{\beta}_{\cdot(\cdot)})^2\right\} = \sigma_\beta^2/q$$

This leads to the expression

$$E_\beta E_\varepsilon\{MS(B)\} = \sigma^2 + r\sigma_\beta^2 \qquad (14.18)$$

$$E_\beta E_\varepsilon\{MS(A)\} = \sigma^2 + r\sigma_\beta^2 + \frac{rq}{(p-1)} \sum_{i=1}^{p} (\alpha_i - \bar{\alpha}) \qquad (14.19)$$

and finally by taking expectations over the distribution of $\alpha$ we have the results shown in Table 14.9.

From this point the methods of analysis follow closely those discussed in Section 14.2. The derivation of formulae for point estimates of $\sigma^2, \sigma_\alpha^2, \sigma_\beta^2$, their variances and certain confidence-interval statements are left as an exercise (see Exercise 14.3.1).

Hypothesis testing is usually not as important as estimation in a compo-

Table 14.9 *Expected mean squares for Table 14.8 (infinite population)*

| Mean square | Expected mean square |
|---|---|
| A | $\sigma^2 + r\sigma_\beta^2 + rq\sigma_\alpha^2$ |
| B | $\sigma^2 + r\sigma_\beta^2$ |
| C | $\sigma^2$ |

nents-of-variance problem, but the procedures are also clear from Table 14.9. In order to test the hypothesis $\sigma_\beta^2 = 0$, we calculate the ratio

$$MS(B)/MS(C)$$

where $B$ and $C$ refer to the mean squares in Table 14.8, and then we look up the $F$-tables for the appropriate numbers of degrees of freedom. Similarly, the hypothesis of $\sigma_\alpha^2 = 0$ is tested using the ratio

$$MS(A)/MS(B)$$

and referring to $F$-tables, whether or not $\sigma_\beta^2$ turns out to be significantly different from zero.

In a components-of-variance problem, care has to be taken to check assumptions made and an analysis of Example 14.3 is discussed in the next section.

---

### Exercise 14.3
1. Obtain point estimates of $\sigma^2$, $\sigma_\alpha^2$ and $\sigma_\beta^2$, and their variances, following the methods used in Section 14.2. Also obtain formulae for confidence intervals for $\sigma_\alpha^2$, by using Equation (14.12).

---

### 14.4 Outline analysis of Example 14.3
One of the first steps in analysing Example 14.3 is to look at the data and see if the assumptions necessary for the analysis we are about to do are seriously in question. We need to graph and tabulate the data in various ways to see what we can learn from it. We shall only do a certain amount of this here.

We start by tabulating the difference (first determination minus the second determination) within each cask, and this is shown in Table 14.10. If all 60 determinations were carried out in random order, and the difference

Table 14.10 *Differences (1st–2nd) observations ($\times$ 10)*

| Batch | Cask 1 | Cask 2 | Cask 3 |
|-------|--------|--------|--------|
| 1     | 2      | −22    | −4     |
| 2     | −14    | 6      | 22     |
| 3     | 12     | 8      | 17     |
| 4     | 7      | −17    | 2      |
| 5     | 0      | 5      | 13     |
| 6     | −15    | 12     | 5      |
| 7     | −1     | 23     | −8     |
| 8     | −2     | −8     | 7      |
| 9     | 0      | 0      | 9      |
| 10    | −10    | 0      | 23     |

between the two determinations in any cell is an analytical error with a homogeneous variance, then we would expect these 30 differences to be consistent with having been sampled from a normal distribution with a variance twice the variance of a determination.

A glance at this table shows a suspiciously large number of zeros (four), and also some differences which appear to be rather extreme. There appears to be some slight evidence of a bias towards positive differences. Some of these features stand out in Fig. 14.2, which is a normal plot of the differences. Ideally one would check back with the experimenters to see if something could account for the features mentioned. For example, if the second observation recorded in any cell was always later than the first, a time drift could account for the slight positive bias in the differences. Certainly the number of zero results and the peculiarities of the positive end of the normal plot need looking into. However, these points do not appear to be serious enough to

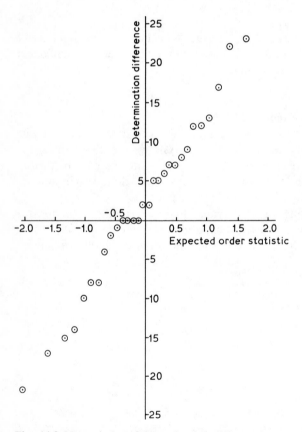

Fig. 14.2 Normal plot of determination differences.

completely invalidate the analysis. In a practical case one would want to investigate before proceeding.

As another illustration of examining the data, we shall look at the batch means. The ordered means are as follows:

Batch:    5        10        9        2        4        7        6        3        1        8
Mean:   55.90    58.58    58.67    59.30    59.70    59.90    61.03    62.05    62.27    63.12

Immediately batch 5 stands out, and there are one or two other large gaps in the ordered sequence of means. Figure 14.3 shows the normal plot of the means. In fact, normal plotting is not a very powerful technique with a small number of observations, but the gaps between batches 5 and 10, 7 and 6, and 1 and 8 show up here. Clearly, some non-randomness in the variation between the batches would not affect the other parts of the analysis. We shall proceed with the analysis as if all of the assumptions held, and then return to consider the variation between the batch means.

Finally, in checking the data we take a brief look at the cask means, as within batches, these should be normally distributed with a constant variance. The cask means are shown in Table 14.11, together with the range and sample variance of the sample means within each batch. The range of the variances

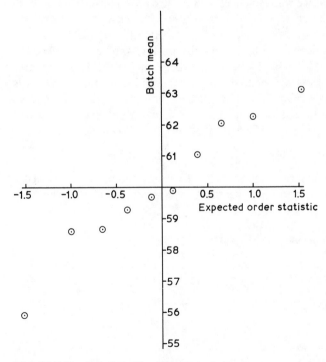

Fig. 14.3 Normal plot of batch means.

Table 14.11  *Cask means for Example* 14.3 *data*

| Batch | Cask 1 | Cask 2 | Cask 3 | Range | Variance |
|-------|--------|--------|--------|-------|----------|
| 1  | 62.70 | 61.20 | 62.90 | 1.70 | 0.863 |
| 2  | 60.70 | 57.20 | 60.00 | 3.50 | 3.430 |
| 3  | 58.10 | 63.50 | 64.55 | 6.45 | 11.977 |
| 4  | 56.75 | 57.75 | 64.60 | 7.85 | 18.257 |
| 5  | 55.10 | 54.45 | 58.15 | 3.70 | 3.902 |
| 6  | 64.15 | 58.70 | 60.25 | 5.45 | 7.886 |
| 7  | 62.55 | 59.85 | 57.30 | 5.25 | 6.892 |
| 8  | 59.30 | 65.60 | 64.45 | 5.15 | 11.256 |
| 9  | 54.80 | 64.00 | 57.25 | 9.20 | 22.701 |
| 10 | 58.80 | 59.20 | 57.75 | 1.45 | 0.561 |

(or ranges) does not look excessive, and if we calculate Hartley's test (Section 13.4(c)), we have

$$H = \max s_i^2 / \min s_i^2 = 22.701/0.561 = 40.46$$

Reference to the tables of percentage points of $H$ in *Biometrika Tables* shows that the 5 per cent point is 550! On this value of $H$ there is no evidence for heterogeneity of variance of the cask means. Clearly this is not a very powerful test of what we want to check – that the cask means within batches are normally distributed with constant variance. We leave it as an exercise for the reader to carry out further checks on this point.

The analysis-of-variance table follows the outline shown in Table 14.8, and results are given in Table 14.12. The expectations of the means squares in Table 14.12 are worked out for an infinite population model, which is something of an approximation in this case. The components of variance between determinations, between casks, and between batches are denoted by $\sigma^2$, $\sigma_\beta^2$, and $\sigma_\alpha^2$ respectively. We readily obtain the following estimates

$$\hat{\sigma}^2 = 0.68 \qquad \hat{\sigma}_\beta^2 = 8.44 \qquad \hat{\sigma}_\alpha^2 = 1.66$$

We see that by far the largest component of variance is that between casks, and this has some important implications for sampling schemes which we shall discuss below.

Table 14.12  *Analysis of variance for Example* 14.3 *data*

| Source | CSS | d.f. | MS | E(MS) |
|--------|-----|------|-----|-------|
| Between batches | 247.402 | 9 | 27.49 | $\sigma^2 + 2\sigma_\beta^2 + 6\sigma_\alpha^2$ |
| Between casks within batches | 350.907 | 20 | 17.55 | $\sigma^2 + 2\sigma_\beta^2$ |
| Between determinations within casks | 20.340 | 30 | 0.68 | $\sigma^2$ |
| Total | 618.649 | 59 | | |

Tests of hypotheses are also straightforward. A test of the hypothesis that $\sigma_\beta^2 = 0$ is made by calculating

$$F = 17.55/0.68 = 25.81$$

and by referring to $F(20, 30)$ tables. We see that the observed $F$-value is very highly significant, and we conclude that it is most unlikely that $\sigma_\beta^2 = 0$.

A test of the hypothesis that $\sigma_\alpha^2 = 0$ is made by calculating

$$F = 27.49/17.55 = 1.57$$

and by referring to $F(9, 20)$ tables. As the 10 per cent point is 1.96 and the observed value is even less than this, the result is not significant and the hypothesis cannot be rejected.

These hypothesis tests are not very informative, and we may be interested in confidence intervals. Confidence intervals for $\sigma^2$ are calculated using Equation (14.10) and this leads to (0.43, 1.21), showing that this component of variance is determined fairly precisely (95 per cent limits are given).

A confidence interval for the ratio $\sigma_\beta^2/\sigma^2$ is obtained using the expression (14.11), and at the 95 per cent level this leads to (5.37, 24.83). The rather wide limits show that there is considerable uncertainty about the true value of this ratio.

In order to obtain confidence intervals for $\sigma_\beta^2$, two methods are available. Firstly, we could take the confidence interval for $\sigma_\beta^2/\sigma^2$, and simply multiply this by $\hat{\sigma}^2$. This leads to (3.65, 16.88). Secondly, we could use the approximation (14.12), by putting $s_b^2 = 17.55$, $s_w^2 = 0.68$, $v_b = 20$, $v_w = 30$, $r = 2$. This leads to a 95 per cent interval as (4.78, 18.00). These results are virtually identical, although the second is to be preferred from a theoretical point of view. In order to contract this interval, a very much larger experiment would be necessary.

Confidence intervals for $\sigma_\alpha^2$ can also be obtained from the interval (14.12), by putting

$$s_b^2 = 27.49, \qquad s_w^2 = 17.55, \qquad v_b = 9, \qquad v_w = 20, \qquad r = 6$$

However, the lower limit is negative, and, the upper limit is 12.4, which shows the poor precision of this component of variance.

Another way of demonstrating the poor precision with which some of the components of variance are estimated is to calculate standard errors by the method leading to the variance given in Equation (14.9). This is not particularly useful for the smaller components of variance owing to the skewness of the distributions.

Now let us return to the questions posed by the batch means. Given the model (14.15), a batch mean is

$$\bar{Y}_{i..} = \mu + \alpha_i + \bar{\beta}_{.(i)} + \bar{\varepsilon}_{i..} \tag{14.20}$$

so that the variance of a batch mean is

$$V(\bar{Y}_{i..}) = V(\alpha_i) + V(\bar{\beta}_{.(i)}) + V(\bar{\varepsilon}_{i..})$$
$$= \sigma_\alpha^2 + \sigma_\beta^2/3 + \sigma^2/6 \qquad (14.21)$$
$$= (\sigma^2 + 2\sigma_\beta^2 + 6\sigma_\alpha^2)/6$$

and the variance of a batch mean is estimated by the mean square between batches, divided by six. For our data this leads to a variance of 4.58, or a standard error of 2.14. If a slope of 2.14 is drawn in on Fig. 14.3, it will be seen that the observed pointed are reasonably consistent with it. There do not appear to be any strong grounds to support the suspicions made earlier about differences between batch means.

Frequently, one of the objectives of an analysis of this type is to compare the effect of alternative sampling schemes, and we can easily do this by repeating the steps leading to Equation (14.21) for different numbers of casks and determinations. For example, if from each batch we sampled four casks and made one determination of each, the variance of a batch mean would be

$$V(\bar{Y}_{i..}) = \sigma_\alpha^2 + \sigma_\beta^2/4 + \sigma^2/4 \qquad (14.22)$$

and on inserting the estimates we get 3.94, which is an improvement on 4.58 obtained above. If six casks are sampled instead of four with one determination made from each, the variance reduces further to 3.18.

---

**Exercises 14.4**

1. In the routine quality control of a chemical, three samples are taken from production every half hour, and duplicate analyses were made on each sample. Some results (in percentages) are shown in Table 14.13. Carry out an appropriate analysis of these data, and report on your conclusions.

Table 14.13

| Time | Sample | | | Time | Sample | | |
|------|--------|------|------|------|--------|------|------|
| 8.30 | 1 | 17.55 | 17.63 | 10.30 | 1 | 17.23 | 17.50 |
|      | 2 | 17.27 | 17.15 |       | 2 | 17.39 | 17.36 |
|      | 3 | 17.43 | 17.55 |       | 3 | 17.28 | 17.31 |
| 9.00 | 1 | 17.25 | 17.35 | 11.00 | 1 | 17.30 | 17.24 |
|      | 2 | 17.44 | 17.29 |       | 2 | 17.59 | 17.41 |
|      | 3 | 17.47 | 17.36 |       | 3 | 17.36 | 17.44 |
| 9.30 | 1 | 17.38 | 17.35 | 11.30 | 1 | 17.31 | 17.55 |
|      | 2 | 17.09 | 17.20 |       | 2 | 17.42 | 17.64 |
|      | 3 | 17.42 | 17.40 |       | 3 | 17.44 | 17.41 |
| 10.00| 1 | 17.39 | 17.39 |       |   |       |       |
|      | 2 | 17.20 | 17.12 |       |   |       |       |
|      | 3 | 17.46 | 17.38 |       |   |       |       |

Table 14.14

| | Sample A | | | | Sample B | | | |
|---|---|---|---|---|---|---|---|---|
| | Chemist 1 | | Chemist 2 | | Chemist 1 | | Chemist 2 | |
| Lot | | | | | | | | |
| 1 | 3.4 | 3.4 | 3.6 | 3.5 | 3.7 | 3.5 | 3.1 | 3.4 |
| 2 | 4.2 | 4.1 | 4.3 | 4.2 | 4.2 | 4.2 | 4.3 | 4.2 |
| 3 | 3.5 | 3.5 | 4.2 | 4.5 | 3.4 | 3.7 | 3.9 | 4.0 |
| 4 | 3.4 | 3.3 | 3.5 | 3.1 | 4.2 | 4.2 | 3.3 | 3.1 |
| 5 | 3.2 | 2.8 | 3.1 | 2.7 | 3.0 | 3.0 | 3.2 | 2.7 |
| 6 | 0.2 | 0.7 | 0.8 | 0.7 | 0.3 | 0.4 | 0.2 | − 1.0 |
| 7 | 0.9 | 0.6 | 0.3 | 0.6 | 1.0 | 1.1 | 0.7 | 1.0 |
| 8 | 3.3 | 3.5 | 3.5 | 3.4 | 3.9 | 3.7 | 3.7 | 3.7 |
| 9 | 2.9 | 2.6 | 2.8 | 2.9 | 3.1 | 3.1 | 2.9 | 2.7 |
| 10 | 3.8 | 3.8 | 3.9 | 3.8 | 3.4 | 3.6 | 4.0 | 3.8 |
| 11 | 3.8 | 3.4 | 3.6 | 3.8 | 3.8 | 3.6 | 3.9 | 4.0 |
| 12 | 3.2 | 2.5 | 3.0 | 3.5 | 4.3 | 3.5 | 3.8 | 3.8 |
| 13 | 3.4 | 3.4 | 3.3 | 3.3 | 3.5 | 3.5 | 3.2 | 3.3 |

2. The data given in Table 14.14 were reported by Bennett (1954). They are part of the results of an experiment designed to study the metal content of a metal oxide produced by a certain process. The lots represent successive production, and two samples were drawn at random from each lot. The samples were submitted to the laboratory where two different chemists, chosen randomly on each occasion, did independent duplicate determinations. The observations recorded are the metal contents, in percentages by weight, less 80. Analyse the results and estimate the components of variance for determinations, chemists, samples and lots.

## 14.5 Nested classifications: Finite population model

When we analysed Example 14.1 we assumed in effect that, for example, the six bobbins chosen were selected from an infinite population of bobbins, so that we could assume a normal distribution for the distribution of the bobbin parameters $\alpha_i$. Clearly, there was only a finite population of bobbins, and the infinite population assumption was an approximation. In some problems the population size is so small relative to the sample size selected that the expectations of the mean squares, etc., must be worked out using results for sampling from finite populations. This section presents a finite population model for the nested-classification arrangement considered in Section 14.3. Some algebra for finite populations will be required, and this is established in Section 14.6; those prepared to assume the results given in the summary at the end of that section may proceed directly with this section. However, some may prefer to read Section 14.6 before proceeding.

We shall assume that we wish to analyse data arising from an experiment similar in design to Example 14.3, but we shall use the general notation and

Table 14.15 *Finite population nested classification*

| Population | Sample size | Population size | Notation for sample value | Index |
|---|---|---|---|---|
| Classes | $p$ | $P$ | $\alpha_i$ | $i = 1, 2, \ldots, p$ |
| Subclasses | $q$ | $Q$ | $\beta_{j(i)}$ | $j = 1, 2, \ldots, q$ |
| Errors | $r$ | Infinite, $N(0, \sigma^2)$ | $\varepsilon_{ijk}$ | $k = 1, 2, \ldots, r$ |

framework of Section 14.3, shown in Table 14.15. Let there be $p$ classes drawn randomly from a population of $P$ such classes. From each class, a random sample of $q$ subclasses is drawn from a population of $Q$ such subclasses.

The model remains (14.15), which we restate for convenience:

$$Y_{ijk} = \mu + \alpha_i + \beta_{j(i)} + \varepsilon_{ijk} \qquad (14.23)$$

However, in this case the parameters $\alpha_i$ and $\beta_{j(i)}$ are selected from finite populations. Much of Section 14.3 remains unaltered by this change of model, including the analysis-of-variance table, Table 14.8, and the expectation of the mean square 'Between determinations within subclasses', labelled $C$ in Table 14.8. The only changes are in the expectations of mean squares $A$ and $B$, and we can start deriving these from Equations (14.16) and (14.17).

Now let us define the measures of location and spread of the populations of the $\beta_{j(i)}$ for given $i$ by $\mu$ and $\tau^2$ as defined in Section 14.6, Equations (14.36) and (14.40). We shall assume that these measures are the same for all of the populations of $\beta$, and equal to $\mu_\beta$, and $\tau_\beta^2$. This last assumption is equivalent to the assumptions of common expectations and variances for the infinite population model. If we take the expectation of Equation (14.16) over the distribution of $\beta_{j(i)}$, and use Equation (14.45), we obtain

$$E_\beta E_\varepsilon \{MS(B)\} = \sigma^2 + r\tau_\beta^2 \qquad (14.24)$$

When we take expectations of Equation (14.17), the second term on the right-hand side is dealt with by an argument parallel to that used before, in Section 14.3. We see from Equation (14.41) that

$$V\{\bar{\beta}_{.(i)}\} = \frac{\tau_\beta^2}{q}\left(1 - \frac{q}{Q}\right) \qquad (14.25)$$

so that we obtain

$$E_\beta E_\varepsilon \{MS(A)\} = \sigma^2 + r(1 - q/Q)\tau_\beta^2 + \frac{rq}{(p-1)}\sum_{i=1}^{p}(\alpha_i - \alpha_.)^2$$

We now let the measures of location and spread of the $\alpha$ population be $\mu_\alpha$ and $\tau_\alpha^2$, and by taking expectations over the distribution of $\alpha$ we obtain the result shown in Table 14.16.

Table 14.16 *Expected mean squares for Table 14.8* (*finite populations*)

| Mean Square | Expected mean square |
|---|---|
| A | $\sigma^2 + r(1 - q/Q)\tau_\beta^2 + rq\tau_\alpha^2$ |
| B | $\sigma^2 + r\tau_\beta^2$ |
| C | $\sigma^2$ |

From this table it is easy to see how to get unbiased estimators of $\sigma^2, \tau_\beta^2$ and $\tau_\alpha^2$. For example, we obtain

$$\hat{\tau}_\alpha^2 = \{MS(A) - (1 - q/Q)MS(B) - (q/Q)MS(C)\}/rq \qquad (14.26)$$

Hypothesis tests are obviously approximate in the finite-population case, since the distributions cannot be normal, except for the distribution of $\varepsilon$. A hypothesis test for $\tau_\beta^2 = 0$ proceeds as in the infinite-population case. However, in order to test the hypothesis $H_0 : \tau_\alpha^2 = 0$, we must calculate the ratio

$$MS(A)/[(1 - q/Q)\{MS(B)\} + (q/Q)\{MS(C)\}] \qquad (14.27)$$

and unless $q/Q$ is either zero or unity, the denominator does not have a $\chi^2$-distribution. In order to carry out an approximate test, the following procedure, due to Welch (1936) and Satterthwaite (1946) can be used. In effect, the denominator is treated as a $\chi^2$-variable with a number of degrees of freedom $v_D$ calculated to give the same variance as the denominator of the expression (14.27). If the composite mean square is

$$MS(D) = b\{MS(B)\} + c\{MS(C)\} \qquad (14.28)$$

where $b$ and $c$ are some constants, and the degrees of freedom of the mean squares are $v_B, v_C$ respectively, then we use $v_D$ satisfying

$$\frac{\{MS(D)\}^2}{v_D} = \frac{b^2\{MS(B)\}^2}{v_B} + \frac{c^2\{MS(C)\}^2}{v_C} \qquad (14.29)$$

Therefore we use Equation (14.28) as a denominator with $v_D$ satisfying Equation (14.29), where

$$b = (1 - q/Q) \qquad c = q/Q$$
$$v_B = p(q - 1) \qquad v_C = pq(r - 1)$$

## Exercises 14.5

1. Show that if the means squares $A, B, C$ are statistically independent, then from Equation (14.26),

$$V(\hat{\tau}_\alpha^2) = \frac{1}{r^2 q^2} [V\{MS(A)\} + (1 - q/Q)^2 \, V\{MS(B)\} + (q/Q)^2 \, V\{MS(C)\}]$$

and hence work out $V(\hat{\tau}_\alpha^2)$.

2. Similarly, find the variance of $\hat{\tau}_\beta^2$.

3. Set up an investigation to determine the conditions under which an infinite population model may be safely used as a basis for inferences, when the finite population model really applies.

---

### 14.6* Sampling from finite populations

This section establishes some results for sampling from finite populations, which are required for Section 14.5. Those prepared to assume the results given in the summary at the end of the section may omit it.

Let the population be denoted $a_1, \ldots, a_N$, and the sample $A_1, \ldots, A_n$, where $0 \le n < N$. For each $i = 1, 2, \ldots, N$, we introduce the variable $\delta_i$, such that

$$\delta_i = \begin{cases} 1 & \text{if } a_i \in (A_1, \ldots, A_n) \\ 0 & \text{otherwise} \end{cases} \tag{14.30}$$

We assume that sampling is random, so that

$$\Pr(\delta_i = 1) = n/N$$

It now follows that

$$E(\delta_i) = \Pr(\delta_i = 1) = n/N \tag{14.31}$$

and

$$V(\delta_i) = E(\delta_i^2) - E^2(\delta_i)$$

but

$$E(\delta_i^2) = \Pr(\delta_i = 1) = n/N$$

so that

$$V(\delta_i) = n/N - (n/N)^2$$

$$= \frac{n}{N}(1 - n/N) \tag{14.32}$$

We shall also need the covariance term,

$$C(\delta_i, \delta_j) = E(\delta_i \delta_j) - E(\delta_i)E(\delta_j)$$

Now we have

$$E(\delta_i \delta_j) = \Pr\{(\delta_i = 1) \cap (\delta_j = 1)\}$$

$$= \frac{n}{N} \frac{(n-1)}{(N-1)} \tag{14.33}$$

Therefore the covariance is

$$C(\delta_i, \delta_j) = \frac{n(n-1)}{N(N-1)} - \frac{n^2}{N^2}$$

$$= -\frac{n(N-n)}{N^2(N-1)} \tag{14.34}$$

We can now use these results to obtain some formulae for sampling from finite populations very easily.

The sample mean is

$$\bar{A} = \sum_1^n A_i/n = \sum_1^N a_i \delta_i/n \tag{14.35}$$

so that

$$E(\bar{A}) = E\left\{\sum_1^N a_i \delta_i/n\right\}$$

$$= \sum_1^N a_i E(\delta_i)/n$$

$$= \sum_1^N a_i/N = \mu \tag{14.36}$$

where the last equation is taken as the definition of $\mu$.

The variance of the sample mean is obtained in the same way, starting from Equation (14.35).

$$V(\bar{A}) = V\left(\sum_1^N a_i \delta_i/n\right)$$

$$= \frac{1}{n^2}\left\{\sum_1^N a_i^2 V(\delta_i) + 2\sum\sum_{i<j} a_i a_j C(\delta_i, \delta_j)\right\}$$

$$= \frac{1}{n^2}\left\{\frac{n(N-n)}{N^2}\sum_1^N a_i^2 - \frac{2n(N-n)}{N^2(N-1)}\sum\sum_{i<j} a_i a_j\right\} \tag{14.37}$$

We now notice that

$$2\sum\sum_{i<j} a_i a_j = \sum\sum_{i\neq j} a_i a_j = \sum_j a_j(N\mu - a_j)$$

$$= N^2\mu^2 - \sum_1^N a_j^2 \tag{14.38}$$

If we insert this into Equation (14.37) we obtain

$$V(\bar{A}) = \frac{1}{n}\left\{\sum_1^N a_i^2\left(\frac{(N-n)}{N^2} + \frac{(N-n)}{N^2(N-1)}\right) - \frac{(N-n)\mu^2}{(N-1)}\right\}$$

$$= \frac{1}{n}\{\Sigma a_i^2 - N\mu^2\}\frac{1}{(N-1)}\left(1 - \frac{n}{N}\right) \tag{14.39}$$

If we now define a measure of the population variability to be

$$\tau^2 = (\sum a_i^2 - N\mu^2)/(N-1) \qquad (14.40)$$

then Equation (14.39) becomes

$$V(\bar{A}) = \frac{\tau^2}{n}\left(1 - \frac{n}{N}\right) \qquad (14.41)$$

and the term $(1 - n/N)$ is known as the *finite population correction*.

Finally in this section we work out the expectation of the sample variance,

$$s^2 = \sum_1^n (A_i - \bar{A})^2/(n-1) \qquad (14.42)$$

First, we write

$$\sum (A_i - \bar{A})^2 = \sum\{A_i - \mu - (\bar{A} - \mu)\}^2$$
$$= \sum (A_i - \mu)^2 - n(\bar{A} - \mu)^2$$

so that

$$E\left\{\sum_1^n (A_i - \bar{A})^2\right\} = E\left\{\sum_1^n (A_i - \mu)^2\right\} - nV(\bar{A}) \qquad (14.43)$$

The second term on the right-hand side of Equation (14.43) is (14.41), and the first term can be written

$$E\left\{\sum_1^n (A_i - \mu)^2\right\} = E\left\{\sum_1^N (a_i - \mu)^2\delta_i\right\}$$
$$= \sum_1^N (a_i - \mu)^2 n/N$$
$$= n(N-1)\tau^2/N \qquad (14.44)$$

Therefore by inserting Equations (14.41) and (14.44) into Equation (14.43) we obtain

$$E\left\{\sum_1^n (A_i - \bar{A})^2\right\} = \tau^2\left\{\frac{n(N-1)}{N} - n\frac{1}{n}\left(1 - \frac{n}{N}\right)\right\}$$
$$= \tau^2(n-1)$$

so that

$$E(s^2) = \tau^2 \qquad (14.45)$$

*Summary of results*
Given

$$\mu = \sum_1^N a_i/N$$

and

$$\tau^2 = \sum_1^N (a_i - \mu)^2/(N-1)$$

then we have

$$E(\bar{A}) = \mu$$

$$V(\bar{A}) = \frac{\tau^2}{n}\left(1 - \frac{n}{N}\right) \right\} \qquad (14.46)$$

and

$$E(s^2) = \tau^2$$

## 14.7 Nested classifications with unequal numbers

Frequently data arise in nested classifications in which the numbers in the groups are not balanced. Davies (1967) points out that a disadvantage of a balanced nested design is the poor accuracy with which the major classification component of variance is estimated. In Example 14.3 a rather large experiment involving 60 observations yielded only nine degrees of freedom between batches. In order to increase the number of degrees of freedom between batches and retain a balanced design, a very much larger experiment would be necessary. Davies suggested using unbalanced designs of the kind illustrated in Fig. 14.4.

We therefore need a modification of the previous analysis to cope with a general (unbalanced) arrangement. We shall use the infinite population model of Section 14.3.

Let our notations be as follows:

Classes $\qquad i = 1, 2, \ldots, p$

Subclasses within classes $\qquad j = 1, 2, \ldots, q_i$

Within subclasses $\qquad k = 1, 2, \ldots, r_{ij}$

$$r_{i.} = \sum_{j=1}^{q_i} r_{ij}, \qquad N = \sum_{i=1}^{p} \sum_{j=1}^{q_i} r_{ij}$$

See the preface for the 'dot' notation.

We shall now derive the least-squares solution for this more general case of a nested classification with unequal subclass numbers. There are two reasons why we do this. Firstly, it will establish in a more rigorous way

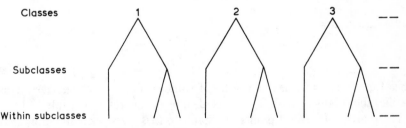

Fig. 14.4 An unbalanced nested design.

the analysis developed rather intuitively in Sections 14.3 and 14.4. Secondly, it will enable us to compare the analysis with a similar but rather different case of cross-classifications with unequal subclass numbers, which we deal with in the next chapter.

We write our model in the general form (10.4), with

$$\boldsymbol{\theta}' = \{\mu, \alpha, \ldots, \alpha_p, \beta_{1(1)}, \ldots, \beta_{q_1(1)}, \ldots, \beta_{1(p)}, \ldots, \beta_{q_p(p)}\} \tag{14.47}$$

and for this we obtain

$$\mathbf{y}'\mathbf{a} = \{y_{\ldots}, y_{1\ldots}, \ldots, y_{p\ldots}, y_{11\ldots}, \ldots, y_{pq_p\ldots}\} \tag{14.48}$$

Following the method of Section 10.8 it is easier to solve the normal equations directly than develop an inverse. We therefore write the sum of squares

$$S = \sum_{i=1}^{p} \sum_{j=1}^{q_i} \sum_{k=1}^{r_{ij}} \{y_{ijk} - \mu - \alpha_i - \beta_{j(i)}\}^2$$

from which the following normal equations are obtained:

$$y_{\ldots} = N\mu + \sum_{i=1}^{p} r_{i\cdot}\alpha_i + \sum_{i=1}^{p} \sum_{j=1}^{q_i} r_{ij}\beta_{j(i)} \tag{14.49}$$

$$y_{i\cdot\cdot} = r_{i\cdot}\mu + r_{i\cdot}\alpha_i + \sum_{j=1}^{q_i} r_{ij}\beta_{j(i)} \tag{14.50}$$

$$y_{ij\cdot} = (\mu + \alpha_i + \beta_{j(i)})r_{ij} \tag{14.51}$$

A solution to these equations is provided by

$$\hat{\mu} = \bar{y}_{\ldots}, \qquad \hat{\alpha}_i = (\bar{y}_{i\cdot\cdot} - \bar{y}_{\ldots}), \qquad \hat{\beta}_{j(i)} = (\bar{y}_{ij\cdot} - \bar{y}_{i\cdot\cdot}) \tag{14.52}$$

where

$$\bar{y}_{\ldots} = y_{\ldots}/N, \bar{y}_{i\cdot\cdot} = y_{i\cdot\cdot}/r_{i\cdot}, \text{ and } \bar{y}_{ij\cdot} = y_{ij\cdot}/r_{ij}.$$

If we insert Equations (14.48) and (14.52) into (10.30) we obtain

$$S_{\min} = \sum_{i=1}^{p} \sum_{j=1}^{q_i} \sum_{k=1}^{r_{ij}} y_{ijk}^2 - N(\bar{y}_{\ldots})^2 - \sum_{i=1}^{p} r_{i\cdot}(\bar{y}_{i\cdot\cdot} - \bar{y}_{\ldots})^2$$

$$- \sum_{i=1}^{p} \sum_{j=1}^{q_i} r_{ij}(\bar{y}_{ij\cdot} - \bar{y}_{i\cdot\cdot})^2 \tag{14.53}$$

and we see that

$$S_{\min} = \sum_{i=1}^{p} \sum_{j=1}^{q_i} \sum_{k=1}^{r_{ij}} (y_{ijk} - \bar{y}_{ij\cdot})^2$$

since Equation (14.52) leads to a fitted value of $\bar{y}_{ij\cdot}$ for each subclass. This justifies the subdivision of the corrected sums of squares as in Table 14.17.

There are now several points to establish. Firstly, we need to show that the mean squares $A$, $B$ and $C$ are statistically independent. One way of doing

Table 14.17 *Nested analysis of variance: Unequal subclass case*

| Source | CSS | d. f. | MS |
|---|---|---|---|
| Between classes | $\sum_{i=1}^{p} r_{i.}(\bar{y}_{i..} - \bar{y}_{...})^2$ | $(p-1)$ | A |
| Between subclasses within classes | $\sum_{i=1}^{p}\sum_{j=1}^{q_i} r_{ij}(\bar{y}_{ij.} - \bar{y}_{i..})^2$ | $\sum_{i=1}^{p}(q_i - 1)$ | B |
| Within subclasses | $\sum_{i=1}^{p}\sum_{j=1}^{q_i}\sum_{k=1}^{r_{ij}} (\bar{y}_{ijk} - \bar{y}_{ij.})^2$ | $\sum_{i=1}^{p}\sum_{j=1}^{q_i}(r_{ij} - 1)$ | C |
| Total | $\sum\sum\sum(y_{ijk} - \bar{y}_{...})^2$ | $(N-1)$ | |

this is to show that the separate terms $(\bar{y}_{i..} - \bar{y}_{...}), (\bar{y}_{ij.} - \bar{y}_{i..})$ and $(y_{ijk} - \bar{y}_{ij.})$ are uncorrelated. This is left as an exercise (see Exercise 14.7.1).

If now we fit the model (14.15) but with all $\beta_{j(i)} = 0$, then the sum of squares and degrees of freedom due to the fitted model are as in line $A$. If we fit the full model (14.15), the sum of squares due to fitting the model is the sum of the corrected sums of squares in lines $A$ and $B$, with degrees of freedom $(\sum q_i - 1)$. If we subtract the corrected sum of squares and degrees of freedom for the restricted model (line $A$), we obtain line $B$. This procedure shows that it is the mean square $B$ which is used for testing the hypothesis $\sigma_\beta^2 = 0$. The number of degrees of freedom for 'Within subclasses' also follows, since the total is $(N-1)$.

So far as the expectations of the mean squares is concerned, it is easy to check that

$$E\{MS(C)\} = \sigma^2 \qquad (14.54)$$

but the results for mean squares $A$ and $B$ are rather more difficult to obtain. The results for the infinite population model are given below; for a derivation see Kempthorne (1975).

$$E\{MS(B)\} = \sigma^2 + \left(N - \sum_{i=1}^{p}\left(\sum_{j=1}^{q_i} r_{ij}^2/r_{i.}\right)\right)\sigma_\beta^2/(\sum q_i - p) \qquad (14.55)$$

$$E\{MS(C)\} = \sigma^2 + \left(\sum_{i=1}^{p}\left(\sum_{j=1}^{q_i} r_{ij}^2/r_{i.}\right) - \sum_{i=1}^{p}\sum_{j=1}^{q_i} r_{ij}^2/N\right)\sigma_\beta^2/(p-1)$$

$$+ \left(N^2 - \sum_{i=1}^{p} r_{i.}^2\right)\sigma_\alpha^2/\{N(p-1)\} \qquad (14.56)$$

For approximate confidence intervals for this case see Davies, (1967).

**Exercise 14.7**
1. Check that the sums of squares of Table 14.17 are statistically independent by the method indicated underneath the table.

2. How would you set about designing a nested classification with unequal numbers?

3. The data given in Example 5.1, and analysed in Chapters 5 and 6, have a nested structure, so there are only 10 crudes and not 32. It is therefore possible to define components of variance between and within crudes. Reanalyse the data.

**Further reading**
A good but advanced treatment of all of the topics in this chapter can be found in either Kempthorne (1975) or Searle (1971). Bennett and Franklin (1954) contains a wealth of practical detail and numerical examples, and is at a slightly lower mathematical level. An extremely good coverage of most of the topics from a practical point of view is given in Snedecor and Cochran (1967).

# CHAPTER FIFTEEN

# Crossed classifications

## 15.1 Crossed classifications and interactions

In this chapter we shall be discussing the analysis of variance for crossed classifications. The following example will be used to illustrate the methods used.

*Example* 15.1    The data given in Table 15.1 were reported by Bennett and Franklin (1954) and we shall base the present discussion on the following description of the experiment (which may differ slightly from the actual conditions). Duplicate determinations were made of the total solid content of wet brewers' yeast, for ten different samples of yeast, using three drying periods: 3 hours, 6 hours and 9 hours. The ten samples were not selected randomly, but they were chosen from different batches of yeast in order to cover the range of variation which naturally occurs. Each sample of yeast so selected was divided into six portions which were randomly allocated to the duplicate determinations for the three drying periods. The 60 observations were carried out in random order.    □ □ □

First of all we introduce some notation. The word 'treatment', as noted before in 13.6, has the important implication that it is under the control of the experimenter, and can be allocated to different experimental units.

Table 15.1 *Percentage total solid content of wet brewers yeast (less 10%)*

| Drying period | Sample | | | | | | | | | |
|---|---|---|---|---|---|---|---|---|---|---|
| | 1 | 2 | 3 | 4 | 5 | 6 | 7 | 8 | 9 | 10 |
| 3 hrs | 3.24 | 3.92 | 9.13 | 8.35 | 5.51 | 6.63 | 9.29 | 8.76 | 8.03 | 6.61 |
| | 3.56 | 3.86 | 9.23 | 8.29 | 5.53 | 6.65 | 9.28 | 8.72 | 8.11 | 6.77 |
| 6 hrs | 3.16 | 3.81 | 8.86 | 8.11 | 5.06 | 6.61 | 8.96 | 8.39 | 7.86 | 6.32 |
| | 3.26 | 3.80 | 8.79 | 8.24 | 5.11 | 6.57 | 9.12 | 8.43 | 7.84 | 6.23 |
| 9 hrs | 2.96 | 3.76 | 8.70 | 7.94 | 4.84 | 6.60 | 8.84 | 8.23 | 7.72 | 6.21 |
| | 3.01 | 3.75 | 8.75 | 7.99 | 4.80 | 6.55 | 9.03 | 8.27 | 7.79 | 6.13 |

In Example 15.1, 'drying periods' is a treatment but 'samples' is not. However, we shall use the term 'factor' to cover treatments and also other relevant methods by which the data are classified, such as 'samples' in Example 15.1. Each factor consists of a number of 'levels' or 'states' and in Example 15.1 these are the length of the drying period (for the drying-periods factor), and the ten different samples (for the samples factor).

In Example 15.1 there are two factors by which we can classify the observations, drying period and sample number, and the data arise only at combinations of the levels of these factors. Each observation is the result of effects due to the drying period, the sample and random variation. One of the chief models used is a simple additive model, of the type already discussed in Chapters 10 and 11 (see below for details). Example 15.1 is called a *crossed classification*, and it differs markedly from the nested classifications discussed in Chapter 14. The crossed-classification concept is illustrated for Example 15.1 in Fig. 15.1. Observations are made at the intersections of the lines, and in this example duplicate determinations were made at each point. The distinction between this and a nested classification is made clear by comparing this with Fig. 14.1 for Example 14.3.

In that experiment there was nothing especially in common between, say, cask 1 of batch 1, and cask 1 of batch 2; the casks were sampled from populations of casks within each batch. The mathematical model used is corres-

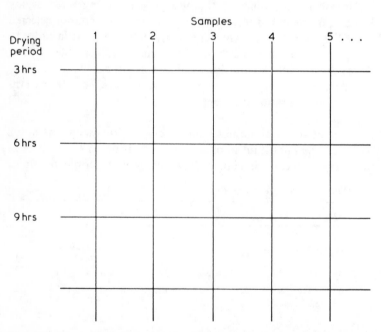

Fig. 15.1 Illustration of crossed classification using Example 15.1.

pondingly different, as in Example 14.3 it would not be appropriate to think of there being cask 1, cask 2 and cask 3 effects which act in a similar way in all batches.

Returning now to cross-classified data, there may be in general several factors which we can label $A, B, C, \ldots$, and with each factor having several 'levels' or 'states' $A_1, A_2, \ldots$, and $B_1, B_2, \ldots$, etc. We assume that observations are taken at intersections of these factors and levels, and in this chapter we assume that the observations are replicated, possibly unequally. For the present we concentrate on two-way classifications, and the factors are labelled $A$ and $B$.

A two-way classification was analysed in Section 10.7, and there we used a simple model in which the expectation of an observation involving $A_i$ and $B_j$ was written in the form

$$E(Y_{ij}) = \text{Constant} + \left(\begin{array}{c}\text{Component due}\\ \text{to} \quad A_i\end{array}\right) + \left(\begin{array}{c}\text{Component due}\\ \text{to} \quad B_j\end{array}\right) \quad (15.1)$$

This expression assumes that the effect of an $A$-classification on an observation is an additive component which is not affected by which of the $B$-classifications also occurs, and similarly for the effect of a $B$-classification. In Chapter 11 we explained that when this simple additive model does not apply, we say that an interaction is present. The terms *main effect* and *interaction* are used frequently in the analysis of cross-classifications, and it will be useful to have some definitions. In these definitions we shall assume that there are an equal number of observations at each combination of the $A$ and $B$ classifications.

*Definition*   The *main effect* of $A$ is defined by

$$ME(A) = \{ME(A_i)\} = \{E(\bar{Y}_{1.} - \bar{Y}_{..}), E(\bar{Y}_{2.} - \bar{Y}_{..}), \ldots\}$$

in standard notation, so that $\bar{Y}_{i.}$ is the mean of all observations involving $A_i$, and $\bar{Y}_{..}$ is the overall mean.   □ □ □

Other main effects can be defined similarly. In this definition we have subtracted the overall mean throughout, and this is admittedly arbitrary, but convenient for our definition of an interaction. The chief point is that $ME(A)$ specifies the relative effect of all of the $A$-levels or states.

*Definition*   An *interaction* between two factors $A$ and $B$ exists if, for any $i, j$,

$$E(Y_{ij}) \neq E(\bar{Y}_{..}) + ME(A_i) + ME(B_j)$$

and we say that there is an $A \times B$ interaction.   □ □ □

If there are other factors, $C, D, \ldots$, then this definition applies either (*a*) by averaging over the other factors and levels, or (b) by speaking of the $A \times B$ interaction at $C_1$, etc. In the latter case we have a component of the $A \times B \times C$ interaction, and we shall discuss this further below.

If an interaction exists, there will not normally be a meaning to the relevant main effects. Finney (1948) put the position well: 'A common situation is that in which the "levels" of factor $B$ represent different methods of performing some operation. If the effect of $A$ depends on which method is chosen, no type of average $A$ effect can be of much interest unless the $q$ methods tested in the experiment either are the only possibilities or are in some way representative of all possible methods. An $A$ effect averaged over an entirely arbitrary selection of levels of $B$ may give a formal completeness to a summary table of results, but is then meaningless or even misleading: the results should instead be interpreted in terms of the effect of $A$ for each $B_j$ separately.' As we shall see later, this has an important effect on the models we consider. Some common forms of interaction are described in the next section.

Let us now return to Example 15.1 and examine Tables 15.2 and 15.3. A glance at the results shows that, for example, the 'percentage total solid content' measurement tends to reduce considerably as the 'drying period' increases. However, samples 2 and 6 do not follow the same pattern as the others, as there is very little change in their results over the three drying periods. When we come to do the analysis we shall see that this feature shows up as a Sample × Drying period interaction.

Finally, we remark that in Section 10.7 the residual sum of squares was estimated from components which we now recognize could contain inter-

Table 15.2 *Totals over duplicates for Example* 15.1

| Drying period | Sample | | | | | | | | | |
|---|---|---|---|---|---|---|---|---|---|---|
| | 1 | 2 | 3 | 4 | 5 | 6 | 7 | 8 | 9 | 10 |
| 3 hrs | 6.80 | 7.78 | 18.36 | 16.64 | 11.04 | 13.28 | 18.57 | 17.48 | 16.14 | 13.38 |
| 6 hrs | 6.42 | 7.61 | 17.65 | 16.35 | 10.17 | 13.18 | 18.08 | 16.82 | 15.70 | 12.55 |
| 9 hrs | 5.97 | 7.51 | 17.45 | 15.93 | 9.64 | 13.15 | 17.87 | 16.50 | 15.51 | 12.34 |

Table 15.3 *Difference of duplicates for Example* 15.1 ($\times$ 100)

| Dryin period | Sample | | | | | | | | | |
|---|---|---|---|---|---|---|---|---|---|---|
| | 1 | 2 | 3 | 4 | 5 | 6 | 7 | 8 | 9 | 10 |
| 3 hrs | 32 | −6 | 10 | −6 | 2 | 2 | −1 | −4 | 8 | 16 |
| 6 hrs | 10 | −1 | −7 | 13 | 5 | −4 | 16 | 4 | −2 | −9 |
| 9 hrs | 5 | −1 | 5 | 5 | −4 | −5 | 19 | 4 | 7 | −8 |

action. In order to separate interaction from error we must have replication, as in Example 15.1, so that there is an independent estimate of error. In Example 15.1 the differences between duplicates shown in Table 15.3 can be used to yield an estimate of error which we can then use in assessing the significance of differences between cells (see below for the details). A statistical analysis of such data usually proceeds on the assumption that the differences between replicated observations within a cell are caused by an error which is independently and normally distributed, with constant variance from cell to cell. Clearly, some checks can be made on how consistent the data are with this assumption. Also, for the error term to be valid there must be some form of randomization, and in Example 15.1 all 60 observations should be carried out in random order.

In this discussion of crossed classifications we have been thinking primarily of a fixed-effects model. However, if, say, the levels of the $A$ factor, $A_1, A_2, \ldots,$ were randomly selected from a population of $A_i$, then this factor is to be thought of in terms of a random-effects model. In this situation, although we can define a main effect for $A$, there is very little meaning or importance to be attached to it, and greater importance is attached to estimating the mean and variance of the effects of the population of the $A_i$. The best procedure is probably to treat the factors as fixed effects to the point where we have completed the analysis-of-variance table, and carried out checks that a random-effects model is reasonable. If we are satisfied about this, we can estimate components of variance, etc., as in the previous chapter. In a $k$-way classification, then, provided there are no interactions, any or all of the factors can be treated in this way as random effects. If there are interactions, problems arise, and we discuss interactions further in the next section.

## 15.2 More about interactions

There are many possible forms of interaction, and Fig. 15.2 shows some of the possibilities for two treatments, with one at three levels and the other at four. Fig. 15.2(a) shows the case of no interaction, and a suitable statistical model has the form

$$\left. \begin{array}{l} E(Y_{ij}) = \mu + \alpha_i + \beta_j \\ V(Y_{ij}) = \sigma^2 \end{array} \right\} \qquad \begin{array}{l} i = 1, 2, 3, 4 \\ j = 1, 2, 3 \end{array} \qquad (15.2)$$

which should be compared with Equation (15.1). This 'parallel lines' picture will occur whenever there is no interaction. In contrast, Fig. 15.2(b) shows a very strong interaction, and there is little we can do except present the separate means.

In Fig. 15.2(c) the interaction is not so drastic as in Fig. 15.2(b), and in Fig. 15.2(d) the interaction affects only one level of $A$. In neither case is it satisfactory to define main effects, in general, but one may be able to make

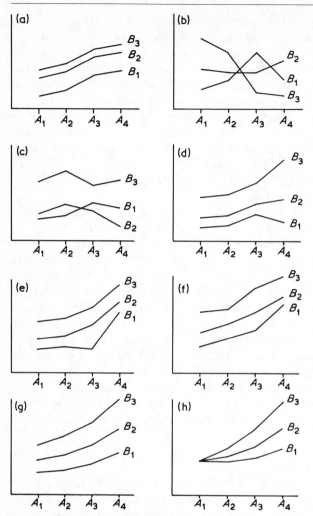

Fig. 15.2 Illustrations of possible interactions.

some cautious use of multiple comparison procedures, for example, to show that $B_3$ gave significantly higher results.

Figure 15.2($e$) shows the sort of results obtained when we have an outlying observation, and an immediate search should be made for one in this case. In Fig. 15.2($f$) the lines are nearly parallel, but seem to be perturbed by some other source of error, and there are several ways in which this might arise. For example, there could be an interaction between the treatments, which though statistically significant has a small effect. In such a case we might, as an approximation, ignore the interaction and summarize the results in terms of main effects.

Another situation in which Fig. 15.2($f$) can arise is in experiments of the following type. We have a set of experimental units, such as plots of land, and there is a two-way system of treatments imposed upon the units ($A$-treatments and $B$-treatments), with replication of the observations within each experimental unit. A suitable model is then sometimes

$$E(Y_{ij}) = \mu + \alpha_i + \beta_j + e_{ij} \tag{15.3}$$

where the expectation is taken over the distribution of errors within a unit. The terms $e_{ij}$ represent unit errors, and might arise because of the position of the unit. For example, in certain animal experiments, animals are grouped into pens, which form the basic experimental units. Here unit errors might arise due to draughts, small variations in temperature, etc. If the treatments related to are feeds, these unit errors are quite different from interactions, although they might affect the experimental results in a similar way. We shall discuss this situation further below.

Sometimes patterns such as that shown in Fig. 15.2($g$) can be simplified by transforming the scale of the observations. For example, a transformation such as $\sqrt{y}$ or $\log_e y$ might remove or reduce the interaction shown. However, the use of such a transformation may not be consistent with other assumptions made in the analysis (see the discussion in Chapter 8).

Another method of dealing with patterns such as those in Fig. 15.2($g$) and ($h$) is to use a simple mathematical model. For example, the results in Fig. 15.2($g$) might be fitted well by the model

$$E(Y_{ij}) = \mu + \alpha_i + \beta_j + \gamma \alpha_i \beta_j \tag{15.4}$$

In the model (15.4), one extra parameter $\gamma$ has been fitted as compared with the model (15.3) and when the experimental results are layed out as in Fig. 15.2($g$) this extra parameter might account for the interaction. This is similar to Tukey's one degree of freedom for non-additivity (Tukey, 1949; Scheffé, 1959), and it can be used when there is some reason to believe that the treatment had a multiplicative effect.

If in any experiment a model can be chosen for a possible interaction on *a priori* grounds, leading to an extra term or terms as in Equation (15.4), then our analysis of variance is easily modified to allow for (and test) these. Apart from models of this type, the discussion above shows that in general there is no meaning to a main effect when interactions are present.

A very good book on the effect and interpretation of interactions is Daniel (1976). Daniel states that the most common form of interaction is that due to a single cell or due to a single level of one treatment. There is also an excellent section on interactions in Cox (1958).

### 15.3 Analysis of a two-way equally replicated design

We now consider the analysis of variance for a set of data such as those in Example 15.1. In general let there be an $A$-classification represented by $r$

rows, a $B$-classification represented by $c$ columns, and suppose that there are $n$ replicated observations for each of the $rc$ cells. In all there are $N = rcn$ observations, and it is assumed that these are carried out in random order.

In writing down models for the data we shall assume that the observations $Y_{ijk}$ ($i = 1, 2, \ldots, r$; $j = 1, 2, \ldots, c$; $k = 1, 2, \ldots, n$) all have constant variance $\sigma^2$. The simplest model is to put

$$E(Y_{ijk}) = \mu + \phi_{ij}$$

$$V(Y_{ijk}) = \sigma^2 \tag{15.5}$$

and if we write this in the form $E(\mathbf{Y}) = \mathbf{a}\,\boldsymbol{\theta}$, we have

$$\boldsymbol{\theta}' = \{\mu, \phi_{11}, \ldots, \phi_{rc}\}$$

This is in effect the case discussed in Section 10.4, so that from Equation (10.44) the least squares estimator of $\boldsymbol{\theta}$ is

$$\hat{\boldsymbol{\theta}}' = \{\bar{Y}_{\ldots} + \delta, \ \bar{Y}_{11.} - \delta, \ \bar{Y}_{12.} - \delta, \ldots, \ \bar{Y}_{rc.} - \delta\} \tag{15.6}$$

where $\delta$ is an arbitrary constant. Again, following the theory of Section 10.4, the analysis-of-variance table for fitting the model (15.5) is that given in Table 15.4. It may be helpful here to restate the argument used in Chapter 10, leading to the analysis-of-variance table, in terms of this example. The bottom line of Table 15.4 gives the minimized sum of squares for fitting the model

$$E(Y_{ijk}) = \mu$$

and since one parameter ($\mu$) is fitted it has ($rcn - 1$) degrees of freedom. The 'Within cells' line is the minimized sum of squares for fitting the model (15.5), and it has $rc(n - 1)$ degrees of freedom since there are $rcn$ observations and $rc$ fitted parameters (one for each cell). The 'Between cells' corrected sum of squares is the difference between the sums of squares for the two models, one fitting only $\mu$ and the other fitting $\{\mu + \phi_{ij}\}$, and it therefore gives a sum of squares due to the $\phi_{ij}$. The numbers of degrees of freedom can also be justified from considerations of rank, as given in Section 10.4.

A test of the hypothesis that there are no cell effects at all is given by

Table 15.4 *Analysis of variance for a crossed classification using model (15.5)*

| Source | CSS | d. f. | Mean square |
|---|---|---|---|
| Between cells | $n \sum \sum (\bar{y}_{ij.} - \bar{y}_{\ldots})^2$ | $rc - 1$ | $A$ |
| Within cells | $\sum \sum \sum (y_{ijk} - \bar{y}_{ij.})^2$ | $rc(n - 1)$ | $B$ |
| Total | $\sum \sum \sum (y_{ijk} - \bar{y}_{\ldots})^2$ | $rcn - 1$ | |

calculating the ratio $A/B$ and referring to $F$-tables on $(rc - 1, rc(n - 1))$ degrees of freedom. There is rarely any interest in this test, and it is not a very powerful one if there is a two-way treatment structure.

The next model to try is one in which the two treatments have a simple additive effect with no interaction, and this can be written

$$E(Y_{ijk}) = \mu + \alpha_i + \beta_j \qquad (15.7)$$

The least-squares estimates and analysis of variance for this model follow from the two-way analysis discussed in Section 10.6, and we arrive at Table 15.5. We have not completed the 'Mean square' column here as there is no appropriate $F$-test, since the 'Residual sum of squares' is not a valid error term. It contains the 'Within cells' variation, and any interaction between the treatments. Table 15.5 is merely a convenient way of showing the calculations to get the residual sum of squares.

We are now in a position to develop a test of the hypothesis that there is no interaction. The 'Within cells' sum of squares of Table 15.4 was obtained by fitting model (15.5), and because this contains a parameter for each cell, it allows for the possible presence of an interaction between row and column factors. We have already seen that the residual sum of squares of Table 15.5 contains any interaction effects. By substraction we see that the sum of squares for interaction is

Residual sum of squares (Table 15.5) — Within-cells CSS (Table 15.4)

$$= n\sum\sum(\bar{y}_{ij.} - \bar{y}_{...})^2 - cn\sum(\bar{y}_{i..} - \bar{y}_{...})^2 - rn\sum(\bar{y}_{.j.} - \bar{y}_{...})^2$$

$$= n\sum\sum(\bar{y}_{ij.} - \bar{y}_{i..} - \bar{y}_{.j.} + \bar{y}_{...})^2 \qquad (15.8)$$

The number of degrees of freedom for this is

$$rcn - r - c + 1 - rc(n - 1) = (r - 1)(c - 1) \qquad (15.9)$$

This seems to be a long way round to get the interaction sum of squares, but it is important to see what we have done. We have fitted two quite distinct models, one which allows an interaction and one which does not; the interaction term is obtained by subtracting the residual sums of squares.

Table 15.5 *Analysis of variance for two-way case using model (15.7)*

| Source | CSS | d.f. |
|---|---|---|
| Due to row effects | $cn\sum(\bar{y}_{i..} - \bar{y}_{...})^2$ | $r - 1$ |
| Due to column effects | $rn\sum(\bar{y}_{.j.} - \bar{y}_{...})^2$ | $c - 1$ |
| Residual sum of squares | (By subtraction) | $rcn - r - c + 1$ |
| Total | $\sum\sum\sum(y_{ijk} - \bar{y}_{...})^2$ | $rcn - 1$ |

This method can be applied when the replication within cells is not equal, but in that case the interaction sum of squares does not simplify to an expression like (15.8). However, for the equally replicated case, the analysis of variance is set out in Table 15.6, and we see that in effect we have separated the mean square $A$ of Table 15.4 into three components $C$, $D$ and $E$.

The expected values of the mean squares are set out in Table 15.7, and this table requires some explanation. The model (15.5) has been used to obtain $E\{MS(E)\}$, but the model (15.7) has been used to obtain $E\{MS(C)\}$, and $E\{MS(D)\}$. That is, we have used *different and alternative models in these cases*. The expected value for $B$ applies for either model, and provides the estimate of error to test the other effects.

Therefore the $F$-tests to be carried out are as follows. First we test the hypothesis of no interaction by calculating the ratio $E/B$. If this is not significant we can proceed to test the row and column effects by calculating $C/B$ and $D/B$. (The sums of squares for 'interaction' and 'Within cells' should not be combined even if $E$ is not significant.) If the interaction term is significant we should make an immediate search for the form of this interaction, by using tables or graphs, and note the comments of Section 15.2 that in general, main effects are meaningless when there is an interaction.

If we have a fixed-effects model, then $F$-tests of row and column effects would be followed by an examination of treatment comparisons such as by

Table 15.6 *Analysis of variance for two-way replicated classification*

| Source | CSS | d.f. | Mean square |
|---|---|---|---|
| Due to row effects | $cn\sum(\bar{y}_{i..} - \bar{y}_{...})^2$ | $r-1$ | $C$ |
| Due to column effects | $rn\sum(\bar{y}_{.j.} - \bar{y}_{...})^2$ | $c-1$ | $D$ |
| Due to interaction | $n\sum\sum(\bar{y}_{ij.} - \bar{y}_{i..} - \bar{y}_{.j.} + \bar{y}_{...})^2$ | $(r-1)(c-1)$ | $E$ |
| Between cells | $n\sum\sum(\bar{y}_{ij.} - \bar{y}_{...})^2$ | $rc-1$ | $A$ |
| Within cells | $\sum\sum\sum(y_{ijk} - \bar{y}_{ij.})^2$ | $rc(n-1)$ | $B$ |
| Total | $\sum\sum\sum(y_{ijk} - \bar{y}_{...})^2$ | $rcn-1$ | |

Table 15.7 *Expected value of the mean squares in Table 15.6*

| Mean square | Expected value |
|---|---|
| $B$ | $\sigma^2$ |
| $C$ | $\sigma^2 + cn\sum(\alpha_i - \bar{\alpha}_.)^2/(r-1)$ |
| $D$ | $\sigma^2 + rn\sum(\beta_j - \bar{\beta}_.)^2/(c-1)$ |
| $E$ | $\sigma^2 + n\sum\sum(\phi_{ij} - \bar{\phi}_{i.} - \bar{\phi}_{.j} + \bar{\phi}_{..})^2/\{(c-1)(r-1)\}$ |

the methods of Chapter 13. If either or both of the row and column effects are thought of as random rather than as a fixed effect, then estimation of components of variance follows.

*Computing formulae*
The algebraic form of the analysis-of-variance table is given in Table 15.6, and the usual computing formulae are as follows.

*Total sum of squares*

$$\sum\sum\sum(y_{ijk} - \bar{y}_{...})^2 = \sum\sum\sum y_{ijk}^2 - \text{(correction)}$$

where

$$\text{correction} = (\sum\sum\sum y_{ijk})^2/rcn$$

*Row sum of squares*

$$\sum_i (y_{i..} - \bar{y}_{...})^2 = \frac{1}{cn}\sum_i\left(\sum_j\sum_k y_{ijk}\right)^2 - \text{(correction)}$$

*Column sum of squares*

$$\sum_j (\bar{y}_{.j.} - \bar{y}_{...})^2 = \frac{1}{rn}\sum_j\left(\sum_i\sum_k y_{ijk}\right)^2 - \text{(correction)}$$

*Between cells*

$$n\sum\sum(\bar{y}_{ij.} - \bar{y}_{...})^2 = \frac{1}{n}\sum\sum\left(\sum_k y_{ijk}\right)^2 - \text{(correction)}$$

The 'Interaction' and 'Within cells' sums of squares are then filled in by subtraction. The 'Between cells' sum of squares calculation involves making a two-way table of cell totals, and this will be useful in interpreting any interaction, as we found earlier in the chapter.

---

**Exercises 15.3**
1. Check the values of the expected mean squares given in Table 15.7.

2. Suggest why it is not correct to combine the CSS for 'Interaction' and 'Within cells' in Table 15.6 even if the mean square $E$ is not significant.

---

**15.4 An analysis of Example 15.1**
In any analysis of Example 15.1 data, an estimate of the error variance uses the cell differences shown in Table 15.3, and we shall start the analysis by looking at these differences more closely. Figure 15.3 shows a normal plot

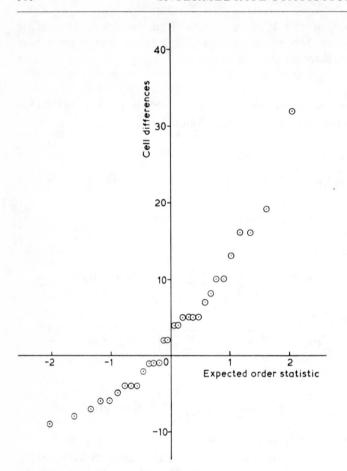

Fig. 15.3 Normal plot of cell differences.

of the differences and two features stand out. Firstly, the distribution of the differences appears to be slightly skew and, secondly, there is one very large difference. In fact, it is possible to show by sampling experiments that normal plots often show skewness as strongly as in Fig. 15.3 even when no skewness is present. Nevertheless, there seems to be something to look into here, and in a live case one would want to check back with the experimenters. If the experimental procedures were properly carried out, including the appropriate randomizations, the distribution of the differences should be approximately symmetrical. As we are dealing with historical data, all we can do is to make the point and proceed.

The one large difference stands out markedly. The average of the differences is 3.5, and the standard deviation just over 9, so that the difference of 32 is more than three standard deviations from the mean (and even more from zero). Again, in a live experiment we would want to check back with the experimenters to see if there was any mis-recording of the data in the appropriate cell, or whether something went unexpectedly wrong with the conditions of the experiment. If, after checking back with the experimenters, some cause is found for the aberrant value, we may decide that it would be better to delete the offending observation before analysis. In the absence of finding any such cause, it is very dangerous to go ahead deleting observations you simply don't like. We shall proceed with the value included but we shall discuss it again later.

The algebraic form of the analysis-of-variance table for Example 15.1 is given in Table 15.6, and the method of calculation is given at the end of Section 15.3. The calculations result in Table 15.8.

One of the important results of any analysis-of-variance table of this sort is the estimate of the error variance, and in Table 15.8 we use the 'Within cells' line for this. Usually we calculate this sum of squares, by subtracting all the other sums of squares from the total. However, there is a simple direct method which can be used whenever there are two observations per cell. Let the observations in cell $(i, j)$ be denoted $y_{ij1}, y_{ij2}$, then the contribution to the within-cells sum of squares is

$$(y_{ij1} - \bar{y}_{ij.})^2 + (y_{ij2} - \bar{y}_{ij.})^2$$

which is easily seen to be $(y_{ij1} - y_{ij2})^2/2$. Therefore the within-cells corrected sum of squares is the sum of these terms, or

$$(\tfrac{1}{2}) \times 10^{-4} \times \text{(Sum of squares of terms in Table 15.3)}$$
$$= (\tfrac{1}{2}) \times 10^{-4} \times 2865$$
$$= 0.1433$$

which checks with the above calculations. This calculation also shows that the one large difference referred to above accounts for

$$\tfrac{1}{2} \times 10^{-4} \times 32^2 = 0.0512$$

of the amount 0.1433 shown in Table 15.8 for 'Within cells'. If the cell contain-

Table 15.8 *Analysis of variance of Example 15.1 data*

| Source | CSS | d. f. | Mean square |
|---|---|---|---|
| Due to samples | 241.8531 | 9 | 26.9114 |
| Due to drying periods | 1.4873 | 2 | 0.7436 |
| Due to interaction | 0.3500 | 18 | 0.0194 |
| Between cells | 243.6904 | 29 | — |
| Within cells | 0.1433 | 30 | 0.0047 |
| Total | 243.8337 | 59 | |

ing the large difference is deleted, the estimate of the error variance is reduced to 0.0032. This is rather a dramatic reduction, and it demonstrates the importance of knowing whether to include or exclude wild observations – especially when estimating variances. (One aberrant value in thirty would have to be wildly out before it made much difference to a mean.)

One outstanding feature of Table 15.8 is that the 'row effect' accounts for a very large percentage of the total sum of squares (241.85 out of 243.83). This indicates large differences between samples, and this feature can be seen in the table of cell totals, Table 15.2.

The next step is to carry out a significance test for the presence of an interaction between samples and drying periods. We calculate the ratio

$$0.0194/0.0047 = 4.13$$

and refer to $F(18, 30)$ tables. We see that the observed result is very highly significant (beyond the 0.1 per cent level), showing very strong evidence of the presence of an interaction. We must, therefore, explore the reason for this interaction, and one way of doing this would be to draw graphs similar to those in Fig. 15.2. However, with 10 samples the graphs are likely to get rather tangled, and it might not be easy to make a simple interpretation. Another method of investigating interactions is to tabulate the array of interaction terms $(\bar{y}_{ij.} - \bar{y}_{i..} - \bar{y}_{.j.} + \bar{y}_{...})$ as in Table 15.9. If there is no interaction, these terms represent error, and with $r$ rows, $c$ columns and $n$ observations per cell, we have

$$r_{ij} = \bar{y}_{ij.} - \bar{y}_{i..} - \bar{y}_{.j.} + \bar{y}_{...}$$

$$V(r_{ij}) = (r-1)(c-1)\sigma^2/rcn \qquad (15.10)$$

but the correlations between these residuals can be large (see Exercise 15.4.1), so that we must take care when attempting to draw conclusions. From the way the residuals are defined, it is clear that they add to zero over each row and column.

In order to examine this table we need an estimate of the standard error of the terms, which we obtain by using Equation (15.10) and the estimate of $\sigma^2$ from Table 15.8. This yields a value

$$\text{Standard error} = \sqrt{\{2 \times 9 \times 0.0047/3 \times 10 \times 2\}} = 0.0376$$

and so we need to use the value 376 in examining the values in Table 15.9. Columns 2 and 6 stand out in the way we had expected, and each contain values equal to several standard errors. It is possible that an investigation at the time might have given adequate grounds for deleting both of these samples from the analysis. However, Table 15.9 shows one further feature which has not been noted so far. Sample 5 also shows very large interaction terms, and the effect is the opposite of the effect in samples 2 and 6. In sample 5

Table 15.9 *Interaction terms for Example* 15.1 *data* ($\times 10^4$)

| Drying period | Sample | | | | | | | | | |
|---|---|---|---|---|---|---|---|---|---|---|
| | 1 | 2 | 3 | 4 | 5 | 6 | 7 | 8 | 9 | 10 |
| 3 hrs | −75 | −1355 | 610 | −425 | 1695 | −1705 | −105 | 660 | −305 | 1030 |
| 6 hrs | 500 | 265 | −470 | 595 | −185 | 265 | −85 | −185 | −35 | −655 |
| 9 hrs | −425 | 1095 | −140 | −185 | −1505 | 1445 | 195 | −455 | 345 | −370 |

Table 15.10 *Analysis of variance of Example* 15.1 *data (deleting samples 2 and 6)*

| Source | CSS | d. f. | Mean square |
|---|---|---|---|
| Due to samples | 182.2985 | 7 | 26.0426 |
| Due to drying periods | 1.6677 | 2 | 0.8338 |
| Due to interaction | 0.1464 | 14 | 0.0105 |
| Between cells | 184.1126 | 23 | — |
| Within cells | 0.1391 | 24 | 0.0058 |
| | 184.2517 | 47 | |

the differences between the values for the three drying periods are much larger than in other samples. Clearly all of these 'interaction' effects, in samples 2, 5 and 6, could be accounted for by some heterogeneity in the state of dryness of the original samples, but it is now impossible to tell.

In order to demonstrate the effect on the analysis of variance of deleting some of the samples, Table 15.10 shows the effect of deleting samples 2 and 6. We see from this table that the significant interaction has now disappeared and the results are interpretable as differences between samples, and an additive effect due to drying periods.

At this point we should recognize that the drying periods chosen were equally spaced levels, 3 hours, 6 hours and 9 hours. We should, therefore, split up the two degrees of freedom for drying periods into a linear and a quadratic component, with one degree of freedom each. It would also be of interest to see the interaction sum of squares split up into a linear by samples and a quadratic by samples component, each on seven degrees of freedom. These terms will enable us to assess the consistency of the linear and quadratic effects of 'drying periods' over samples. These calculations, and the remainder of the analysis, will be left as an exercise (see Exercise 15.4.2). A full analysis should include tables of means of the drying periods and of the samples, checks of assumptions, and a conclusion written up for the original experimenters (not for statisticians).

## Exercise 15.4

1. Check the derivation of Equation (15.10), assuming the model (15.7) to hold. Also check that $C(r_{ij}, r_{hk}) = \sigma^2/rcn$, $i \neq h$, $j \neq k$.

2. Continue the analysis of Example 15.1 from the point reached in Table 15.10. Include appropriate graphs or tables so that the results are demonstrated clearly, and also include a written up form of the conclusions suitable for presentation to the original experimenters.

3. The data given in Table 15.11 were reported by Smith (1961, 1969), and they are coded production rates for a chemical production process. Four reagents and three catalysts were used, and each combination was run twice. The 24 observations were taken in random orders. Make a report on the data for the manufacturer.

## 15.5 Unit errors

We now return to emphasize a point we touched on briefly in Section 15.2. The question which arises is whether in an analysis-of-variance table such as that shown in Table 15.10, the 'within cells' or the 'interaction' mean square is taken to estimate the error variance.

The model we have decided upon for Example 15.1 can be written in the form

$$Y_{ijk} = \mu + \alpha_i + \beta_j + \varepsilon_{ijk} \begin{cases} i = 1, 2, 3 \\ j = 1, 2, \dots, 10 \\ k = 1, 2 \end{cases}$$

where $\varepsilon$ is an independently and normally distributed error of expectation zero and variance $\sigma^2$. The term $\alpha_i$ could be written as an orthogonal poly-

Table 15.11

| | Catalyst | | |
|---|---|---|---|
| Reagent | X | Y | Z |
| A | 4 | 11 | 5 |
| | 6 | 7 | 9 |
| B | 6 | 13 | 9 |
| | 4 | 15 | 7 |
| C | 13 | 15 | 13 |
| | 15 | 9 | 13 |
| D | 12 | 12 | 7 |
| | 12 | 14 | 9 |

nomial in drying time. If we had decided that there was an interaction a term $\theta_{ij}$ would replace $(\alpha_i + \beta_j)$. The terms $\varepsilon$ represent the errors made in the 60 chemical determinations, and the 'within cells' line provides an estimate of $\sigma^2$. We call the $\varepsilon$ terms *technical errors*.

Suppose now that duplicate determinations were made together, but that the order of the 30 duplicate determinations was randomized. In this situation it would be difficult to rule out the possibility of small variations in the measurements over the period in which the determinations were made. The model would therefore have to be modified to the form

$$y_{ijk} = \mu + \alpha_i + \beta_j + e_{ij} + \varepsilon_{ijk} \tag{15.11}$$

where the $e_{ij}$ represent errors applying to both duplicate determinations for the cells $(i, j)$, and we call them *unit errors*. Now, in general, unit errors are likely to be correlated; for example, there may be a time trend over the period in which the determinations were made. However, if the 30 duplicate determinations were carried out in random order it would probably be plausible to assume the $e_{ij}$ to be independently distributed $N(0, \tau^2)$. The effect of this is that the expected values of mean squares $C$, $D$ and $E$ of Table 15.6 have terms $(\sigma^2 + n\tau^2)$ replacing the existing terms $\sigma^2$ (in the general case). This has an obvious effect on standard errors and $F$-tests. The 'interaction' mean square has to be used as our estimate of the error variance, and an analysis can only proceed by assuming the interaction to be zero. In effect, we are back to the model and situation of section 10.6. As a final comment on this case, we remark that if the 30 duplicates are not carried out in random order, the analysis can only proceed by making an unlikely set of assumptions, and a completely different analysis may be necessary.

A similar effect to the presence of unit errors is produced when the observations within a cell are correlated. Suppose we have

$$c(y_{ijk}, y_{ije}) = \rho\sigma^2 \tag{15.12}$$

but zero correlation between all other observations, then the variance of cell means is

$$\sigma^2/n + \rho\sigma^2(1 - 1/n) \tag{15.13}$$

so that in the absence of correlation the expected value of the 'interaction' mean square is $n$ times this. Under the same conditions, the expected value of the 'within cells' mean square (see Equation (1.16)) is

$$\sigma^2(1 - 2\rho/n) \tag{15.14}$$

If the experiment is an agricultural one, and the experimental units are plots of land, the correlations are often positive, and the 'interaction' mean square tends to be inflated with respect to the 'within cells' mean square. However, in certain animal experiments the correlation can be negative, due to competition, and the reverse occurs.

Consider as an example an experiment on feedstuffs for pigs. Suppose that 20 pens, each of 8 pigs are used. Let there be two treatments, $A$ and $B$, forming a two-way classification on the 20 pens. However, it is not practicable to feed pigs individually, and all one can do is to apply the treatments to the feeds put into the pens. Competition for food between pigs then produces a greater variation of pig weights within pens than one would expect from the variation of the averages of the pen weights. That is, we have the situation just set out but with negative $\rho$.

At this point we should emphasize that the 'unit errors' model (15.11), and the correlation model with positive correlation are alternatives. (There is no such equivalent when correlations are negative.)

In summing up, this discussion means that some care has to be taken in writing down a model and in deciding which term represents error.

## Exercise 15.5
1. Check formula (15.13) as follows. Let there be $n$ random variables $Y_1$, $\ldots, Y_n$, such that $E(Y_i) = \mu$, $V(Y_i) = \sigma^2$, and $C(Y_i, Y_j) = \rho\sigma^2$, then apply Equation (A.3) of Appendix A to obtain the result.

## 15.6 Random-effects models
So far we have discussed the analysis of variance of crossed classifications, in terms of fixed-effects models for the factors, but sometimes they must be regarded in the random-effects sense. Again it will be helpful to discuss this by using Example 15.1 to illustrate the points.

In Example 15.1 we stated that the ten samples were not chosen randomly, but in fact this statement was not in the original description of the experiment. (The statement was inserted into Example 15.1 to postpone discussions of random-effects models.) It seems clear in fact that the samples were chosen to represent a population of possible samples. Let us suppose that we have reached the stage in the analysis at which we have ruled out the possibility of an interaction, but we will leave in a unit error term in our model, so that we are working with Equation (15.11). All we need to do now for an infinite population model is to take the $\alpha_i$ to be normally dis-

Table 15.12 *Expected values of the mean squares in Table 15.6*

| Mean square | Expected value |
|---|---|
| B | $\sigma^2$ |
| C | $\sigma^2 + n\tau^2 + cn\sigma_\alpha^2$ |
| D | $\sigma^2 + n\tau^2 + rn\sum(\beta_j - \bar{\beta})^2/(c-1)$ |
| E | $\sigma^2 + n\tau^2$ |

tributed with expectation and variance $\mu_\alpha$ and $\sigma_\alpha^2$. If we have $r$ rows, $c$ columns and $n$ observations per cell, we have the analysis of variance as in Table 15.6, and the expectations of the mean squares are as in Table 15.12.

From this table we can easily see how to obtain estimates of $\sigma^2$, $\tau^2$ and $\sigma_\alpha^2$, and the methods of Chapter 14 can be applied to obtain estimates of the variances of these estimates and also to obtain confidence intervals, etc.

These methods can be applied to any or all the factors in a crossed classification provided there are no interactions. If we do have interactions then we cannot proceed by this route, but we must first investigate the reason for the interaction. Example 15.1 provides a good illustration of the sort of difficulty which arises. If we regard samples as a random effect, and drying periods as a fixed effect, the data do show that there is an interaction between these two. Our investigations have shown that samples 2, 5 and 6 show a pattern of results which is rather different from that of the other samples. If we had been able to check with the experimenters, we may have been able to determine:

(a) Whether the samples 2, 5 and 6 were or were not subject to misrecording or some other error. If they were, they could be adjusted or deleted from the analysis.
(b) Whether samples giving an untypical pattern do occur from time to time, in which case the model has to be completely revised.
(c) Whether we can find a plausible model for the interaction.

Once an interaction arises, investigations of this sort must be carried out, and the further direction of the analysis depends on the outcome.

### 15.7* Analysis of a two-way unequally replicated design
We now return to the situation of Section 15.3 but with unequal replication within cells. The observations are written $y_{ijk}$, where $i = 1, 2, \ldots, r$, $j = 1, 2, \ldots, c$, and $k = 1, 2, \ldots, n_{ij}$. For example, this might arise in the analysis of Example 15.1 if some of the duplicate observations had to be discarded in certain cells. One of the effects of unequal replication is (usually) that the row and column effects are no longer orthogonal, and the analysis is much more complicated. As before, we assume that all observations are statistically independent and have a constant variance $\sigma^2$.

We start by fitting model (15.5),

$$E(y_{ijk}) = \mu + \phi_{ij} \tag{15.15}$$

and from Section 10.8 (Exercise 10.8.1) we see that this leads to the analysis-of-variance table given in Table 15.13. In this table we define $\delta(n_{ij})$ to be unity if $n_{ij} \neq 0$, and zero otherwise; clearly cells with no observations do not contribute to the degrees of freedom.

The model (15.15) allows for the possible presence of an interaction

Table 15.13 *Analysis of variance for unequally replicated design*

| Source | CSS | d. f. |
|---|---|---|
| Due to the $\phi_{ij}$ in (15.15) | $\sum_i \sum_j n_{ij} (\bar{y}_{ij.} - y_{...})^2$ | $\sum_i \sum_j \delta(n_{ij}) - 1$ |
| Residual | $\sum_i \sum_j \sum_k (y_{ijk} - \bar{y}_{ij.})^2$ | $\sum_i \sum_j (n_{ij} - \delta(n_{ij}))$ |
| Total | $\sum_i \sum_j \sum_k (y_{ijk} - \bar{y}_{...})^2$ | $\sum_i \sum_j n_{ij} - 1$ |

between the row and column effects, and following the method of Section 15.3, we now wish to compare the fit obtained by this model with the fit obtained by using the simple additive model. We shall follow the method used in Section 10.6, and consider a sequence of models:

Model 1 $\qquad\qquad E(Y_{ijk}) = \mu$

Model 2 $\qquad\qquad E(Y_{ijk}) = \mu + \alpha_i$

Model 3 $\qquad\qquad E(Y_{ijk}) = \mu + \beta_j$

Model 4 $\qquad\qquad E(Y_{ijk}) = \mu + \alpha_i + \beta_j$

The next step is to obtain the sums of squares due to fitting these models, and then the sums of squares to be used for testing various hypotheses are obtained by differencing these.

The sum of squares due to model 1 is easily seen to be

$$S_{\text{par}}^{(1)} = N\bar{y}_{...}^2 \qquad\qquad (15.16)$$

where

$$N = \sum_{i=1}^{r} \sum_{j=1}^{c} n_{ij}$$

The sums of squares due to models 2 and 3 can be written down from Section 10.8 (Exercise 10.8.1), and they are as follows:

$$S_{\text{par}}^{(2)} = N\bar{y}_{...}^2 + \sum_{i=1}^{r} n_{i.} (\bar{y}_{i..} - \bar{y}_{...})^2 \qquad\qquad (15.17)$$

$$S_{\text{par}}^{(3)} = N\bar{y}_{...}^2 + \sum_{j=1}^{c} n_{.j} (\bar{y}_{.j.} - \bar{y}_{...})^2 \qquad\qquad (15.18)$$

where

$$n_{i.} = \sum_{j=1}^{c} n_{ij} \quad \text{and} \quad n_{.j} = \sum_{i=1}^{r} n_{ij}$$

So far, the results follow those in Section 10.6, but a major difference comes when we try to fit model 4. The sum of squares to be minimized is

$$S = \sum_{i=1}^{r} \sum_{j=1}^{c} \sum_{k=1}^{n_{ij}} (y_{ijk} - \mu - \alpha_i - \beta_j)^2$$

and the normal equations are given by

$$
\left.
\begin{aligned}
N\bar{y}_{...} - N\hat{\mu} - \sum_{i=1}^{r} n_{i.}\hat{\alpha}_i - \sum_{j=1}^{c} n_{.j}\hat{\beta}_j &= 0 \\
n_{i.}\bar{y}_{i..} - n_{i.}\hat{\mu} - n_{i.}\hat{\alpha}_i - \sum_{j=1}^{c} n_{ij}\hat{\beta}_j &= 0 \qquad \text{for} \quad i = 1, 2, \dots, r \\
n_{.j}\bar{y}_{.j.} - n_{.j}\hat{\mu} - \sum_{i=1}^{r} n_{ij}\hat{\alpha}_j - n_{.j}\hat{\beta}_j &= 0 \qquad \text{for} \quad j = 1, 2, \dots, c
\end{aligned}
\right\} \quad (15.19)
$$

and

The method given in Section 10.8 is to find a solution for $(\hat{\mu} + \hat{\alpha}_i + \hat{\beta}_j)$ from these equations, and then find the relevant sum of squares. Unfortunately, this procedure is not straightforward for Equations (15.19), and in general it does *not* result in an expression like (10.64). This means that in going from the equally replicated to the unequally replicated case we have lost orthogonality. We would find, for example, that in general the adjusted sum of squares due to the $\beta_j$, given that the terms $\mu$ and $\alpha_i$ are included in the model, is not equal to the sum of squares due to the $\beta_j$ when the $\alpha_i$ terms are ignored, that is, we have

$$
S_{\text{par}}^{(4)} - S_{\text{par}}^{(2)} \neq S_{\text{par}}^{(3)} - S_{\text{par}}^{(1)}
$$

It is therefore necessary to revert to the methods of Chapter 6 for obtaining sums of squares for testing hypotheses, and the skeleton analysis-of-variance table in Table 15.14 illustrates the method.

Alternatively, we could calculate all the sums of squares adjusted for fitting $\mu$, and the numbers of degrees of freedom in the first two rows of Table 15.14 would be reduced by one each. The numbers of the degrees of freedom can be justified either by writing out the design matrix to see the rank, or else from physical considerations.

Suppose, for example, that there are two rows and three columns, and the numbers of observations in each cell is as follows:

| row | column | | |
|-----|--------|---|---|
|     | 1 | 2 | 3 |
| 1 | 1 | 1 | 2 |
| 2 | 2 | 1 | 1 |

then for the parameter vector

$$
\theta' = (\mu, \alpha_1, \alpha_2, \beta_1, \beta_2, \beta_3)
$$

Table 15.14 *Method of obtaining the adjusted sum of squares due to the* $\beta_j$

| Source | d.f. |
|---|---|
| Due to $\mu$, $\alpha_i$, $\beta_j$ | $r + c - 1$ |
| Due to $\mu$, $\alpha_i$ (ignoring $\beta_j$) | $r$ |
| Due to $\beta_j$ (adjusting for $\mu$, $\alpha_i$) | $c - 1$ |

the design matrix is as follows:

$$\mathbf{a} = \begin{bmatrix} 1 & 1 & 0 & 1 & 0 & 0 \\ 1 & 1 & 0 & 0 & 1 & 0 \\ 1 & 1 & 0 & 0 & 0 & 1 \\ 1 & 1 & 0 & 0 & 0 & 1 \\ 1 & 0 & 1 & 1 & 0 & 0 \\ 1 & 0 & 1 & 1 & 0 & 0 \\ 1 & 0 & 1 & 0 & 1 & 0 \\ 1 & 0 & 1 & 0 & 0 & 1 \end{bmatrix}$$

Here columns 2 and 3 add to column 1, and also columns 4, 5 and 6 add together to form column 1. The rank is therefore four. Alternatively, we see that having inserted $\mu$, we only need one further parameter to fix two independent row means, and two others to fix independent column means. With these considerations in mind, the degrees of freedom listed in Table 15.14 should be obvious.

In the two-way unequally replicated design, it is necessary to calculate two separate tables similar to Table 15.14, one to obtain the adjusted sum of squares due to the $\beta_j$ and another to obtain the adjusted sum of squares due to the $\alpha_i$. As a denominator for the $F$-tests we use the mean square for 'Residual' in Table 15.13.

In order to test whether there is an interaction present, we need to calculate the sum of squares

$$\{S_{par}(\text{model }(15.15)) - S_{par}^{(4)}\} \tag{15.20}$$

and test this against the mean square 'Residual' in Table 15.13.

The full procedure for analysis should be as follows:

(i) A preliminary examination of the data.
(ii) Calculation of Table 15.13 in order to get the residual mean square for testing hypotheses.
(iii) A test of whether or not an interaction is present. If there is an interaction, the table of cell means needs to be examined to explore this.
(iv) Tests of whether both row and column parameters are necessary.
(v) Checks of the assumptions using residuals.

The main disadvantage of the unequally replicated case is that although it is possible to calculate sums of squares for testing various hypotheses, it is not usually possible to list them all neatly in an analysis-of-variance table (see below for the exception).

The above procedures require the calculation of $S_{par}^{(4)}$, and in general this will be done on a computer; a simple formula for calculation of this sum of squares does not exist in the general case, but see, for example, Kempthorne (1975) for a detailed discussion of this point.

*Special case of proportional cell numbers*
There is one special case which simplifies, and this is when the cell frequencies $n_{ij}$ satisfy the restriction

$$n_{ij} = n_{i.} n_{.j} / N \qquad (15.21)$$

This means that along each row or column the frequencies are proportional to the marginal totals. An example is as follows:

|        |   | Column |    |        |
|--------|---|--------|----|--------|
| Row    | 1 | 2      | 3  | Totals |
| 1      | 2 | 6      | 4  | 12     |
| 2      | 3 | 9      | 6  | 18     |
| Totals | 5 | 15     | 10 | 30     |

If Equation (15.21) holds then Equations (15.19) reduce to the following:

$$
\left.
\begin{aligned}
N\bar{y}_{...} - N\hat{\mu} - \sum_{i=1}^{r} n_{i.}\hat{\alpha}_i - \sum_{j=1}^{c} n_{.j}\hat{\beta}_j &= 0 \\
N\bar{y}_{i..} - N\hat{\mu} - N\hat{\alpha}_i - \sum_{j=1}^{c} n_{.j}\hat{\beta}_j &= 0 \\
N\bar{y}_{.j.} - N\hat{\mu} - \sum_{i=1}^{r} n_{i.}\hat{\alpha}_i - N\hat{\beta}_j &= 0
\end{aligned}
\right\} \qquad (15.22)
$$

From these equations we see that the fitted values are given by

$$\hat{\mu} + \hat{\alpha}_i + \hat{\beta}_j = \bar{y}_{i..} + \bar{y}_{.j.} - \bar{y}_{...}$$

so that the sum of squares is

$$
\begin{aligned}
S_{par}^{(4)} &= \sum_{i=1}^{r} \sum_{j=1}^{c} \sum_{k=1}^{n_{ij}} \{(\bar{y}_{i..} - \bar{y}_{...}) + (\bar{y}_{.j.} - \bar{y}_{...}) + \bar{y}_{...}\}^2 \\
&= \sum_{i=1}^{r} n_{i.}(\bar{y}_{i..} - \bar{y}_{...})^2 + \sum_{j=1}^{c} n_{.j}(\bar{y}_{.j.} - \bar{y}_{...})^2 + N\bar{y}_{...}^2 \qquad (15.23)
\end{aligned}
$$

Table 15.15 *Analysis of variance for a two-way classification with proportional frequencies*

| Source | CSS | d.f. |
|---|---|---|
| Due to rows | $\displaystyle\sum_{i=1}^{r} n_{i.}(\bar{y}_{i..} - \bar{y}_{...})^2$ | $r - 1$ |
| Due to columns | $\displaystyle\sum_{j=1}^{c} n_{.j}(\bar{y}_{.j.} - \bar{y}_{...})^2$ | $c - 1$ |
| Interaction | $\displaystyle\sum_{i=1}^{r}\sum_{j=1}^{c} n_{ij}(\bar{y}_{.j.} - \bar{y}_{i..} - \bar{y}_{.j.} + \bar{y}_{...})^2$ | $\displaystyle\sum_i\sum_j \delta(n_{ij}) - r - c + 1$ |
| Residual | $\displaystyle\sum_{i=1}^{r}\sum_{j=1}^{c}\sum_{k=1}^{n_{ij}} (y_{ijk} - \bar{y}_{ij.})^2$ | $\displaystyle\sum_i\sum_j (n_{ij} - \delta(n_{ij}))$ |
| Total | $\displaystyle\sum_{i=1}^{r}\sum_{j=1}^{c}\sum_{k=1}^{n_i} (y_{ijk} - \bar{y}_{...})^2$ | $\displaystyle\sum_i\sum_j n_{ij} - 1$ |

We observe that we again have the orthogonal case here, since

$$S_{\text{par}}^{(4)} - S_{\text{par}}^{(2)} = S_{\text{par}}^{(3)} - S_{\text{par}}^{(1)}$$

and

$$S_{\text{par}}^{(4)} - S_{\text{par}}^{(3)} = S_{\text{par}}^{(2)} - S_{\text{par}}^{(1)}$$

That is, given we fit the parameter $\mu$, the row and column parameters are fitted independently. The interaction sum of squares can also be obtained easily, and the analysis of variance is as shown in Table 15.15.

The value of the orthogonal arrangement is not simply the ease of calculation, but rather the ease of interpretation. For example, the rows and columns sums of squares in Table 15.15 are statistically independent. This means that there is some advantage in having a proportional frequencies arrangement if an unequally replicated design is appropriate.

One great advantage of using a design with some replication within cells is that an independent estimate of $\sigma^2$ is provided, which is free from any possible contamination by interactions. However, equal replication can be very expensive in terms of observations, and the 'proportional frequencies' design offers a method of obtaining an independent estimate of error without losing orthogonality of the row and column effects.

**Further reading**
For reading on interactions see Daniel (1976) and Cox (1958); see also Brown (1975) and Bradu and Gabriel (1975).

# CHAPTER SIXTEEN

# *Further analysis of variance*

## 16.1* Three-way crossed classification

In Chapter 15 we considered two-way crossed classifications, and in this section we discuss briefly the theory of the three-way crossed classification with one observation per cell. The theory is based upon principles discussed in earlier chapters. As an example we shall use some data from an experiment discussed by Lemus (1960), slightly modified.

*Example* 16.1   The data given in Table 16.1 were obtained in an experiment performed to determine the force necessary to separate electrical connectors. The connectors were for use in the electrical connections between a satellite or missile with its booster, and they disconnect when the vehicles separate. Lack of homogeneity of the force needed to disconnect could impart some instability to the vehicle. Five connectors were tested on a special machine in the laboratory, and the connectors were pulled apart at four different angles. Five trials were made at each combination of connector and angle of pull.   □ □ □

There were two factors in this experiment which arise directly from the treatments applied, angle of pull and connector number, but as the order number of the trial seems to be relevant this is brought into the analysis as a third factor, even though it does not arise directly from the treatments applied. This type of experiment, with two or more factors, each at several levels, is called a *factorial experiment*. This experiment, with the three factors at 5, 5 and 4 levels, is referred to as a $5 \times 5 \times 4$ factorial experiment. The treatment here is a general one; certain special types of factorial experiment, particularly the $2^n$ series, have certain special methods of analysis available, for which we refer the reader to the list of further reading at the end of the chapter.

There is a large literature on factorial experiments, and, in particular, some very important material about the design of such experiments is contained in Cox (1958). We emphasize here a point which arises out of

Table 16.1 *Force (in lbs, less* 40) *to disconnect electrical connectors*

| Angle | Connector | Trial | 1 | 2 | 3 | 4 | 5 | Total |
|-------|-----------|-------|-----|-----|-----|-----|-----|-------|
| 0° | 1 | | 6.5 | 3.0 | 6.5 | 4.0 | 6.5 | 26.5 |
| | 2 | | − 1.0 | 2.0 | 4.0 | 4.0 | 2.0 | 11.0 |
| | 3 | | − 2.5 | 0.0 | 1.5 | 0.0 | − 1.0 | − 2.0 |
| | 4 | | − 4.0 | − 2.0 | − 2.5 | − 4.0 | − 3.5 | − 16.0 |
| | 5 | | 6.0 | 4.5 | 4.0 | 8.0 | 6.5 | 29.0 |
| | Total | | 5.0 | 7.5 | 13.5 | 12.0 | 10.5 | 48.5 |
| 2° | 1 | | 2.5 | 3.5 | 5.0 | 5.0 | 4.5 | 20.5 |
| | 2 | | 2.5 | 4.0 | 5.5 | 4.5 | 4.0 | 20.5 |
| | 3 | | − 0.5 | − 0.5 | − 2.5 | − 2.0 | − 2.5 | − 8.0 |
| | 4 | | − 5.0 | − 4.0 | − 2.0 | − 1.0 | 2.0 | − 10.0 |
| | 5 | | 4.5 | 6.0 | 8.0 | 8.0 | 9.5 | 36.0 |
| | Total | | 4.0 | 9.0 | 14.0 | 14.5 | 17.5 | 59.0 |
| 4° | 1 | | 3.5 | 3.0 | 3.0 | 2.0 | 2.0 | 13.5 |
| | 2 | | 3.5 | 2.0 | 2.5 | 2.5 | 3.0 | 13.5 |
| | 3 | | 2.0 | 2.5 | 2.5 | 2.0 | 4.0 | 13.0 |
| | 4 | | 3.5 | 3.0 | 4.0 | 5.5 | − 5.0 | 11.0 |
| | 5 | | 9.5 | 11.0 | 6.0 | 7.5 | 10.5 | 44.5 |
| | Total | | 22.0 | 21.5 | 18.0 | 19.5 | 14.5 | 95.5 |
| 6° | 1 | | 3.5 | 5.0 | 1.5 | 4.0 | 3.6 | 17.5 |
| | 2 | | 4.0 | 4.0 | 4.0 | 7.0 | 10.0 | 29.0 |
| | 3 | | 7.0 | 8.5 | 7.0 | 7.5 | 9.5 | 39.5 |
| | 4 | | − 5.0 | − 3.0 | − 2.0 | − 0.5 | 0.0 | − 10.5 |
| | 5 | | 12.0 | 13.0 | 17.0 | 19.0 | 21.0 | 82.0 |
| | Total | | 21.5 | 27.5 | 27.5 | 37.0 | 44.0 | 157.5 |

comments made in Section 13.6, that when analysing any experiment, factorial or otherwise, it is important to examine the design carefully. Errors in design, such as a lack of the appropriate randomizations, often show up in the data. This is one reason why statisticians should be consulted at the design stage of an experiment wherever this is possible. In this example some of the design details are not given, so that we must proceed without checking.

Next, we shall examine the data briefly before starting to construct a model for it. If we take ranges of the observations across the trials, there is some evidence of the need for a transformation in order to stabilize variance. Larger ranges tend to go with observations of larger mean. However, the evidence is not very strong, and in this experiment there is probably some pressure to carry out an analysis in the original scale.

Table 16.2 shows the three two-way tables of marginal totals. The response increases with angle and trial number, and there seem to be some large

Table 16.2 *Two-way tables of means for Example* 16.1

|            | Angle |      |      |      |       |
|------------|-------|------|------|------|-------|
| Connector  | 0     | 2    | 4    | 6    | Total |
| 1          | 26.5  | 20.5 | 13.5 | 17.5 | 78.0  |
| 2          | 11.0  | 20.5 | 13.5 | 29.0 | 74.0  |
| 3          | − 2.0 | − 8.0| 13.0 | 39.5 | 42.5  |
| 4          | − 16.0| − 10.0| 11.0| − 10.5| − 25.5|
| 5          | 29.0  | 36.0 | 44.5 | 82.0 | 191.5 |
| Total      | 48.5  | 59.0 | 95.5 | 157.5| 360.5 |

|       | Angle |      |      |      |       |
|-------|-------|------|------|------|-------|
| Trial | 0     | 2    | 4    | 6    | Total |
| 1     | 5.0   | 4.0  | 22.0 | 21.5 | 52.5  |
| 2     | 7.5   | 9.0  | 21.5 | 27.5 | 65.5  |
| 3     | 13.5  | 14.0 | 18.0 | 27.5 | 73.0  |
| 4     | 12.0  | 14.5 | 19.5 | 37.0 | 83.0  |
| 5     | 10.5  | 17.5 | 14.5 | 44.0 | 86.5  |
| Total | 48.5  | 59.0 | 95.5 | 157.5| 360.5 |

|           | Trial |      |      |      |      |       |
|-----------|-------|------|------|------|------|-------|
| Connector | 1     | 2    | 3    | 4    | 5    | Total |
| 1         | 16.0  | 14.5 | 16.0 | 15.0 | 16.5 | 78.0  |
| 2         | 9.0   | 12.0 | 16.0 | 18.0 | 19.0 | 74.0  |
| 3         | 6.0   | 10.5 | 8.5  | 7.5  | 10.0 | 42.5  |
| 4         | − 10.5| − 6.0| − 2.5| 0.0  | − 6.5| − 25.5|
| 5         | 32.0  | 34.5 | 35.0 | 42.5 | 47.5 | 191.5 |
| Total     | 52.5  | 65.5 | 73.0 | 83.0 | 86.5 | 360.5 |

differences between connectors. The increase with trials is a bit unusual, but apparently not impossible, with the sort of metals and conditions under which such trials are done. In spite of these large effects, the two-way table of angle versus connector shows evidence of interaction. A good way to see this is to use Daniel's (1976) 'eyeball' test for interactions. If the effects of angle and connector are purely additive, then row differences will be constant down the rows (not across). Table 16.3 shows very large departures from this which will cause trouble when we are trying to interpret the analysis of variance. Another way of examining this interaction is to graph the two-way table totals, as in Fig. 16.1. Although connector 5 gives the highest total responses for all angles, and connector 4 the lowest total responses for all totals, the interaction is too strong to make any sense of a main effect due to

Table 16.3 *Table of angle differences for*
*connectors, Example* 16.1

| Connector | | | |
|---|---|---|---|
| 1 | −  6.0 | −  7.0 | 4.0 |
| 2 | 9.5 | −  7.0 | 15.5 |
| 3 | 6.0 | 21.0 | 26.5 |
| 4 | 6.0 | 21.0 | − 21.5 |
| 5 | 7.0 | 8.5 | 37.5 |

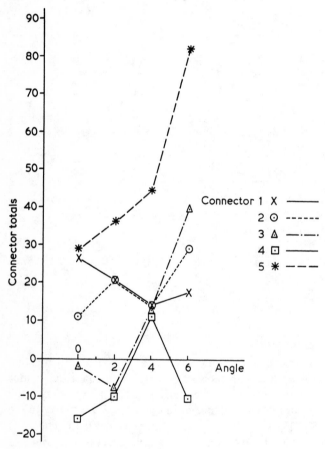

Fig. 16.1 Connectors by angles, two-way table.

connectors. The mean response depends on which particular angle and connector occur together.

There are some similar evidences of an interaction in the angles by trials two-way table, but this is nowhere near as strong, and may not be statistically significant.

We shall now start modelling the data, but we shall do this in a general context, and return later to a detailed consideration of Example 16.1. We shall suppose that there are three factors, $A$, $B$ and $C$, at $r$, $c$ and $d$ levels respectively, and we write the data $y_{ijk}$, $i = 1, 2, \ldots, r$; $j = 1, 2, \ldots, c$; and $k = 1, 2, \ldots, d$.

The first step is to fit model 1,

$$\text{Model 1:} \quad E(Y_{ijk}) = \mu \tag{16.1}$$

which leads to a minimized sum of squares

$$S_{\min}^{(1)} = \sum_i \sum_j \sum_k (y_{ijk} - \bar{y}_{...})^2 \tag{16.2}$$

on $(rcd - 1)$ degrees of freedom. The next model to consider is one which allows for all three possible interactions between pairs of factors. Following the method in Chapter 15, this is written

$$\text{Model 2:} \quad E(Y_{ijk}) = \mu + \phi_{ij}^{AB} + \phi_{ik}^{AC} + \phi_{jk}^{BC} \tag{16.3}$$

In this model, the $\phi_{ij}^{AB}$ term represents the fact that the mean response depends upon the combination of the $i$th level of factor $A$ and the $j$th level of factor $B$, as in the discussion about angle and connector for Example 16.1 above. We cannot introduce a term $\phi_{ijk}^{ABC}$ here, representing a three-factor interaction, as there is only one observation per cell, and there would be no degrees of freedom left for error. In the unreplicated three-factor experiment, we can go no further than two-factor interactions.

Model 2 is easily fitted by the method of Section 10.8. The easiest way to obtain the least-squares estimators is to recognize that the sufficient statistics are the marginal means,

$$\bar{Y}_{i..}, \bar{Y}_{.j.}, \bar{Y}_{..k}, \bar{Y}_{ij.}, \bar{Y}_{i.k}, \bar{Y}_{.jk}$$

for all $i$, $j$, $k$, and we need a linear combination of the expectations of those which has an expectation equal to the right-hand side of Equation (16.3). For example, we have

$$E(\bar{Y}_{i..}) = \mu + \bar{\phi}_{i.}^{AB} + \bar{\phi}_{i.}^{AC} + \bar{\phi}_{..}^{BC} \tag{16.4}$$

$$E(\bar{Y}_{i.k}) = \mu + \bar{\phi}_{i.}^{AB} + \bar{\phi}_{ik}^{AC} + \bar{\phi}_{.k}^{BC} \tag{16.5}$$

and it is readily checked that this procedure leads to the estimator

$$E(\widehat{Y_{ijk}}) = \bar{Y}_{ij.} + \bar{Y}_{i.k} + \bar{Y}_{.jk} - \bar{Y}_{i..} - \bar{Y}_{.j.} - \bar{Y}_{..k} + \bar{Y}_{...} \tag{16.6}$$

so that the minimized sum of squares for fitting model 2 is given by

$$S_{\min}^{(2)} = \sum_i \sum_j \sum_k (y_{ijk} - \bar{y}_{ij.} - \bar{y}_{i.k} - \bar{y}_{.jk} + \bar{y}_{i..} + \bar{y}_{.j.} + \bar{y}_{..k} - \bar{y}_{...})^2 \tag{16.7}$$

This sum of squares has $(r - 1)(c - 1)(d - 1)$ degrees of freedom, and it provides us with an error term for testing and estimation purposes.

Suppose we now drop one of the interaction terms in the model (16.3), say the term $\phi_{ij}^{AB}$, then we have a new model,

$$\text{Model 3:} \quad E(Y_{ijk}) = \mu + \phi_{ik}^{AC} + \phi_{jk}^{BC} \qquad (16.8)$$

The least-squares estimator can be found as before. We can readily check that in terms of model 3, we have

$$E(\bar{Y}_{i.k} + \bar{Y}_{.jk} - \bar{Y}_{..k}) = \mu + \phi_{ik}^{AC} + \phi_{jk}^{BC}$$

so that the minimized sum of squares is

$$S_{\min}^{(3)} = \sum_i \sum_j \sum_k (y_{ijk} - \bar{y}_{i.k} - \bar{y}_{.jk} + \bar{y}_{..k})^2 \qquad (16.9)$$

The difference between models 2 and 3 is that there is no two-way interaction $AB$ in model 3. The difference between the corresponding minimized sums of squares therefore gives a sum of squares for testing this. In fact, we see that

$$S_{\min}^{(3)} = S_{\min}^{(2)} + d \sum_i \sum_j (\bar{y}_{ij.} - \bar{y}_{i..} - \bar{y}_{.j.} + \bar{y}_{...})^2 \qquad (16.10)$$

so that the last term in Equation (16.10) is the sum of squares for testing the presence of the two-way interaction $AB$. In this way we can generate three two-way interaction sums of squares.

We now consider the model

$$\text{Model 4:} \quad E(Y_{ijk}) = \mu + \alpha_i + \phi_{jk}^{BC} \qquad (16.11)$$

In this model we have just one two-factor interaction and a simple additive term representing the main effect of $A$. The minimized sum of squares for fitting this model is found to be

$$S_{\min}^{(4)} = \sum_i \sum_j \sum_k (y_{ijk} - \bar{y}_{i..} - \bar{y}_{.jk} + \bar{y}_{...})^2 \qquad (16.12)$$

The difference between models 3 and 4 is the absence of an $AB$ interaction term in model 4. In fact, we find

$$S_{\min}^{(4)} = S_{\min}^{(3)} + c \sum_i \sum_k (\bar{y}_{i.k} - \bar{y}_{i..} - \bar{y}_{..k} + \bar{y}_{...})^2 \qquad (16.13)$$

That is, the same interaction terms arise as in Equation (16.10) when comparing models such as 2 and 3. Clearly, in an unbalanced arrangement, such as if some responses were missing, this would not necessarily happen.

Finally, we consider the model

$$\text{Model 5:} \qquad E(Y_{ijk}) = \mu + \alpha_i + \beta_j + \gamma_k \qquad (16.14)$$

This model is appropriate if there are no interactions present, and the least-squares estimator of Equation (16.14) is

$$\hat{\mu} + \hat{\alpha}_i + \hat{\beta}_j + \hat{\gamma}_k = \bar{Y}_{i..} + \bar{Y}_{.j.} + \bar{Y}_{..k} - 2\bar{Y}_{...}$$

Table 16.4 *Three-way crossed classification analysis of variance*

| Source | CSS | d. f. |
|---|---|---|
| A | $cd\sum_i(\bar{y}_{i..} - \bar{y}_{...})^2$ | $(r-1)$ |
| B | $rd\sum_j(\bar{y}_{.j.} - \bar{y}_{...})^2$ | $(c-1)$ |
| C | $rc\sum_k(\bar{y}_{..k} - \bar{y}_{...})^2$ | $(d-1)$ |
| AB | $d\sum_i\sum_j(\bar{y}_{ij.} - \bar{y}_{i..} - \bar{y}_{.j.} + \bar{y}_{...})^2$ | $(r-1)(c-1)$ |
| AC | $c\sum_i\sum_k(\bar{y}_{i.k} - \bar{y}_{i..} - \bar{y}_{..k} + \bar{y}_{...})^2$ | $(r-1)(d-1)$ |
| BC | $r\sum_j\sum_k(\bar{y}_{.jk} - \bar{y}_{.j.} - \bar{y}_{..k} + \bar{y}_{...})^2$ | $(c-1)(d-1)$ |
| Residual | $\sum_i\sum_j\sum_k(y_{ijk} - \bar{y}_{ij.} - \bar{y}_{i.k} - \bar{y}_{.jk} + \bar{y}_{i..} + \bar{y}_{.j.} + \bar{y}_{..k} - \bar{y}_{...})^2$ | $(r-1)(c-1)(d-1)$ |
| Total | $\sum_i\sum_j\sum_k(y_{ijk} - \bar{y}_{...})^2$ | $rcd - 1$ |

In this case it is easier to get the corrected sum of squares due to the model, and this is

$$S^{(5)}_{par} = cd\sum_i(\bar{y}_{i..} - \bar{y}_{...})^2 + rd\sum_j(\bar{y}_{.j.} - \bar{y}_{...})^2 + rc\sum_k(\bar{y}_{..k} - \bar{y}_{...})^2 \quad (16.15)$$

If we drop, say, $\gamma_k$ from the model, then the last contribution of Equation (16.15) is dropped. We have therefore arrived at the analysis-of-variance table shown in Table 16.4.

The calculation of the corrected sums of squares will usually be done by computer. However, if it is done by hand, the two-way interaction terms are calculated from the appropriate two-way marginal table in the same way as the interaction term in a two-way equally replicated classification (see the formulae in Section 15.3). The 'Residual' term is then obtained by subtraction.

Some of the steps of the analysis follow from the way we constructed the model. Tests of significance are first carried out for the two-factor interactions. If none are significant, we can proceed to test the main effects $A$, $B$ and $C$. If any of the interaction terms are significant, we must investigate closely, along the lines indicated in Chapter 15. Sometimes a model of the form (16.11) might be usable. Sometimes, the interactions though real, may be small enough compared to the main effects for us to proceed to test these. However, the main message should be clear – if there are significant interactions, *Stop and look at the data*!

## Exercises 16.1

1. Justify the numbers of degrees of freedom listed in Table 16.4.

2. Check the derivations leading to the minimized sums of squares given in this section.

## 16.2 An analysis of Example 16.1

We have already had a brief look at the data in Example 16.1, and our next step is to calculate the anova table. The total sum of squares is obtained in the usual way from Table 16.1, and the main effects and two-factor interactions are obtained from the two-way tables in Table 16.2. The result is shown in Table 16.5.

The angles by connectors interaction is very highly significant, since the $F$-ratio is 10.62 while the 0.1 per cent point of the $F(12, 48)$ distribution is only about 3.5. The reason for this can be seen from Table 16.3, and the appropriate two-way table in Table 16.2. Connector 5 gives results rather different from the others, especially at $6°$. The results for some other connectors at $6°$ also seem to be out of the general pattern. The rather unexpected effect of increase in force necessary with trial number also appears to depend rather heavily on the $6°$ results.

If the residuals are calculated, just two of them stand out. For trial 5, connector 4, at $4°$, we get $-5.21$ when the standard error of the residuals is only 1.31. It seems that this observation may have been misread as 35.0 instead of 45.0. The other large residual is $+3.93$ for trial 5, connector 4, at $2°$, and it is difficult to see why this occurred. Neither of these residuals will account for the large interaction just noted.

The two-way table of interaction terms (means) is shown in Table 16.6. The sum of squares of these terms is $455.02/5 = 91.004$; the $6°$ column contains the three largest terms, and accounts for 54 per cent of the total interaction.

It is difficult to know how to proceed in this example. It may well be that some combinations of angle and connector produce large disconnecting forces, and this would be an important result. Misrecording seems unlikely, and a few misrecorded observations will not account for the strength of the

Table 16.5 *Anova of Example* 16.1 *data*

| Source | CSS | d. f. | MS |
|---|---|---|---|
| Angles | 290.79 | 3 | 96.93 |
| Connectors | 1234.84 | 4 | 308.71 |
| Trials | 37.74 | 4 | 9.44 |
| Angles × connectors | 455.02 | 12 | 37.92 |
| Angles × trials | 65.42 | 12 | 5.45 |
| Trials × connectors | 42.01 | 16 | 2.63 |
| Residual | 171.33 | 48 | 3.57 |
| Total | 2297.15 | 99 | |

Table 16.6 *Interaction terms for angles × connectors*

| Connector | Angle | | | |
|---|---|---|---|---|
| | 0° | 2° | 4° | 6° |
| 1 | 3.065 | 1.445 | − 1.416 | − 3.095 |
| 2 | 0.165 | 1.645 | − 1.215 | − 0.595 |
| 3 | − 0.86 | − 2.48 | 0.26 | 3.08 |
| 4 | − 0.26 | 0.52 | 3.26 | − 3.52 |
| 5 | − 2.11 | − 1.13 | − 0.89 | 4.13 |

Table 16.7 *Anova of Example* 16.1 *omitting the* 6° *results*

| Source | CSS | d.f. | MS |
|---|---|---|---|
| Angles | 48.69 | 2 | 24.34 |
| Connectors | 644.51 | 4 | 161.13 |
| Trials | 10.38 | 4 | 2.60 |
| Angles × connectors | 129.25 | 8 | 16.16 |
| Angles × trials | 29.08 | 8 | 3.64 |
| Trials × connectors | 32.99 | 16 | 2.06 |
| Residual | 123.65 | 32 | 3.86 |
| Total | 1018.55 | 74 | |

interactions seen. It would be nice to know more about the conditions and conduct of the experiment, as for example, there may be environmental effects, such as temperature, or there may be some 'unit error' effect, to do with the setting up of the apparatus at different times. As we don't know any of the details, there is not much we can do.

Table 16.7 above shows the analysis of variance omitting all of the results at 6°. The trials effect has now disappeared, and although the angles by connectors interaction is still significant at the 0.5 per cent level, it is drastically reduced. However, if the two-way table of interaction terms for angles × connectors is examined, such as by drawing a graph, it will be seen that they are still too large to make any sense of main effects.

In the original experiment, 'connectors' was considered as a random effect. If we had been able to discount all interactions, a component of variance for connectors could have been estimated, but it would be difficult to justify making any such estimates is Example 16.1. A main effect due to trials can be estimated, but this is not very strong, and it is of little interest in the experiment.

If the interaction had not been so strong then, for example, we might have gone ahead and fitted main effects due to connectors and angles, and merely noted the interaction with the estimates.

## Exercises 16.2

1. The data given in Table 16.8 were discussed by Wilkie (1962). In an experiment, a roughened rod was placed in a smooth pipe, and air was blown down the annular space between them. The roughness of the rod consisted of isolated ribs of various heights and pitches. At a position downstream in the airflow, the position of maximum velocity was noted. The response measured was the distance $r$ in inches from the centre of the pipe. Three factors were involved:

$A$: rib height (0.010, 0.015 and 0.020 inches, denoted levels 0, 1 and 2 respectively)

$B$: rib pitch (0.125 and 0.250 inches, denoted levels 0 and 1 respectively)

$C$: Reynolds number (six levels equally spaced logarithmically, denoted levels 0 to 5 respectively)

The numbers recorded below are values of $(r - 1.4)10^3$. Analyse the data and report on your conclusions.

2. Voltage regulators fitted to private motor cars are required to operate within the range 15.8 to 16.4 volts, and the following experiment was conducted to estimate the components of variance. The normal procedure was for a regulator from the production line to be passed to one of a number of setting stations, where an operative adjusted the regulator on a test rig. These regulators then passed to one of four testing stations, where the regulator was

Table 16.8

| Levels of | | | | Levels of | | | |
|---|---|---|---|---|---|---|---|
| $A$ | $B$ | $C$ | Response | $A$ | $B$ | $C$ | Response |
| 0 | 0 | 0 | − 24 | 0 | 0 | 3 | 8 |
| 1 | 0 | 0 | 33 | 1 | 0 | 3 | 57 |
| 2 | 0 | 0 | 37 | 2 | 0 | 3 | 95 |
| 0 | 1 | 0 | − 13 | 0 | 1 | 3 | 24 |
| 1 | 1 | 0 | 39 | 1 | 1 | 3 | 51 |
| 2 | 1 | 0 | 96 | 2 | 1 | 3 | 95 |
| 0 | 0 | 1 | − 23 | 0 | 0 | 4 | 29 |
| 1 | 0 | 1 | 28 | 1 | 0 | 4 | 74 |
| 2 | 0 | 1 | 79 | 2 | 0 | 4 | 101 |
| 0 | 1 | 1 | − 3 | 0 | 1 | 4 | 29 |
| 1 | 1 | 1 | 36 | 1 | 1 | 4 | 50 |
| 2 | 1 | 1 | 76 | 2 | 1 | 4 | 103 |
| 0 | 0 | 2 | 1 | 0 | 0 | 5 | 23 |
| 1 | 0 | 2 | 45 | 1 | 0 | 5 | 80 |
| 2 | 0 | 2 | 79 | 2 | 0 | 5 | 111 |
| 0 | 1 | 2 | 4 | 0 | 1 | 5 | 27 |
| 1 | 1 | 2 | 44 | 1 | 1 | 5 | 62 |
| 2 | 1 | 2 | 96 | 2 | 1 | 5 | 106 |

Table 16.9

| Setting station | Regulator | Testing stations | | | |
|---|---|---|---|---|---|
| | | 1 | 2 | 3 | 4 |
| A | 1 | 16.5 | 16.5 | 16.6 | 16.6 |
| | 2 | 15.8 | 16.7 | 16.2 | 16.3 |
| | 3 | 16.2 | 16.5 | 15.8 | 16.1 |
| | 4 | 16.3 | 16.5 | 16.3 | 16.6 |
| | 5 | 16.2 | 16.1 | 16.3 | 16.5 |
| | 6 | 16.9 | 17.0 | 17.0 | 17.0 |
| | 7 | 16.0 | 16.2 | 16.0 | 16.0 |
| | 8 | 16.0 | 16.0 | 16.1 | 16.0 |
| B | 1 | 16.0 | 16.1 | 16.0 | 16.1 |
| | 2 | 15.4 | 16.4 | 16.8 | 16.7 |
| | 3 | 16.1 | 16.4 | 16.3 | 16.3 |
| | 4 | 15.9 | 16.1 | 16.0 | 16.0 |
| C | 1 | 16.0 | 16.0 | 15.9 | 16.3 |
| | 2 | 15.8 | 16.0 | 16.3 | 16.0 |
| | 3 | 15.7 | 16.2 | 15.3 | 15.8 |
| | 4 | 16.2 | 16.4 | 16.4 | 16.6 |
| | 5 | 16.0 | 16.1 | 16.0 | 15.9 |
| | 6 | 16.1 | 16.1 | 16.1 | 16.1 |
| | 7 | 16.1 | 16.0 | 16.1 | 16.0 |
| D | 1 | 16.1 | 16.0 | 16.0 | 16.1 |
| | 2 | 16.0 | 15.9 | 16.2 | 16.0 |
| | 3 | 15.7 | 15.8 | 15.7 | 15.7 |
| | 4 | 15.6 | 16.4 | 16.1 | 16.2 |
| | 5 | 16.0 | 16.2 | 16.1 | 16.1 |
| | 6 | 15.7 | 15.7 | 15.7 | 15.7 |
| | 7 | 16.1 | 16.1 | 16.1 | 16.0 |

tested, and if found to be unsatisfactory, it was passed down the production line to be reset.

In the experiment, a random sample of the setting stations took part, and a number of regulators from each setting station were passed through each testing station. The results are shown in Table 16.9. Carry out the analysis of variance. If the mean could be kept constant at 16.1, what percentage of regulators would be unsatisfactory? [Data modified from an experiment reported by Desmond (1954). This is *not* a three-way crossed classification.]

**Further reading**
For the design aspects of factorial experiments, see Cox (1958). For detailed numerical illustrations, see Davies (1963).

# CHAPTER SEVENTEEN

# The generalized linear model

## 17.1 Introduction
In Chapters 3 and 4 the method of maximum likelihood was introduced as a general method by which a model could be fitted to data. In Chapter 5 we specialized by restricting ourselves to normally distributed random variables, and to cases where the model is linear in the unknown parameters. Most of the rest of the text has been taken up with the applications of the theory of Chapter 5, and with certain practical points related to these applications. The methods of transformation discussed in Chapter 8 showed how the theory could be applied to a slightly wider class of models than those immediately satisfying the restrictions of Chapter 5. This chapter introduces some important work, started by Nelder and Wedderburn (1972), by which the ideas behind the theory of the model linear in the parameters can be applied to a very much wider class of models. The range of possibilities is vast, and we can do no more here than introduce the topic, and leave the reader to explore for himself through the 'Further reading' list. In particular, there is a computer package available, called 'GLIM', which is written to facilitate use of the generalized linear model. The exposition here is written to link directly into the further reading list, particularly to GLIM, but readers are advised to study this chapter carefully first. However, the concept of the generalized linear model is important in its own right.

Some of the underlying ideas can be put as follows. Suppose we have observations $y_i, i = 1, 2, \ldots, n$, and if, for example, we take the continuous case, we assume that these observations are drawn independently from a distribution with a p.d.f. $f(y_i, \eta_i, \phi)$ belonging to a special class to be defined in Section 17.3. The parameter $\phi$ is usually a nuisance parameter, such as $\sigma^2$ in the theory discussed in Chapter 5, but the parameter $\eta$ is taken to have the form

$$\eta_i = \sum_{j=1}^{p} \beta_j x_{ji} \qquad (17.1)$$

*Example* 17.1 Suppose that the distribution $f$ is normal, with a variance

$\phi = \sigma^2$ and expectation

$$E(Y_i) = \eta_i = \alpha + \beta x_i + \gamma x_i^2$$

where the $x_i$ are known, and the $y_i$ are response variables. This is simply the case of polynomial regression, which was discussed in Chapter 7.

□ □ □

*Example* 17.2  Suppose that groups of $n_1, n_2, \ldots, n_g$, insects are sprayed with an insecticide of strength $x_1, x_2, \ldots, x_g$, for the different groups. The probability of observing $r_i$ deaths in the $i$th group is

$$^nC_r \theta_i^r (1 - \theta_i)^{n-r}$$

where we need a model to relate $\theta$ to $x$. One such model is the logistic model

$$\theta(x) = e^{\eta i}/(1 + e^{\eta i})$$

where

$$\eta_i = \alpha + \beta x_i.$$

□ □ □

In these examples the parameter $\eta_i$ of the distribution of the response variables is related to known quantities $x_i$, through a model linear in the parameters, but $\eta_i$ is not necessarily equal to the mean, and the distribution is not necessarily normal.

In general, therefore, the model (17.1) has $(p + 1)$ unknown parameters, $\beta_1, \ldots, \beta_p$, and $\phi$, which can be estimated by maximum likelihood, and an iterative method of solving the likelihood equations will usually be necessary.

The terms of Equation (17.1) could be such that the $x$'s are explanatory variables of the kind we meet with in regression analysis, or the $x$'s could be $(1, 0)$ variables so that the corresponding $\beta_j$'s represent row, column or treatment effects as in Chapter 10, or a mixture of both types of variable.

We now need a method of testing whether or not some of the parameters can be deleted from the model, or be given special values, parallel to the methods using the 'extra sum of squares' principle and analysis of variance, which have been used throughout the book. The answer is given by an asymptotic method called the maximum likelihood ratio test, which leads to a procedure equivalent to the analysis of variance when the model satisfies the conditions of Chapter 5. This is described in the next section.

## 17.2  The maximum likelihood ratio test

The maximum likelihood ratio test can be described as follows. Let $\mathbf{x} = (x_1, \ldots, x_n)$ have a joint p.d.f. $f(\mathbf{x}, \boldsymbol{\theta})$, and suppose that we are considering two hypotheses,

$$H_0 : \boldsymbol{\theta} \in \omega \qquad \text{and} \qquad H_1 : \boldsymbol{\theta} \in \Omega - \omega \qquad (17.2)$$

where $\Omega$ is thought of as specifying a 'full' model. For example, for the p.d.f.

outlined at the end of Section 17.1 we could have

$$\Omega = (\beta_1, \ldots, \beta_p, \phi) \qquad (17.3)$$

$$\omega = (\beta_1, \ldots, \beta_s, 0, \ldots, 0, \phi) \qquad (17.4)$$

where $s < p$. In this case the hypothesis $H_0$ is a hypothesis that $(p - s)$ of the $p$ parameters $(\beta_1, \ldots, \beta_p)$ are zero, whereas under $\Omega$ any parameter may take any value. We now write the likelihood

$$l(\mathbf{x}, \boldsymbol{\theta}) = \prod_{i=1}^{n} f(x_i, \boldsymbol{\theta})$$

and we consider the statistic

$$W = \left\{ \max_{\boldsymbol{\theta} \in \omega} l(\mathbf{x}, \boldsymbol{\theta}) \right\} \bigg/ \left\{ \max_{\boldsymbol{\theta} \in \Omega} l(\mathbf{x}, \boldsymbol{\theta}) \right\} \qquad (17.5)$$

which falls in the range $0 < W < 1$.

The statistic $W$ is clearly a reasonable test statistic for testing the hypotheses (17.2) since it will tend to be close to unity if $H_0$ is true, and close to zero if $H_1$ is true. We now rely on a result that, provided the conditions for Theorem 4.2 hold, the asymptotic distribution of

$$R = -2 \log W$$

is $\chi_v^2$, where $v$ is the difference between the number of parameters for $\omega$ and the full model $\Omega$.

*Example* 17.3   Suppose we observe $(y_i, x_i)$, for $i = 1, 2, \ldots, n$, where the $x$'s are observed without error and the $y$'s are drawn independently from normal distributions with expectations and variances

$$E(Y) = \alpha + \beta(x - \bar{x})$$

$$V(Y) = \sigma^2.$$

First let us suppose that $\sigma^2$ is known, and that we wish to consider the hypothesis $H_0 : \beta = 0$. This can be put in the above form by writing

$$\Omega = (\alpha, \beta) \qquad \omega = (\alpha) \qquad \text{for } -\infty < \alpha, \beta < \infty.$$

The likelihood is

$$l(\alpha, \beta) = \frac{1}{(2\pi)^{n/2} \sigma^n} \exp \left\{ -\frac{1}{2\sigma^2} \sum [y_i - \alpha - \beta(x_i - \bar{x})]^2 \right\}.$$

Under the full model $\Omega$ we maximize this for choice of $\hat{\alpha}, \hat{\beta}$, and this leads to

$$\hat{\alpha} = \bar{y}, \quad \hat{\beta} = CS(y, \ x) / CS(x, x) \qquad (17.6)$$

which are the estimates used in Chapter 2. The maximized likelihood is

$$\max_{\theta \in \Omega} l(\alpha, \beta) = \frac{1}{(2\pi)^{n/2}\sigma^n} \exp\left\{-\frac{1}{2\sigma^2} S_{\min}^{(\Omega)}\right\} \qquad (17.7)$$

where

$$S_{\min}^{(\Omega)} = CS(y, y) - \{CS(x, y)\}^2 / CS(x, x) \qquad (17.8)$$

Similarly, under the model $\omega$ we obtain $\hat{\alpha} = \bar{y}$, and

$$\max_{\theta \in \omega} l(\alpha) = \frac{1}{(2\pi)^{n/2}\sigma^n} \exp\left\{-\frac{1}{2\sigma^2} CS(y, y)\right\}.$$

Therefore the ratio $W$ is

$$W = \exp\left\{-\frac{1}{2\sigma^2} [CS(y, y) - S_{\min}^{(\Omega)}]\right\}$$

so that we find

$$R = \frac{\{CS(x, y)\}^2}{\sigma^2 CS(x, x)}$$

Our general result tells us that the distribution of $R$ is asymptotically $\chi_1^2$ if $H_0$ is true, since the difference between the number of parameters in the models $\omega$ and $\Omega$ is one. However, we know from Chapter 2 that the distribution is exactly $\chi_1^2$ in this case. □ □ □

As a further illustration of the use of the maximum likelihood ratio test, we now repeat Example 17.3, but treat $\sigma^2$ as unknown.

*Example* 17.4  If, in Example 17.3, $\sigma^2$ is unknown, we define

$$\Omega' = (\alpha, \beta, \sigma^2) \qquad \omega' = (\alpha, \sigma^2).$$

Then under $\Omega'$, maximum likelihood estimates of $\alpha$ and $\beta$ are as given in Equation (17.6) and we find that

$$\hat{\sigma}^2 = S_{\min}^{(\Omega')}/n$$

where $S_{\min}^{(\Omega')}$ is again defined by Equation (17.8). Therefore the maximized likelihood is

$$\max_{\theta \in \Omega} l(\alpha, \beta, \sigma^2) = \frac{1}{(2\pi)^{n/2}[S_{\min}^{(\Omega')}/n]^{n/2}} \exp\left(-\frac{n}{2}\right) \qquad (17.9)$$

Under $\omega$ the maximum likelihood estimates are

$$\hat{\alpha} = \bar{y} \qquad \hat{\sigma}^2 = CS(y, y)/n$$

so that we find

$$\max_{\theta\in\omega'} l(\alpha, \sigma^2) = \frac{1}{(2\pi)^{n/2}[CS(y,y)/n]^{n/2}} \exp\left(-\frac{n}{2}\right)$$

Therefore the ratio $\omega$ is

$$W = \{S_{\min}^{(\Omega')}/CS(y,y)\}^{n/2}$$

and our general result tells us that $R = -2 \log W$ is asymptotically distributed as $\chi_2^2$ if $H_0$ is true. Again, we have exact distribution theory for the sums of squares in this case, given in Theorems 5.7 and 5.8, and we do not need to rely on the asymptotic result. $\qquad \square\square\square$

For a detailed discussion of the properties of the maximum-likelihood ratio test, see Cox and Hinkley (1974); it has optimal asymptotic properties under the same sort of conditions under which Theorem 4.2 applies. It is a method which is used a great deal in testing composite hypotheses of the sort we deal with below.

### 17.3 The family of probability distributions permitted
The general linear model which we now describe has a structure a little more complex than the discussion of Section 17.1 implies, in order to leave plenty of flexibility. We write the observations – assumed independent – as $y_i$, for $i = 1, 2, \ldots, n$, as before.

The probability distribution is taken to have the form

$$f(y) = \exp\{[y\,a(\mu) - b(\mu)]/\phi + c(y, \phi)\} \qquad \text{where} \qquad \mu = E(Y) \qquad (17.10)$$

This is a particular case of a family of distributions called the *exponential* family. For a detailed discussion of this family of distributions, see Cox and Hinkley (1974). For our purposes here, all we need to note is that by choice of $a(\mu), b(\mu), c(y, \phi)$ and $\phi$, Equation (17.10) can represent a very wide range of distributions; see Table 17.1 for illustrations.

In all of the applications below, the means of the observations differ, so that $\mu_i$ depends on $i$, as in linear regression. In principle the method to be outlined allows a different distribution to be specified for each observation.

Some properties of the distribution (17.10) can now be worked out. For one observation, the log-likelihood is

$$L(y, \mu, \phi) = \{y\,a(\mu) - b(\mu)\}/\phi + c(y, \phi)$$

so that

$$\frac{\partial L}{\partial \mu} = \{y\,a'(\mu) - b'(\mu)\}/\phi \qquad (17.11)$$

Table 17.1 *Illustrations of distributions of the form* (17.7)

| No. | Range of y | $a(\mu)$ | $b(\mu)$ | $\phi$ | $c(y, \phi)$ | $f(y)$ |
|-----|-----------|----------|----------|--------|--------------|--------|
| 1 | $-\infty < y < \infty$ | $\mu$ | $\frac{1}{2}\mu^2$ | $\sigma^2$ | $-\frac{1}{2}[\log(2\pi\sigma^2) + y^2/\sigma^2]$ | $N(\mu, \sigma^2)$ |
| 2 | $y = 0, 1, 2, \ldots$ | $\log \mu$ | $\mu$ | 1 | $-\log y!$ | Poisson $e^{-\mu}\mu^y/y!$ |
| 3 | $y = 0, 1, \ldots, n$ | $\log\{p/(1-p)\}$ | $-n\log(1-p)$ | 1 | $\log\binom{n}{y}$ | Binomial $\binom{n}{y}p^y(1-p)^{n-y}$ |
| 4 | $0 < y < \infty$ | $-\alpha/\beta$ | $\log \alpha$ | $1/\beta$ | $(\beta - 1)\log y - \log \Gamma(\beta)$ | Gamma $\alpha(\alpha y)^{\beta-1}e^{-\alpha y}/\Gamma(\beta)$ (with $\beta$ known) |

However, it follows from an obvious extension of Equation (3.23) that

$$E\left(\frac{\partial L}{\partial \mu}\right) = 0$$

so that from Equation (17.11), and by definition of $\mu$, we have

$$E(Y) = b'(\mu)/a'(\mu) = \mu \tag{17.12}$$

This equation is a condition on the functions $a(\mu)$ and $b(\mu)$ in Equation (17.10), resulting from the definition of $\mu$. It can be readily checked that Equation (17.12) is satisfied for the distributions of Table 17.1.

Further, by a small extension of Equation (3.24) we have

$$E\left(\frac{\partial L}{\partial \mu}\right)^2 = -E\left(\frac{\partial^2 L}{\partial \mu^2}\right) \tag{17.13}$$

and by using this it is possible to get an expression for $V(Y)$ in terms of the functions $a(\mu)$ and $b(\mu)$. The parameter $\phi$ determines the scale, and for most applications of the GLIM this is assumed to be estimable from replicated observations, or else known separately.

If we have $n$ independent observations from the distribution in (17.10), the log-likelihood is

$$L = \sum_i \{y_i a(\mu_i) - b(\mu_i)\}/\phi + \text{constant}$$

The expectations $\mu_i$ are related to a set of parameters $\beta_j$ through: (*a*) an expression $\eta_i$, defined in Equation (17.1), which is linear in the parameters; and (*b*) a link function relating $\mu_i$ to $\eta_i$. The link function is discussed in the next section.

**Exercise 17.3**
1. Obtain $V(Y)$ of distributions of the form (17.10), as a function of $a(\mu)$, $b(\mu)$ and their derivatives.

## 17.4 The generalized linear model

The final step in constructing a generalized linear model is to link the probability distribution (17.10) with (17.1). This is done through a *link function*, which links the mean $\mu_i$ to the parameter $\eta_i$ of the distribution, in the form

$$\eta_i = g_i(\mu_i) \tag{17.14}$$

where $\mu_i$ is the expectation (17.12) of the $i$th observation, and where $\eta_i$ is given by Equation (17.1). Usually we shall use one link function $g$, for all observations, rather than specifying one for each observation. Sometimes we can use the *identity link*, $\eta = \mu$, as in Example 17.5 below. However, since $\eta$ can take values in the range $(-\infty, \infty)$, and for example the Poisson mean has the range $(0, \infty)$, some non-identity link is required. Some examples will help to explain the ideas.

*Example* 17.5   If we use case (1) of Table 17.1, with the identity link function,

$$\eta_i = \mu_i$$

and

$$\mu_i = \alpha + \beta(x_i - \bar{x})$$

then we generate the normal distribution theory regression model which was discussed in Examples 17.3 and 17.4.                                    ☐ ☐ ☐

*Example* 17.6   In Example 13.4 we discussed a randomized block experiment on the weed control of cereals, in which the response variable on each cell was expected to have a Poisson distribution, and we therefore have case (2) of Table 17.1. In this problem we can use the link function

$$\eta_{ij} = \log \mu_{ij} \tag{17.15}$$

where $\eta_i$ has the form (17.1), with constants to represent the treatment and the block effects. Therefore we might have, for example,

$$\mu_{ij} = \exp(\eta_{ij}) = \exp\{\mu + \beta_i + \gamma_j\} \tag{17.16}$$

where $\beta_i$ are parameters representing the block effects, and the $\gamma_j$ are parameters representing the treatment effects.                              ☐ ☐ ☐

*Example* 17.7   In the testing of plastic conduit pipe, one-foot lengths were put in a rig, and a striker of a given weight was dropped from 2 metres. The numbers of lengths which failed on impact are noted below:

| Weight of striker (kg) | 5.0 | 5.2 | 5.5 | 5.7 | 6.0 |
|---|---|---|---|---|---|
| No. of lengths tested | 12 | 12 | 12 | 12 | 12 |
| No. failed | 2 | 6 | 9 | 11 | 12 |

In this problem the distribution of the number of pieces of pipe failing is binomial, with $n = 12$, but with the probability of failure $p$ varying with the weight of the striker. We therefore have case (3) of Table 17.1, and a link function which is frequently used (see Cox, 1970) is

$$p_i = e^{\eta_i}/(1 + e^{\eta_i}) \qquad (17.17)$$

where

$$\eta_i = \alpha + \beta x_i \qquad (17.18)$$

and $x_i$ is the weight of the striker. The expression (17.17) is called the *logistic link function*.                                               □ □ □

These examples will suffice to show the flexibility of the method. The form (17.10) for the probability distribution can be made to represent a variety of distributions by suitable choice of the functions $a(\mu_i), b(\mu_i), c(y, \phi)$, and the constants, while having fixed the distribution, the link function enables us to represent a variety of structures using a model linear in the parameters for $\eta$.

Two classes of models are not covered by the generalized linear model:

(i) Problems where the probability distribution cannot be represented by the form (17.10), such as the Cauchy or logistic distributions.
(ii) Problems where the underlying model is intrinsically non-linear, such as models including relationships (ix) to (xi) of Section 8.2.

Nevertheless, Examples 17.5 to 17.7 show that a very large range of models can be covered, and a more detailed example is given below.

Once we have set up a possible model, we wish to investigate whether or not some of the parameters are zero, or take on other special values. For this we apply the asymptotic theory of Section 17.2, and the next section gives a discussion of how this works out.

### 17.5 The analysis of deviance
First we will deal with those models where the scale parameter is known although the principles involved do not depend on this restriction. Following the method introduced in Section 17.2, we can test hypotheses about parameters in our model by analysing the differences of the quantity

$$- 2 L_{max}$$

where $L$ is the log-likelihood. We call this quantity the *deviance*, and in order to test hypotheses we compare differences of the deviance with the relevant $\chi^2$-distribution.

There are two points on the deviance scale which are of special note. If there are $n$ observations, and a model is fitted with $n$ linearly independent parameters, we call this the *full* model, and denote the appropriate deviance

$$- 2\, L_{\max}^{(f)}$$

In most problems there will be a *minimal* model which we are willing to consider, such as one having block effects only in a randomized block experiment. We denote the deviance of the minimal model

$$- 2\, L_{\max}^{(m)}$$

When we are at the stage of formulating our model, the deviance of the current model under consideration will lie between these two extremes.

If we denote the maximum-likelihood estimates of the $\mu_i$ under the full and current models as $\hat{\mu}_i^{(f)}$ and $\hat{\mu}_i^{(c)}$ respectively, then from Equation (17.10) the difference in the deviance between the current and full models is

$$2 \sum_{i=1}^{n} \{y_i(a(\hat{\mu}_i^{(f)}) - a(\hat{\mu}_i^{(c)}) + b(\hat{\mu}_i^{(c)}) - b(\hat{\mu}_i^{(f)})\}/\phi \qquad (17.19)$$

If $\phi$ is known, we can multiply through by this and consider the terms

$$D(c,f) = 2\sum \{y_i(a(\hat{\mu}_i^{(f)}) - a(\hat{\mu}_i^{(c)})) + b(\hat{\mu}_i^{(c)}) - b(\hat{\mu}_i^{(f)})\} \qquad (17.20)$$

For Example 17.5 the full model yields $\hat{\mu}_i = y_i$, and $D(c,f)$ is the residual sum of squares under the current model. The computer package GLIM enables one to calculate the quantities (17.20), which can be used instead of $-2\,L_{\max}$ to test hypotheses.

If the scale parameter $\phi$ is not known, then an independent estimate of it can be used if this is available. However, the principle outlined in Section 17.2 can be used for testing hypotheses about the model in any case.

A final note here is that little is known about the approximation of the $\chi^2$-distribution to the distribution of differences of deviance. For some discussion of this see Baker and Nelder (1978).

### 17.6 Illustration using the radiation experiment data

In order to illustrate the techniques described in this chapter, we use the radiation-experiment data given in Exercise 10.7.4. The GLIM package was used to do the calculations. The first step in any analysis is to look at the data, and Fig. 17.1 shows a plot of the percentages of the embryos reaching mid-mitosis. From this figure we see that there are very clear differences between experiments 1 and 3, and 2 and 4. Further, the control groups reach much higher percentages than the X-ray or beta-ray treated groups of

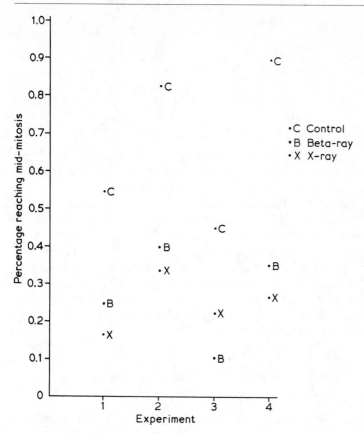

Fig. 17.1 The data for the radiation experiment.

embryos. A peculiarity in the results is that the order of the X-ray and beta-ray results is different in experiment 3 to the order in the other experiments.

In starting the model building, we take the response variable to be the number of embryos reaching mid-mitosis. If we denote the treatments control, X-ray, beta-ray as 1, 2 and 3 respectively, then the data can be denoted $y_{ij}$, for $i = 1, 2, 3, 4$ and $j = 1, 2, 3$. We shall assume the distribution of $y_{ij}$ to be binomial with parameters $(n_{ij}, p_{ij})$, where $n_{ij}$ is the number of embryos observed, and $p_{ij}$ is the probability of reaching mid-mitosis, in cell $(i, j)$. We shall use the link function of the form (17.17):

$$p_{ij} = e^{\eta_{ij}}/(1 + e^{\eta_{ij}})  \tag{17.21}$$

where

$$\eta_{ij} = \mu + \beta_i + \gamma_j  \tag{17.22}$$

and where $\beta_i$ and $\gamma_j$ are experiment and treatment parameters respectively. Table 17.2 shows the deviances for some models based on Equation (17.22).

Table 17.2 *Deviances for radiation experiment*

|   | Model for $\eta_{ij}$ | Deviance | d. f. | Difference deviance | d. f. |
|---|---|---|---|---|---|
| 1 | $\mu$ | 54.84 | 11 | — | — |
| 2 | $\mu + \beta_i$ | 43.18 | 8 | 11.66 | 3 |
| 3 | $\mu + \beta_i + \gamma_j$ | 4.96 | 6 | 38.22 | 2 |

The deviance differences shown in Table 17.2 can be regarded as approximately distributed as $\chi^2$ on the specified numbers of degrees of freedom. Line 2 shows that the experiments effect is highly significant, on a one-sided test, and line 3 shows that the treatments effect is also highly significant.

If now we fit a model including parameters for the experiments and the control treatment, but only one parameter for the X-ray and beta-ray results, we obtain a deviance of 5.04 on 7 d.f., which is a fit as good as that of model 3 of Table 17.2. The difference between the deviances is not significant, showing that there is no significant difference between the X-ray and beta-ray results. In a similar way, it can be shown that there is no significant difference between experiments 1 and 3, or between 2 and 4.

The fitted parameters for model 3 are as follows:

$$\hat{\mu} = \quad 0.367 \qquad \hat{\beta}_1 = \quad 0 \qquad \hat{\beta}_2 = 1.039$$
$$\hat{\beta}_3 = -0.295 \qquad \hat{\beta}_4 = \quad 0.955 \qquad \hat{\gamma}_1 = 0$$
$$\hat{\gamma}_2 = -2.940 \qquad \hat{\gamma}_3 = -1.849$$

(The GLIM package sets the first parameter to zero when we have a model of less than full rank, and so we obtain $\hat{\beta}_1 = \hat{\gamma}_1 = 0$.) Therefore, for example, the fitted value for cell $(2,2)$ would be Equation (17.21) with

$$\hat{\eta}_{22} = 0.367 + 1.039 - 2.940 = -1.534$$

This leads to $\hat{p}_{11} = 0.5907$, and an expected value for the response variable of

$$n_{11}\hat{p}_{11} = 22 \times 0.5907 = 13.0$$

It is now possible to calculate *standardized residuals*,

$$r_{11} = \{y_{11} - n_{11}\hat{p}_{11}\}/\sqrt{(n\hat{p}_{11}(1 - \hat{p}_{11}))} \qquad (17.23)$$

and then analyse these by methods similar to those used in previous chapters. It is important to notice when using formulae like (17.23) that the $\hat{p}_{ij}$ are correlated with $y_{ij}$, and the $r_{ij}$ are correlated with each other. The denominator of Equation (17.23) is not quite the standard error of the numerator. However, these standardized residuals are likely to show up some of the departures from the model.

Asymptotic standard errors for the parameters can also be calculated, using the asymptotic variance for maximum-likelihood estimators given by Equation (4.19).

This will suffice to illustrate the technique. The generalized linear model is indeed a very powerful tool, especially when it is coupled with a computer package which enables the calculations and plotting to be done painlessly. The model covers logit (or probit) analysis, which is illustrated in Example 17.7, models for the analysis of contingency tables, and a large range of models. More details of the method are given in the further-reading list, to which the reader is now invited to turn, to explore the field opened up by this technique.

---

**Exercise 17.6**
1. Show how to use the approach of this chapter to analyse a two-way contingency table. Assume that the distributions of all counts are all Poisson, and use the link function

$$\eta_{ij} = \log \mu_{ij}$$

where $\mu_{ij}$ is the expectation for cell $(i,j)$. How does the analysis proceed if there are polynomial effects possible in both rows and columns?

---

**Further reading**
For further reading on the generalized linear model see Nelder and Wedderburn (1972), and the GLIM manual, which is Baker and Nelder (1978). For some rather similar work see Cox (1970).

# References

Anderson, T.W. (1958) 'An Introduction to Multivariate Analysis'. Wiley, New York.

Anscombe, F.J., (1960) 'Rejection of outliers', *Technometrics*, **2**, 123–47.

Anscombe, F.J. & Tukey, J.W. (1963) 'Examination and analysis of residuals', *Technometrics*, **5**, 141–60.

Armitage, P. (1959) 'The comparison of survival curves', *J.R. Stat. Soc.* A, **122** (3), 279–300.

Bain, W.A. & Batty, J.E. (1956) 'Inactivation of adrenaline and noradrenaline by human and other mammalian liver in vitro', *Br.J. Pharmacol.*, **11**, 52–7.

Baker, R.J. & Nelder, J.A. (1978) 'The GLIM System; Release 3*, Royal Statist. Soc., London.

Barnard, G.A. Jenkins, G.M., & Winsten, C.B. (1962) 'Likelihood inference and time series' (with discussions), *J.R. Stat Soc.* A, **125** (3), 321–72.

Barnett, V.D. (1966) 'Evaluation of the maximum-likelihood estimator where the likelihood equation has multiple roots', *Biometrika*, **53** (1 & 2), 151–68.

Bartlett, M.S. (1935) 'The effect of non-normality on the $t$-distribution', *Proc. Camb. Phil. Soc.*, **31**, 223–31.

Bartlett, M.S. (1936) 'The square root transformation in analysis of variance', *Suppl. J.R. Statist. Soc.*, **3**, 68–78.

Bartlett, M.S. (1937) 'Properties of sufficiency and statistical tests', *Proc. R. Soc.* A, **160**, 268–82.

Bartlett, M.S. (1947) 'The use of transformations', *Biometrics*, **3**, 39–52.

Beale, E.M.L., Kendall, M.G. & Mann, D.W. (1967) 'The discarding of variables in multivariate analysis', *Biometrika*, **54**, 357–66.

Ben Israel, A. & Greville, T.N.E. (1974) *Generalised Inverses: Theory and Applications*, Wiley, New York.

Bennett, C.A. (1954) 'Effect of measurement error on chemical process control, *Ind. Qual. Control*, **10**(4), 17–20.

Bennett, C.A. & Franklin, N.L. (1954) *Statistical Analysis in Chemistry and the Chemical Industry*, Wiley, New York.

Berk, K.N. & Francis, I.S. (1978) 'A review of the manuals for BMDP and SPSS', *J. Amer. Stat. Ass.*, **73**, 65–70 (plus discussion to p. 98).

Berkson, J. (1953) 'A statistically precise and relatively simple method of estimating the bio-assay with quantal response, based on the logistic function', *J. Amer. Stat. Ass.*, **48**, 565–99.

Berkson, J. (1955) 'Maximum likelihood and minimum $\chi^2$ estimates of the logistic function'. *J. Amer. Stat. Ass.*, **50**, 130–62.

Box, G.E.P. (1954) 'Some theorems on quadratic forms applied in the study of analysis of variance problems: I. Effect of inequality of variance in the one-way classification', *Ann. Math. Statist.*, **25**, 290–302.

Box, G.E.P. & Andersen, S.L. (1975) 'Permutation theory in the derivation of robust criteria and the study of departures from assumption', *J.R. Stat. Soc.*, B, **17**, 1–34.

Box, G.E.P. & Cox, D.R. (1964) 'An analysis of transformations' (with discussion), *J.R. Stat, Soc.* B, **26**, (2) 211–52.

Box, G.E.P. & Newbold, P. (1971) 'Some comments on a paper by Coen, Gomme and Kendall, *J.R. Stat. Soc.* A., **134**, 229–40.

Box, G.E.P., & Tidwell, P.W. (1962) 'Transformations of the independent variables', *Technometrics*, **4**, 531–50.

Bradu, D. & Gabriel, K.R. (1974) 'Simultaneous statistical inference on interactions in two-way analysis of variance', *J. Amer. Stat. Ass.*, **69**, 428–36.

Brown, M.R. (1975) 'Exploring interaction effects in the Anova', *Appl. Statist.*, **24**, 288–98.

Cameron, J.M. (1951) 'The use of components of variance in preparing schedules for sampling of baled wool', *Biometrics*, **7**, 83–96.

Chambers, J.M. (1977) *Computational Methods for Data Analysis*, Wiley, New York.

Chambers, J.M. & Ertel, J.E. (1974) 'Remark AS R11', *Appl. Statist.*, **23**, 250–1.

Chernoff, H. & Lieberman, G.J. (1954) 'Use of normal probability paper', *J. Amer. Stat. Ass.*, **49**, 778–85.

Chew, V. (1976) 'Comparing treatment means: a compendium', *Hortscience*, **11**(4), 348–57.

Church, A.E.R. (1925) 'On the moments of the distribution of squared standard deviations for samples of $N$ drawn from an indefinitely large population', *Biometrika*, **17**, 79–83.

Cochran, W.G. (1941) 'The distribution of the largest of a set of estimated variances as a fraction of their total', *Ann. Eugenics London*, **11**, 47–52.

Cochran, W.G. & Cox, G.M. (1957) *Experimental Design*, Wiley, New York.

Coen, P.J., Gomme, E.D. & Kendall, M.G. (1969) 'Lagged relationships in economic forecasting' (with discussion), *J.R. Stat. Soc.* A, **132** (2), 133–63.

Cox, D.R. (1958) *The Planning of Experiments*, Wiley, New York.

Cox, D.R. (1968) 'Notes on some aspects of regression analysis', *J.R. Stat. Soc.* A, **131**, 265–79.

Cox, D.R. (1970) *Analysis of Binary Data*, Methuen, London.

Cox, D.R. & Hinkley, D.V. (1974) *Theoretical Statistics*, Chapman & Hall, London.

Cox, D.R. & Snell, E.J. (1968) 'A general definition of residuals' (with discussion), *J.R. Stat. Soc.* B, **30** (2), 248–75.

Cramér, H. (1952) *Mathematical Methods of Statistics*, Princeton U.P.

Daniel, C. (1976) *Application of Statistics to Industrial Experiments*, Wiley, New York.

David, F.N. (1938) *Tables of the Ordinates and Probability Integral of the Distribution of the Correlation Coefficient in Small Samples*, Cambridge U.P.

Davies, O.L. (1963) *Design and Analysis of Industrial Experiments*, Oliver & Boyd, Edinburgh.

Davies, O.L. (1967) *Statistical Methods in Research and Production, with Special Reference to the Chemical Industry*, Oliver & Boyd, Edinburgh.

Dempster, P., Schatsoff, M., & Wermuth, D. (1977) 'A simulation study of alternatives to ordinary least sequares', *J. Amer. Statist. Ass.*, **72–91**.

Desmond, D.J. (1954) 'Quality control on the setting of voltage regulators', *Appl. Statist.*, **3**, 65–73.

Dixon, W. (1972) *B.M.D. Biomedical Computer Programs*, Univ. California Press.

Dolby, J.L. (1963) 'A quick method for choosing a transformation', *Technometrics*, **5**, 317–26.

Draper, N.R. & Smith, H. (1966) *Applied Regression Analysis*, Wiley, New York.

Duncan, A.J. (1955) 'Multiple range and multiple $F$-tests', *Biometrics*, **11**, 1–42.

Dyer, A.R. (1974) 'Comparison of tests for normality with a cautionary note', *Biometrika*, **61**, 185–9.

Enrick, N.L. (1962) 'Variations flow analysis', *Ind. Qual. Control* (July), 23–9.

Ezekiel, M. & Fox, K.A. (1959) *Methods of Correlation and Regression Analysis*, Wiley, New York.

Feigl, P. and Zelen, M. (1965) 'Estimation of exponential survival probabilities with concomitant information', *Biometrics*, 826–38.

Finney, D.J. (1948) 'Main effects and interactions', *J. Amer. Stat. Ass.*, **43**, 566–71.

Finney, D.J. (1960) *An Introduction to the Theory of Experimental Design*, Univ. Chicago Press, Chicago.

Fisher, R.A. (1956) *Statistical Methods for Research Workers*, Oliver & Boyd, Edinburgh.

Fisher, Sir R.A. & Yates, F. (1963) *Statistical Tables* (6th Edn), Oliver & Boyd, Edinburgh.

Galton, F. (1888) *Proc. Roy. Soc., London*, **45**, 135.

Gardner, M.J. (1972) 'On using an estimated regression line in a second sample', *Biometrika*, **59**, 263–74.

Garside, M.J. (1965) 'The best subset in multiple regression analysis', *Appl. Statist.*, **14**, 196–200.

Garside, M.J. (1971) 'Some computational procedures for the best subset problem', *Appl. Statist.*, **20**, 8–15.

Gayen, A.K. (1949) 'The distribution of student's $t$ in random samples of any size drawn non-normal universes', *Biometrika*, **36**, 353–69.

Gayen, A.K. (1950a) 'The distribution of the variance ratio in random samples of any size drawn from non-normal universes', *Biometrika*, **37**, 236–55.

Gayen, A.K. (1950b) 'Significance of the difference between the means of two non-normal samples', *Biometrika*, **37**, 399–408.

Gayen, A.K. (1951) 'The frequency distribution of the product moment correlation coefficient in random samples of any size drawn from non-normal universes', *Biometrika*, **38**, 219–47.

Geary, R.C. (1936) 'The distribution of student's ratio for non-normal samples', *J.R. Stat. Soc.*, Suppl., **3**, 178–84.

Geary, R.C. (1947) 'Testing for normality', *Biometrika*, **34**, 209–42.

Gerson, M. (1975) 'The techniques and uses of probability plotting', *The Statistician*, **24**, 235–57.

Ghurye, S.G. (1949) 'On the use of student's $t$-test in an asymmetrical population', *Biometrika*, **36**, 426–30.

Gnedenko, B.V. (1963) *The Theory of Probability* (2nd edn), Chelsea, New York.

Grubbs, F.E. (1969) 'Procedures for detecting outlying observations in samples', *Technometrics*, **11**, 1–21.

Hartley, H.O. (1950) 'The maximum $F$-ratio as a short-cut test for heterogeneity of variance', *Biometrika*, **37**, 308–12.

Hocking, R.R. (1972) 'Criteria for selection of a subset regression: which one should be used?', *Technometrics*, **14**, 967–70.

Hocking, R.R. (1976) 'The analysis and selection of variables in linear regression', *Biometrics*, **32**, 1–49.

Huff, D. (1973) *How to Lie with Statistics*, Pelican, Harmondsworth.

Janacek, G. & Negus, B. (1974) 'Some observations on design and accuracy of the Biomedical Computer Programs package', *Bull. Inst. Math. and Appl.*, **10**, 166–72.

Jeffers, J.N.R., Howard, D.M. & Howard, P.J.A. (1976) 'An analysis of litter respiration at different temperatures', *Appl. Statist.*, **35**, 139–46.

John, J.A. & Quenouille, M.H. (1977) *Experiments: Design and Analysis*, Griffin, London.

Kempthorne, O. (1975) *The Design and Analysis of Experiments*, Kreiger, Huntington, N.Y.

Kendall, M.G. & Stuart, A. (1973) *The Advanced Theory of Statistics*. Vol. 2, *Inference and Relationship* (3rd edn), Griffin, London.

Kuratori, I.S. (1966) 'Experiments with mixtures of components having lower bounds', *Ind. Qual. Control*, **22**, 592–6.

Lemus, F. (1960) 'A mixed model factorial experiment in testing electrical connectors', *Ind. Qual. Control*, **17**, 2–16.

Mansfield, E. & Wein, H.H. (1958) 'A regression control chart for costs', *Appl. Statist.*, **7**, 48–57.

Miller, R.G. Jr. (1966) *Simultaneous Statistical Inference*, McGraw-Hill, New York.

Moran, M.A. & Wright, N. (1974) 'The performance of a multiple regression equation with and without selection of variables', *Proc. 8th, Internat. Biom. Conf.*, Constantsa, Romania.

Morrison, D.F. (1975) *Multivariate Statistical Methods*, (2nd edn), McGraw-Hill, New York.

Muxworth, D.T. (1974) 'A review of some statistical packages', *Bull. Inst. Math. & Appl.*, **10**, 171–4.

Nelder, J.A. & Mead, R. (1965) 'A simplex method for function minimisation, *Computer J.*, **7**(4).

Nelder, J.A. & Wedderburn, R.M.M. (1972) 'Generalised linear models', *J.R. Stat. Soc.* A, **135**, 370–84.

Norden, R.H. (1972) 'A survey of maximum likelihood estimation, Part 1', *Int. Stat. Rev.*, **40**(3), 329–54.

Norden, R.H. (1973) 'A survey of maximum likelihood estimation, Part 2', *Int. Stat. Rev.*, **41**(1), 329–54.

Olkin, I. & Pratt, J.W. (1958) 'Unbiased estimation of certain correlation coefficients', *Ann Math. Statist.*, **29**, 201.

O'Neill, R. (1971) 'Function minimisation using a simplex procedure. Algorithm AS27', *Appl. Statist.*, **20**, 388–45.

O'Neill R. & Wetherill, G.B. (1971) 'The present state of multiple comparison methods', *J.R. Stat. Soc.* B., **33**, 218–50.

Parzen, E. (1960) *Modern Probability Theory and its Applications*, Wiley, New York.

Pearce, S.C. (1965) *Biological Statistics; an Introduction*, McGraw-Hill, New York.

Pearson, E.S. (1931) 'The analysis of variance in cases of non-normal variation', *Biometrika*, **23**, 114–33.

Pearson, E.S. & Hartley, H.O. (1970) *Biometrika Tables for Statisticians*, Vol. 1. (3rd edn), Cambridge U.P.

Pearson, E.S. & Please, N.W. (1975) 'Relations between the shape of population distribution and the robustness of four simple test statistics', *Biometrika*, **62**, 223–41.

Pearson, K. & Lee, A. (1902–3), *Biometrika*, **2**, 357–462.

Prater, N.H. (1956) 'Estimate gasoline yields from crudes', *Petroleum Refiner*, Vol. 35, No. 5.

Rao, C.R. (1965) *Linear Statistical Inference and its Applications*, Wiley, New York.

Rao, C.R. (1971) *Generalised Inverse of Matrices and its Applications*, Wiley, New York.

Sampford, M.R. & Taylor J. (1959) 'Censored observations in randomised block experiments', *J.R. Stat. Soc.* B, **21**(1), 214–37.

Sarhan, A.E. & Greenberg, B.G. (1956) 'Estimation of location and scale parameters by order statistics from singly and doubly censored samples, Part I. The normal distribution up to samples of size 10', *Ann. Math. Statist.*, **27**, 427–51.

Satterthwaite, F.E. (1946) 'An approximate distribution of estimates of variance components', *Biometrics Bull.*, **2**, 110–14.

Scheffé, H. (1959) *The Analysis of Variance*, Wiley, New York.

Searle, S.R. (1971) *Linear Models*, Wiley, New York.

Seber, G.A.F. (1977) *Linear Regression Analysis*, Wiley, New York.

Shapiro, S.S. & Francia, R.S. (1972) 'An approximate analysis of variance test for normality', *J. Amer. Stat. Ass.*, **67**, 215–6.

Shapiro, S.S. & Wilk, M.B. (1965) 'An analysis of variance test, for normality and complete samples', *Biometrika*, **52**, 592–611.

Shapiro, S.S., Wilk, M.B. & Chen, H.J. (1968) 'A comparative study of various tests for normality', *J. Amer. Stat. Ass.*, **63**, 1343–72.

Silvey, S.D. (1975) *Statistical Inference*, Chapman & Hall, London.

Smith, H. (1961) 'Statistical applications manual', Proctor and Gamble Company.

Smith, H. (1969) 'The analysis of data from a designed experiment', *J. Qual. Tech.*, **1**, 259–63.

Snedecor, G.W. & Cochran, W.G. (1967) *Statistical Methods*, Iowa State U.P., Ames, Iowa.

Snee, R.D. (1971) 'Design and analysis of mixture experiments', *J. Qual. Tech.*, **3**, 159–69.

Snoke, L.R. (1956) 'Specific studies on soil-block procedure for bioassay of wood preservatives', *Applied Microbiology*, **4**, 21–31.

Sprent, P. (1969) *Models in Regression and Related Topics*, Methuen, London.

Sprott, D.A. & Kalbefleisch, J.D. (1969) 'Examples of likelihoods and comparison with point estimates and large sample approximations, *J. Amer. Stat. Ass.*, **64**, 468–84.

Stefansky, W. (1972a) 'Rejecting outliers in factorial designs', *Technometrics*, **14**, 469–79.

Stefansky, W. (1972b) 'Rejecting outliers by maximum normal residual', *Ann. Math. Statist.*,**42**, 35–45.

Stigler, S.M. (1977) 'Do robust estimators work with real data?', *Ann. Stat.*, **5**, 1055–98.

Stone, M. (1974) 'Cross validatory choice and assessments of statistical predictions', *J.R. Stat. Soc.* B, **35**, 111–47.

Tiku, M.L. (1963) 'Approximation to student's *t*-distribution in terms of Hermite and Laguerre polynomials', *J. Indian Math. Soc.*, **27**, 91–102.

Tukey, J.W. (1949) 'One degree of freedom for non-additivity', *Biometrics*, **5**, 232–42.

Tukey, J.W. (1962) 'The future of data analysis', *Ann. Math. Statist.*, **33**, 1–67.

Tukey, J.W. & Moore, P.G. (1954) 'Answer to query 112', *Biometrics*, **10**, 562–8.

Ury, H.K. (1976) 'A comparison of four procedures for multiple comparisons among means (pairwise contrasts) for arbitrary sample sizes', *Technometrics*, **18**, 89–97.

Wald, A. (1949) 'Note on the consistency of the maximum likelihood estimate', *Ann. Math. Statist.*, **20**, 595–601

Welch, B.L. (1936) 'The specification of rules for rejecting too variable a product', *J.R. Stat. Soc*, Suppl., **3**, 29.

Wernimont, G. (1947) 'Quality control in the chemical industry II', *Ind. Qual. Control*, **3**, 5.

Wetherill, G.B. (1960) 'The Wilcoxon text and non-null hypothesis', *J.R. Stat. Soc.* B, **22**, 402–18.

Wetherill, G.B. (1969) *Sampling Inspection and Quality Control*, Methuen, London.

Wetherill, G.B. (1972) *Elementary Statistical Methods*, Chapman & Hall, London.

Wilkie, D. (1962) 'A method of analysis of mixed level factorial experiments', *Appl. Statist.*, **11**, 184–95.

Wilks, S.S. (1962) *Mathematical Statistics*, Wiley, New York.

Williams, E.J. (1959) *Regression Analysis*, Wiley, New York.

Wolfowitz, J. (1949) 'On Wald's proof of the consistency of the maximum likelihood estimate', *Ann. Math. Statist.*, **20**, 601–2.

Wooding, W.M. (1969) 'The computation and use of residuals in the analysis of experimental data', *J. Qual. Tech.*, **1**, 175–88.

# Appendix A: Some important definitions and results

The purpose of this appendix is to give some definitions and results which are used in the text. Precise mathematical statements will not be given; such statements, together with proofs, are available in many textbooks on probability, such as Gnedenko (1963), Parzen (1960), and Rao (1965).

It is important that students have a clear intuitive understanding of the points below, rather than an ability to reproduce certain mathematical proofs. For those willing to accept without proof the central-limit theorem (see below), a discussion of other points is given in Wetherill (1972) and other similar books.

We assume that readers are familiar with the concepts of random variable and probability distribution. Some important material on multivariate probability distributions is given in Section 3.7. We shall denote probability density functions by $f(x), f(x, y)$ for the univariate and bivariate cases. Some definitions are quoted for continuous random variables only, leaving the definitions for discrete random variables to be understood.

*Definition* 1   The *expectation* of a random variable $X$ is defined

$$E(X) = \int x f(x) \, dx$$

and the *variance* is defined as

$$V(X) = E\{X - E(X)\}^2$$
$$= E(X^2) - E^2(X)$$

where
$$E(X^2) = \int x^2 f(x) \, dx$$

*The normal distribution*
The normal distribution has a probability density function (p.d.f.)

$$\frac{1}{\sqrt{(2\pi)}\sigma} \exp\left\{ -\frac{1}{2}\left(\frac{x-\mu}{\sigma}\right)^2 \right\}$$

for $\sigma > 0$, $-\infty < \mu < \infty$, $-\infty < x < \infty$, and $E(x) = \mu$, $V(x) = \sigma^2$. (See Fig.

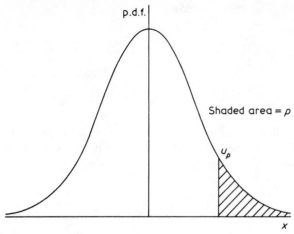

Fig. A.1 The normal distribution.

A.1). This distribution is denoted $N(\mu, \sigma^2)$. In order to refer to the percentage points of the normal distribution we shall use the notation $u_p$, where

$$p = \int_{u_p}^{\infty} \frac{1}{\sqrt{(2\pi)}} \exp\left(-\frac{1}{2}x^2\right) dx$$

Thus from tables we find that $u_{0.025} = 1.96$, etc.

*Result 1*    For any random variables $X_i, i = 1, 2, \ldots, n$, whether statistically independent or not, and for any fixed constants $a_i$, then if we define a random variable $Y$ such that

$$Y = \sum_{i=1}^{n} a_i X_i$$

we have

$$E(Y) = \sum_{i=1}^{n} a_i E(X_i) \tag{A.1}$$

*Definition 2*    The *covariance* of two random variables $X_1$ and $X_2$ is defined

$$
\begin{aligned}
C(X_1, X_2) &= E\{(X_1 - E(X_1))(X_2 - E(X_2))\} \\
&= E(X_1 X_2) - E(X_1)E(X_2)
\end{aligned}
\tag{A.2}
$$

where

$$E(X_1 X_2) = \iint x_1 x_2 f(x_1, x_2) dx_1 dx_2$$

*Result 2*    For any random variables $X_i, i = 1, 2, \ldots, n$, and for any fixed constants $a_i$, then if we define a random variable $Y$ such that

$$Y = \sum_{i=1}^{n} a_i X_i$$

We have

$$V(Y) = \sum_{i=1}^{n} a_i^2 V(X_i) + 2 \sum\sum_{i<j} a_i a_j C(X_i, X_j) \qquad (A.3)$$

*Result* 3    Any linear combination of normal variables is normally distributed. Specifically, if $X_i$ are independently and normally distributed random variables, $N(\mu_i, \sigma_i^2)$, for $i = 1, 2, \ldots$, then for any $a_i$, if we write

$$Y = \sum a_i X_i$$

then $Y$ is normally distributed with

$$E(Y) = \sum a_i \mu_i$$
and
$$V(Y) = \sum a_i^2 \sigma_i^2 \qquad (A.4)$$

(These last results follow from Results 1 and 2.)

*Result* 4    (Central-limit theorem) If $X_i$ are random variables with almost any distributions (not necessarily identical), with $E(X_i) = \mu_i$, $V(X_i) = \sigma_i^2$, for $i = 1, 2, \ldots, n$, then under certain very general conditions,

$$Z = (\sum X_i - \sum \mu_i)/\sqrt{(\sum \sigma_i^2)}$$

is approximately $N(0, 1)$ for large $n$.

*Result* 5    It follows from Results 3 and 4 that if $X_i$ is a random variable with almost any distribution having $E(X_i) = \mu$, $V(X_i) = \sigma^2$, then the distribution of

$$\bar{X} = \sum X_i/n$$

is approximately normal for large enough $n$, with $E(\bar{X}) = \mu$ and $V(\bar{X}) = \sigma^2/n$. If the distribution of the $X_i$ is normal, then $\bar{X}$ is exactly normally distributed.

We comment here that sums tend to normality so quickly that whatever the original distribution of the $X_i$, the distribution of $\bar{X}$ will usually be nearly normal for sample sizes as low as 20. One of the rare exceptions to Results 4 and 5 is if the original random variables have a Cauchy distribution, with p.d.f.

$$\frac{1}{\pi(1 + (x - \mu)^2)}$$

In this case $E(X)$ and $V(X)$ do not exist, and the distribution of $\bar{X}$ does not tend to normality.

*Definition* 3   The *coefficient of correlation* is defined as

$$\rho = C(X, Y)/\sqrt{[V(X)V(Y)]}$$

where $C(X, Y)$ is the covariance between $X$ and $Y$.

*Definition* 4   Given two random variables $X_1$ and $X_2$ with a cumulative distribution function (c.d.f.)

$$G(X_1, X_2) = \Pr\{x_1 \le X_1, \quad x_2 \le X_2\}$$

the random variables are said to be statistically independent if and only if $G(X_1, X_2)$ factorizes into a product of the two univariate c.d.f.'s, so that

$$G(X_1, X_2) = G_1(X_1)G_2(X_2)$$

where

$$G_i(X_i) = \Pr\{x_i \le X_i\} \qquad i = 1, 2.$$

*Result* 6   Random variables which are jointly normally distributed and uncorrelated are independent.

*The $\chi^2$-distribution*

*Definition* 5   If $Z_1, Z_2, \ldots, Z_\nu$ are independently and normally distributed

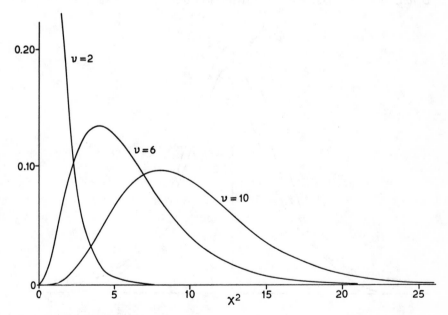

Fig. A.2 The p.d.f. of $\chi^2$.

random variables with expectation zero and unit variance, the distribution of

$$W = Z_1^2 + Z_2^2 + \dots + Z_v^2 \tag{A.5}$$

is said to be a $\chi^2$-distribution with $v$ degrees of freedom.

The p.d.f. of the $\chi^2$-distribution is shown in Fig. A.2. It can be verified from Equation (A.4) that

$$\begin{aligned} E(\chi_v^2) &= v \\ V(\chi_v^2) &= 2v \end{aligned} \tag{A.6}$$

In order to refer to the percentage points of the $\chi_v^2$ distribution we shall use the notation $\chi_v^2(p)$, where

$$p = \Pr\{\chi^2 < \chi_v^2(p)\}$$

*Result 7* If $X_i, i = 1, 2, \dots, n$ are independently distributed $N(\mu, \sigma^2)$, then

$$s^2 = \sum(X_i - \bar{X})^2/(n-1)$$

is such that $\{(n-1)s^2/\sigma^2\}$ has a $\chi_{(n-1)}^2$ distribution. Further, $\bar{X}$ and $s^2$ are statistically independent.

*The t-distribution*

*Definition 6* If $X$ has a $N(0, \sigma^2)$ distribution and $(vs^2/\sigma^2)$ has a $\chi_v^2$ distribution,

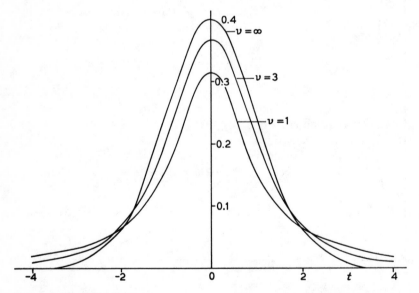

Fig. A.3 The *t*-distribution.

where $X$ and $s$ are statistically independent then

$$t_v = X/s$$

is said to have a $t$-distribution on $v$ degrees of freedom.

The p.d.f. of the $t$-distribution is symmetrical, and tends to normality as $v \to \infty$, as shown in Fig. A.3. We find that

$$E(t_v) = 0 \qquad\qquad\qquad\qquad\qquad\qquad (A.7)$$

$$V(t_v) = v/(v-2) \qquad\qquad\qquad\qquad\qquad (A.8)$$

In order to refer to percentage points of the $t$-distribution we shall use the notation $t_v(p)$, where

$$p = \Pr\{t_v > t_v(p)\}$$

Thus from tables we find that the upper two-sided 5 per cent point for $v = 10$ is $t_{10}(2\frac{1}{2}\%) = 2.23$.

*Result 8*   If $X_i, i = 1, 2, \dots, n$, are independently distributed $N(\mu, \sigma^2)$, then $\sqrt{n}(\bar{X} - \mu)$ is distributed $N(0, \sigma^2)$, and it follows from Results 5 and 7 and Definition 6 that

$$t = \sqrt{n}(\bar{X} - \mu)/s$$

has a $t$-distribution with $(n - 1)$ degrees of freedom.

*The F-distribution*

*Definition 7*   If $W_1$ has a $\chi^2_{v_1}$-distribution and $W_2$ has a $\chi^2_{v_2}$-distribution then the random variable $F$, where

$$F = \frac{W_1/v_1}{W_2/v_2}$$

is said to have an $F$-distribution on $(v_1, v_2)$ degrees of freedom, and this distribution is written $F(v_1, v_2)$.

The p.d.f. of the $F$-distribution has a shape rather similar to that of the $\chi^2$-distribution. It can be shown that

$$E\{F(v_1, v_2)\} = v_2/(v_2 - 2) \qquad v_2 > 2 \qquad\qquad (A.9)$$

and

$$V\{F(v_1, v_2)\} = \frac{2v_2^2(v_1 + v_2) - 2}{v_1(v_2 - 2)^2(v_2 - 4)} \simeq 2\left(\frac{1}{v_1} + \frac{1}{v_2}\right) \qquad v_2 > 4. \quad (A.10)$$

The percentage points of the $F(v_1, v_2)$ distribution are written $F_p(v_1, v_2)$, where

$$p = \Pr\{F < F_p(v_1, v_2)\}.$$

*Result* 9    It follows from Definition 7 that the distribution of $W_2 v_1/W_1 v_2$ is $F(v_2, v_1)$. For this reason, $F$-tables are usually tabulated for $F > 1$ only.

*Result* 10    If $X_1, X_2, \ldots, X_n$, are independently distributed $N(\mu_X, \sigma_X^2)$ and $Y_1, Y_2, \ldots, Y_m$ are independently distributed $N(\mu_Y, \sigma_Y^2)$, then if we write

$$s_X^2 = \sum (X_i - \bar{X})^2/(n-1)$$

and

$$s_Y^2 = \sum (Y_i - \bar{Y})^2/(m-1)$$

it follows from Result 7 and Definition 7 that

$$F = (s_X^2/\sigma_X^2)/(s_Y^2/\sigma_Y^2) = s_X^2 \sigma_Y^2/(s_Y^2 \sigma_X^2)$$

has an $F((n-1), (m-1))$ distribution.

---

**Exercises**

1. Show that for a normally distributed random variable $X$ with parameters $\mu, \sigma^2$, then

$$E(X) = \mu, \qquad V(X) = \sigma^2.$$

2. Show that for any random variables $X_i, i = 1, 2, \ldots$, and for any fixed constants $a_i, i = 1, 2, \ldots, n$, and $b_j, j = 1, 2, \ldots, m$, if we define

$$Y_1 = \sum_{i=1}^{n} a_i X_i$$

$$Y_2 = \sum_{j=1}^{m} b_j X_j$$

then

$$C(Y_1, Y_2) = \sum_{i=1}^{n} \sum_{j=1}^{m} a_i b_j C(X_i, X_j)$$

3. If $W_1$ and $W_2$ have $\chi^2$-distributions with $v_1$ and $v_2$ degrees of freedom respectively, what is the distribution of $Z = W_1 + W_2$?

# Appendix B: Some published data sets

The references listed below all contain data sets which can be used in various ways. They can be used to illustrate applications of statistical methods covered in the theory; they can be used as a basis for further exercises; or they can be used for seminars or discussions on methods of analysis. Some need to be read critically, asking such questions as to whether the analysis is correct, whether transformations are needed, whether the analysis of residuals is complete, etc. The selection of material for this collection is rather arbitrary except that most of the references are readily available. At any rate the supply of data sets below seems to be quite sufficient for most purposes. The following classifications will be used:

ML    maximum likelihood
LR    simple linear regression
MR    multiple regression
PR    polynomial regression
RE    response surfaces
AN    analysis of variance
AC    components of variance
RB    randomized blocks
LA    latin squares or higher squares
CO    analysis of covariance
FA    factorial experiments
OT    other problems

I.      Good problems to start looking at
II.     Problems of moderate difficulty
III.    Problems a bit beyond the material in the book

Ahamad, B. (1967) 'An analysis of crimes by the method of principal components', *Applied Statistics*, **16**, 17–35. [OT, III]
Anderson, R.L. (1946) 'Missing-plot techniques', *Biometrics*, **2**, 41–7. [OT, III]
Anderson, R.L. & Manning, H.L. (1948) 'An experimental design to estimate the optimum planting date for cotton', *Biometrics*, **4**, 171–96. [AN, III]

Bainbridge, T.R. (1965) 'Staggered, nested designs for estimating variance components'. *Ind. Qual. Control*, **22** (1), 12–20. [FA,II]

Baird, H.R. & Kramer, C.Y. (1960) 'Analysis of variance of a balanced incomplete block design with missing observations', *Applied Statistics*, **9**, 189–98. [FA,III]

Barnett, N.K. & Mead, F.C. (1956) 'A $2^4$ factorial experiment in blocks of eight–a study in radioactive decontamination', *Applied Statistics*, **5**, 122–31. [RB,FA,III]

Bartlett, M.S. (1937) 'Some examples of statistical methods of research in agriculture and applied biology', *Supp. J. Royal Stat. Soc.*, **4**, 137–83. [AN,III]

Barlett, M.S. (1947) 'The use of transformations', *Biometrics*, **3**, 39–52. [RB,II]

Beech, D.G. (1953) 'Experiences of correlation analysis', *Applied Statistics*, **2**, 73–85. [LR, I]

Bennett, C.A. (1954) 'Effect of measurement error on chemical process control', *Ind. Qual. Control*, **10**(4), 17–20. [AN,I]

Bicking, C.A. (1952) 'Statistical methods in chemical development', *Ind. Qual. Control*, **8**(4), 9–15. [FA,II]

Box, G.E.P. (1950) 'Problems in the analysis of growth and wear curves', *Biometrics*, **6**, 362–89. [FA,I]

Box, GEP. (1954) The exploration and exploitation of response surfaces', *Biometrics*, **10**, 16–60. [FA,RE,III]

Brown, M.B. (1975) 'Exploring interaction effects in the analysis of variance', *Applied Statistics*, **24**, 288–98. [AN,II]

Brumbaugh, M.A. (1954) 'Principles of sampling in the chemical field', *Ind. Qual. Control*, **10**(4), 6–14. [AC,II]

Burt, C. (1947) 'A comparison of factor analysis and analysis of variance', *British J. of Psychology*, **1**, 3–26. [OT,III]

Calvin, L.D. (1954) 'Doubly balanced incomplete block designs for experiments in which the treatment effects are correlated', *Biometrics*, **10**, 61–88. [OT,III]

Cochran, W.G. (1939) 'Long-term agricultural experiments', *Suppl. J. Royal Stat. Soc.*, **6**, 104–48. [AN,LA,II]

Cochran, W.G. (1947) 'Some consequences when the assumptions for the analysis of variance are not satisfied', *Biometrics*, **3**, 22–38. [RB,II]

Corlett, E.N. & Gregory, G. (1960) 'Consistency of setting of a machine tool hand-wheel', *Applied Statistics*, **9**, 92–102. [FA,III]

Delury, D.B. (1948) 'The analysis of covariance', *Biometrics*, **4**, 153–70. [CO,III]

Desmond, D.J. (1954) 'Quality control on the setting of voltage regulators', *Applied Statistics*, **3**, 65–73. [AC,II]

Doehlert, D.H. (1970) 'Uniform shell designs', *Applied Statistics*, **19**, 231–9. [MR,I]

Dutka, A.F. & Ewens, F.J. (1971) 'A method for improving the accuracy of polynomial regression analysis', *J. Qual. Tech.*, **3**, 149–55. [MR,I]

Feder, P.I. (1974) 'Some differences between fixed, mixed and random effects analysis of variance models', *J. Qual. Tech.*, **6**, 98–106. [AC,II]

Federer, W.T. (1949) 'The general theory of prime-power lattice designs', *Biometrics*, **5**, 144–64. [FA,III]

Federer, W.T. (1950) 'The general theory of prime-power lattice designs'. *Biometrics*, **6**, 34–58. [FA,III]

Fisher, R.A. (1949) 'A biological assay of tuberculins', *Biometrics*, **5**, 300–16. [LA,III]

Freund, R.A. (1974) 'Contrast analysis of experiments', *J. Qual. Tech.*, **6**, 2–21. [FA,III]

Gibson, W.M. & Jowett, G.H. (1957) ' "Three group" regression analysis. Part I. Simple regression analysis', *Applied Statistics*, **6**, 114–22. [LR,I]

Gibson, W.M. & Jowett, G.H. (1957) '"Three group" regression analysis. Part II. Multiple regression analysis', *Applied Statistics*, **6**, 189–97. [MR,I]

Glenn, W.A. & Kramer, C.Y. (1958) 'Analysis of variance of a randomised block design with missing observations', *Applied Statistics*, 7, 173–85. [RB,III]

Gutsell, J.S. (1951) 'The effect of sulfamerazine on the erythroiyte and haemoglobin content of the trout blood', *Biometrics*, 7, 171–9. [AN,I]

Harshbarger, B. (1949) 'Triple rectangular lattices', *Biometrics*, 5, 1–13. [OT,III]

Hill, W.J. & Wiles, R.A. (1975) 'Plant experimentation', *J. Qual. Tech.*, 7, 115–22. [FA,II]

Ineson, J.L. (1939) 'The application of elementary statistical methods to problems encountered in the operation of generating stations', *Supp. J. Royal Stat. Soc.*, 6, 149–68. [MR,I]

Inkson, R.H.E. (1961) 'Analysis of a $3^2 \times 2^2$ factorial experiment with confounding', *Applied Statistics*, 10, 98–107. [FA,III]

Irwin, J.O. & Cheeseman, E.A. (1939) 'On the maximum-likelihood method of determining dosage–response curves and approximations to the median-effective dose in cases of a quantal response', *Supp. J. Royal Stat. Soc.*, 6, 174–85. [ML,II]

Jacob, W.C. (1953) 'Split-plot half plaid squares for irrigation experiments', *Biometrics*, 9, 157–75. [FA,III]

Jeffers, J.N.R. (1976) 'An analysis of litter respiration at different temperatures', *Applied Statistics*, 25, 139–46. [AN,II]

Jennett, W.J. & Dudding, B.P. (1936) 'The application of statistical principles to an industrial problem', *Supp. J. Royal Stat. Soc.*, 3, 1–28. [LR,MR,I]

John, P.W.M. (1962) 'Testing two treatments when there are three experimental units in each block', *Applied Statistics*, 11, 164–9. [AN,II]

Jowett, G.H. (1952) 'The accuracy of systematic sampling from conveyor belts', *Applied Statistics*, 1, 50–9. [AC,II]

Kamat, S. (1976) 'A smoothed Bayes control procedure for the control of a variable quality characteristic with linear shift', *J. Qual. Tech.*, 8, 98–104. [AN, I]

Kramer, C.Y. & Glass, S. (1960) 'Analysis of variance of a latin square design with missing observations', *Applied Statistics*, 9, 43–50. [LA, III]

Longley, J.W. (1967) 'An appraisal of least squares programs for the electronic computer from the point of view of the user', *J. American Stat. Ass.*, 62, 819–41. [MR,I]

Mandel, B.J. (1969) 'The regression control chart', *J. Qual. Tech.*, 1, 1–9. [LR,I]

Mandel, B.J. (1976) 'Models, transformations of scale and weighting', *J. Qual. Tech.*, 8, 86–97. [AN. II].

Marcuse, S. (1949) 'Optimum allocation and variance components in nested sampling with an application to chemical analysis', *Biometrics*, 5, 189–206. [AC,I]

Marquardt, D.W. & Snee, R.D. (1975) 'Ridge regression in practice', *American Statistician*, 29, 3–20. [MR,II]

Mendenhall, W. & Ott, L. (1971) 'A method for the calibration of an on-line density meter', *J. Qual. Tech.*, 3, 80–6. [MR,AN,II]

Nelson, W. (1969) 'Hazard plotting for incomplete failure data', *J. Qual. Tech.*, 1, 27–52. [OT,III]

Peake, R.E. (1953) 'Planning an experiment in a cotton mill', *Applied Statistics*, 2, 184–95. [RB,LA,II]

Pearson, J.C.G. & Sprent, P. (1968) 'Trends in hearing loss associated with age or exposure to noise', *Applied Statistics*, 17, 205–15. [PR,II]

Quenouille, M.H. (1948) 'The analysis of covariance and non-orthogonal comparisons', *Biometrics*, 4, 240–6. [CO,III]

Read, D.R. (1954) 'The design of chemical experiments', *Biometrics*, 10, 1–15. [AN,III]

Rowland, H.A.K. (1966) 'The relationship between the dose of oral iron and the response in severe iron deficiency anaemia', *Applied Statistics*, 15, 50–55.

Sandon, F. (1956) 'A regression control chart for use in personnel selection', *Applied Statistics*, **5**, 20–31. [LR,I]

Sarkar, A.D. (1974) 'A factorial experiment with greens and mixes' The Quality Engineer, **38**(5), 112–17. [FA, II]

Schilling, E.G. (1974) 'The relationship of analysis of variance to regression', *J. Qual. Tech.*, **6**, 74–83. [CO,III]

Smith, H. (1969) 'The analysis of data from a designed experiment', *J. Qual. Tech.*, **1**, 259–63. [AN,I]

Snedecor, G.W. (1948a) 'Query 57 and Answer', *Biometrics*, **4**, 132–4. [AN, II]

Snedecor, G.W. (1948b) 'Query 59 and Answer', *Biometrics*, **4**, 211–13. [AN, III]

Snedecor, G.W. & Haber, E.S. (1946) 'Statistical methods for an incomplete experiment on a perennial crop', *Biometrics*, **2**, 61–7. [AN,II]

Snee, R.D. (1971) 'Design and analysis of mixture experiments', *J. Qual. Tech.*, **3**, 159–69. [MR,II]

Snee, R.D. (1974) 'Computation and use of expected mean squares in analysis of variance', *J. Qual. Tech.*, **6**, 128–37. [AC,II]

Stevens, W.L. (1951) 'Asymptotic regression', *Biometrics* **7**, 247–67. [LA,III]

Wilkie, D. (1961) 'Confounding in $3^3$ factorial experiments in nine blocks', *Applied Statistics*, **10**, 83–92. [FA, III]

Wilkie, D. (1962) 'A method of analysis of mixed level factorial experiments', *Applied Statistics*, **11**, 184–95. [AN,II]

Wilm, H.G. (1945) 'Notes on analysis of experiments replicated in time', *Biometrics*, **1**, 16–20. [RB,II]

Wooding, W.M. (1969) 'The computation and use of residuals in the analysis of experimental data', *J. Qual. Tech.*, **1**, 175–88. [RB,II]

Woods, S.R. & Hartvigsen, D.E. (1964) 'Statistical design and analysis of qualification Test program for a small rocket engine', *Ind. Qual. Control*, **20**(12), 14–18 [FA,II]

Yates, F. 1935 'Complex experiments', *Supp. J. Royal Stat. Soc.*, **2**, 181–247. [AN,III]

Yates, F. & Hale, R.W. (1939) 'The analysis of latin squares when two or more rows, columns or treatments are missing', *Supp. J. Royal Stat. Soc.*, **6**, 67–79. [LA,III]

# Appendix C: Statistical Tables

Table C.1 *Cumulative distribution function of the standard normal distribution*
(a) *For x in 0.1 intervals*

| $x$ | $\Phi(x)$ | $x$ | $\Phi(x)$ | $x$ | $\Phi(x)$ |
|-----|-----------|-----|-----------|-----|-----------|
| 0.0 | 0.5000 | 1.3 | 0.9032 | 2.6 | 0.9953 |
| 0.1 | 0.5398 | 1.4 | 0.9192 | 2.7 | 0.9965 |
| 0.2 | 0.5793 | 1.5 | 0.9332 | 2.8 | 0.9974 |
| 0.3 | 0.6179 | 1.6 | 0.9452 | 2.9 | 0.9981 |
| 0.4 | 0.6554 | 1.7 | 0.9554 | 3.0 | 0.9987 |
| 0.5 | 0.6915 | 1.8 | 0.9641 | 3.1 | 0.9990 |
| 0.6 | 0.7257 | 1.9 | 0.9713 | 3.2 | 0.9993 |
| 0.7 | 0.7580 | 2.0 | 0.9772 | 3.3 | 0.9995 |
| 0.8 | 0.7881 | 2.1 | 0.9821 | 3.4 | 0.99966 |
| 0.9 | 0.8159 | 2.2 | 0.9861 | 3.5 | 0.99977 |
| 1.0 | 0.8413 | 2.3 | 0.9893 | 3.6 | 0.99984 |
| 1.1 | 0.8643 | 2.4 | 0.9918 | 3.7 | 0.99989 |
| 1.2 | 0.8849 | 2.5 | 0.9938 | 3.8 | 0.99993 |

(*Contd.*)

(b) *For x in* 0.01 *intervals*

| x | $\Phi(x)$ | x | $\Phi(x)$ | x | $\Phi(x)$ |
|------|--------|------|--------|------|--------|
| 1.60 | 0.9452 | 1.87 | 0.9693 | 2.14 | 2.9838 |
| 1.61 | 0.9463 | 1.88 | 0.9699 | 2.15 | 0.9842 |
| 1.62 | 0.9474 | 1.89 | 0.9706 | 2.16 | 0.9846 |
| 1.63 | 0.9484 | 1.90 | 0.9713 | 2.17 | 0.9850 |
| 1.64 | 0.9495 | 1.91 | 0.9719 | 2.18 | 0.9854 |
| 1.65 | 0.9505 | 1.92 | 0.9726 | 2.19 | 0.9857 |
| 1.66 | 0.9515 | 1.93 | 0.9732 | 2.20 | 0.9861 |
| 1.67 | 0.9525 | 1.94 | 0.9738 | 2.21 | 0.9865 |
| 1.68 | 0.9535 | 1.95 | 0.9744 | 2.22 | 0.9868 |
| 1.69 | 0.9545 | 1.96 | 0.9750 | 2.23 | 0.9871 |
| 1.70 | 0.9554 | 1.97 | 0.9756 | 2.24 | 0.9875 |
| 1.71 | 0.9564 | 1.98 | 0.9761 | 2.25 | 0.9878 |
| 1.72 | 0.9573 | 1.99 | 0.9767 | 2.26 | 0.9881 |
| 1.73 | 0.9582 | 2.00 | 0.9772 | 2.27 | 0.9884 |
| 1.74 | 0.9591 | 2.01 | 0.9778 | 2.28 | 0.9887 |
| 1.75 | 0.9599 | 2.02 | 0.9783 | 2.29 | 0.9890 |
| 1.76 | 0.9608 | 2.03 | 0.9788 | 2.30 | 0.9893 |
| 1.77 | 0.9616 | 2.04 | 0.9793 | 2.31 | 0.9896 |
| 1.78 | 0.9625 | 2.05 | 0.9798 | 2.32 | 0.9898 |
| 1.79 | 0.9633 | 2.06 | 0.9803 | 2.33 | 0.9901 |
| 1.80 | 0.9641 | 2.07 | 0.9808 | 2.34 | 0.9904 |
| 1.81 | 0.9649 | 2.08 | 0.9812 | 2.35 | 0.9906 |
| 1.82 | 0.9656 | 2.09 | 0.9817 | 2.36 | 0.9909 |
| 1.83 | 0.9664 | 2.10 | 0.9821 | 2.37 | 0.9911 |
| 1.84 | 0.9671 | 2.11 | 0.9826 | 2.38 | 0.9913 |
| 1.85 | 0.9678 | 2.12 | 0.9830 | 2.39 | 0.9916 |
| 1.86 | 0.9686 | 2.13 | 0.9834 | 2.40 | 0.9918 |

The function tabulated is

$$\Pr(x < X) = \int_{-\infty}^{x} \frac{1}{\sqrt{(2\pi)}} e^{-(1/2)x^2} dx = \Phi(X)$$

Table C.2 *Percentiles of the standard normal distribution*

| P | x | P | x | P | x |
|----|--------|-----|--------|-----|--------|
| 20 | 0.8416 | 5 | 1.6449 | 2 | 2.0537 |
| 15 | 1.0364 | 4 | 1.7507 | 1 | 2.3263 |
| 10 | 1.2816 | 3 | 1.8808 | 0.5 | 2.5758 |
| 6 | 1.5548 | 2.5 | 1.9600 | 0.1 | 3.0902 |

This table gives one-sided percentage points,

$$P/100 = \int_{x}^{\infty} \frac{1}{\sqrt{(2\pi)}} e^{-(1/2)x^2} dx$$

The two-sided percentage appropriate to any x is 2P.

Table C.3 *Percentage points of the t-distribution*

| | Probability in per cent | | | | | |
| | 20 | 10 | 5 | 2 | 1 | 0.1 |
|---|---|---|---|---|---|---|
| 1 | 3.08 | 6.31 | 12.71 | 31.82 | 63.66 | 636.62 |
| 2 | 1.89 | 2.92 | 4.30 | 6.96 | 9.92 | 31.60 |
| 3 | 1.64 | 2.35 | 3.18 | 4.54 | 5.84 | 12.92 |
| 4 | 1.53 | 2.13 | 2.78 | 3.75 | 4.60 | 8.61 |
| 5 | 1.48 | 2.01 | 2.57 | 3.36 | 4.03 | 6.87 |
| 6 | 1.44 | 1.94 | 2.45 | 3.14 | 3.71 | 5.96 |
| 7 | 1.42 | 1.89 | 2.36 | 3.00 | 3.50 | 5.41 |
| 8 | 1.40 | 1.86 | 2.31 | 2.90 | 3.36 | 5.04 |
| 9 | 1.38 | 1.83 | 2.26 | 2.82 | 3.25 | 4.78 |
| 10 | 1.37 | 1.81 | 2.23 | 2.76 | 3.17 | 4.59 |
| 11 | 1.36 | 1.80 | 2.20 | 2.72 | 3.11 | 4.44 |
| 12 | 1.36 | 1.78 | 2.18 | 2.68 | 3.05 | 4.32 |
| 13 | 1.35 | 1.77 | 2.16 | 2.65 | 3.01 | 4.22 |
| 14 | 1.34 | 1.76 | 2.14 | 2.62 | 2.98 | 4.14 |
| 15 | 1.34 | 1.75 | 2.13 | 2.60 | 2.95 | 4.07 |
| 20 | 1.32 | 1.72 | 2.09 | 2.53 | 2.85 | 3.85 |
| 25 | 1.32 | 1.71 | 2.06 | 2.48 | 2.79 | 3.72 |
| 30 | 1.31 | 1.70 | 2.04 | 2.46 | 2.75 | 3.65 |
| 40 | 1.30 | 1.68 | 2.02 | 2.42 | 2.70 | 3.55 |
| 60 | 1.30 | 1.67 | 2.00 | 2.39 | 2.66 | 3.46 |
| 120 | 1.29 | 1.66 | 1.98 | 2.36 | 2.62 | 3.37 |
| ∞ | 1.28 | 1.64 | 1.96 | 2.33 | 2.58 | 3.29 |

This table gives two-sided percentage points,

$$P/100 = 2 \int_{t}^{\infty} f(x|v)\,d\acute{x}$$

where $f(x|v)$ is the p.d.f. of the *t*-distribution.

For one-sided percentage points the percentages shown should be halved.

Table C.4 *Percentage points of the $\chi^2$-distribution*

| Degrees of freedom ($v$) | 1 | 5 | 90 | 95 | 99 | 99.9 |
|---|---|---|---|---|---|---|
| 1 | 0.03157 | 0.00393 | 2.71 | 3.84 | 6.63 | 10.83 |
| 2 | 0.0201 | 0.103 | 4.61 | 5.99 | 9.21 | 13.81 |
| 3 | 0.115 | 0.352 | 6.25 | 7.81 | 11.34 | 16.27 |
| 4 | 0.297 | 0.711 | 7.78 | 9.49 | 13.28 | 18.47 |
| 5 | 0.554 | 1.15 | 9.24 | 11.07 | 15.09 | 20.52 |
| 6 | 0.872 | 1.64 | 10.64 | 12.59 | 16.81 | 22.46 |
| 7 | 1.24 | 2.17 | 12.02 | 14.07 | 18.48 | 24.32 |
| 8 | 1.65 | 2.73 | 13.36 | 15.51 | 20.09 | 26.12 |
| 9 | 2.09 | 3.33 | 14.68 | 16.92 | 21.67 | 27.88 |
| 10 | 2.56 | 3.94 | 15.99 | 18.31 | 23.21 | 29.59 |
| 11 | 3.05 | 4.57 | 17.28 | 19.68 | 24.73 | 31.26 |
| 12 | 3.57 | 5.23 | 18.55 | 21.03 | 26.22 | 32.91 |
| 14 | 4.66 | 6.57 | 21.06 | 23.68 | 29.14 | 36.12 |
| 16 | 5.81 | 7.96 | 23.54 | 26.30 | 32.00 | 39.25 |
| 18 | 7.01 | 9.39 | 25.99 | 28.87 | 34.81 | 42.31 |
| 20 | 8.26 | 10.85 | 28.41 | 31.41 | 37.57 | 45.31 |
| 22 | 9.54 | 12.34 | 30.81 | 33.92 | 40.29 | 48.27 |
| 24 | 10.86 | 13.85 | 33.20 | 36.42 | 42.98 | 51.18 |
| 26 | 12.20 | 15.38 | 35.56 | 38.89 | 45.64 | 54.05 |
| 28 | 13.56 | 16.93 | 37.92 | 41.34 | 48.28 | 56.89 |
| 30 | 14.95 | 18.49 | 40.26 | 43.77 | 50.89 | 59.70 |

The table gives the percentage points $\chi^2$, where

$$P = 100 \int_0^{x^2} g(y|v)\,dy$$

where $g(y|v)$ is the probability density function of the $\chi^2$-distribution.

For $v > 30$, $\sqrt{(2\chi^2)}$ is approximately normally distributed with mean $(2v - 1)$ and unit variance.

Table C.5 *Percentage points of the F-distribution*
(*a*) 95% points

| | | Degrees of freedom of numerator ($v_1$) | | | | | | | | |
|---|---|---|---|---|---|---|---|---|---|---|
| | | 1 | 2 | 3 | 4 | 5 | 6 | 8 | 12 | 24 | ∞ |

|  | | 1 | 2 | 3 | 4 | 5 | 6 | 8 | 12 | 24 | ∞ |
|---|---|---|---|---|---|---|---|---|---|---|---|
| *Degrees of freedom of denominator ($v_2$)* | 1 | 161.4 | 199.5 | 215.7 | 224.6 | 230.2 | 234.0 | 238.9 | 243.9 | 249.0 | 254.3 |
| | 2 | 18.51 | 19.00 | 19.16 | 19.25 | 19.30 | 19.33 | 19.37 | 19.41 | 19.45 | 19.50 |
| | 3 | 10.13 | 9.55 | 9.28 | 9.12 | 9.01 | 8.94 | 8.85 | 8.74 | 8.64 | 8.53 |
| | 4 | 7.71 | 6.94 | 6.59 | 6.39 | 6.26 | 6.16 | 6.04 | 5.91 | 5.77 | 5.63 |
| | 5 | 6.61 | 5.79 | 5.41 | 5.19 | 5.05 | 4.95 | 4.82 | 4.68 | 4.53 | 4.36 |
| | 6 | 5.99 | 5.14 | 4.76 | 4.53 | 4.39 | 4.28 | 4.15 | 4.00 | 3.84 | 3.67 |
| | 7 | 5.59 | 4.74 | 4.35 | 4.12 | 3.97 | 3.87 | 3.73 | 3.57 | 3.41 | 3.23 |
| | 8 | 5.32 | 4.46 | 4.07 | 3.84 | 3.69 | 3.58 | 3.44 | 3.28 | 3.12 | 2.93 |
| | 9 | 5.12 | 4.26 | 3.86 | 3.63 | 3.48 | 3.37 | 3.23 | 3.07 | 2.90 | 2.71 |
| | 10 | 4.96 | 4.10 | 3.71 | 3.48 | 3.33 | 3.22 | 3.07 | 2.91 | 2.74 | 2.54 |
| | 11 | 4.84 | 3.98 | 3.59 | 3.36 | 3.20 | 3.09 | 2.95 | 2.79 | 2.61 | 2.40 |
| | 12 | 4.75 | 3.89 | 3.49 | 3.26 | 3.11 | 3.00 | 2.85 | 2.69 | 2.51 | 2.30 |
| | 14 | 4.60 | 3.74 | 3.34 | 3.11 | 2.96 | 2.85 | 2.70 | 2.53 | 2.35 | 2.13 |
| | 16 | 4.49 | 3.63 | 3.24 | 3.01 | 2.85 | 2.74 | 2.59 | 2.42 | 2.24 | 2.01 |
| | 18 | 4.41 | 3.55 | 3.16 | 2.93 | 2.77 | 2.66 | 2.51 | 2.34 | 2.15 | 1.92 |
| | 20 | 4.35 | 3.49 | 3.10 | 2.87 | 2.71 | 2.60 | 2.45 | 2.28 | 2.08 | 1.84 |
| | 25 | 4.24 | 3.39 | 2.99 | 2.76 | 2.60 | 2.49 | 2.34 | 2.16 | 1.96 | 1.71 |
| | 30 | 4.17 | 3.32 | 2.92 | 2.69 | 2.53 | 2.42 | 2.27 | 2.09 | 1.89 | 1.62 |
| | 40 | 4.08 | 3.23 | 2.84 | 2.61 | 2.45 | 2.34 | 2.18 | 2.00 | 1.79 | 1.51 |
| | 60 | 4.00 | 3.15 | 2.76 | 2.53 | 2.37 | 2.25 | 2.10 | 1.92 | 1.70 | 1.39 |
| | ∞ | 3.84 | 3.00 | 2.60 | 2.37 | 2.21 | 2.10 | 1.94 | 1.75 | 1.52 | 1.00 |

(*b*) 97.5% points

| | | Degrees of freedom of numerator ($v_1$) | | | | | | | | |
|---|---|---|---|---|---|---|---|---|---|---|
| | | 1 | 2 | 3 | 4 | 5 | 6 | 8 | 12 | 24 | ∞ |

|  | | 1 | 2 | 3 | 4 | 5 | 6 | 8 | 12 | 24 | ∞ |
|---|---|---|---|---|---|---|---|---|---|---|---|
| *Degrees of freedom of denominator ($v_2$)* | 1 | 648 | 800 | 864 | 900 | 922 | 937 | 957 | 977 | 997 | 1018 |
| | 2 | 38.5 | 39.0 | 39.2 | 39.2 | 39.3 | 39.3 | 39.4 | 39.4 | 39.5 | 39.5 |
| | 3 | 17.4 | 16.0 | 15.4 | 15.1 | 14.9 | 14.7 | 14.5 | 14.3 | 14.1 | 13.9 |
| | 4 | 12.2 | 10.6 | 9.98 | 9.60 | 9.36 | 9.20 | 8.98 | 8.75 | 8.51 | 8.26 |
| | 5 | 10.0 | 8.43 | 7.76 | 7.39 | 7.15 | 6.98 | 6.76 | 6.52 | 6.28 | 6.02 |
| | 6 | 8.81 | 7.26 | 6.60 | 6.23 | 5.99 | 5.82 | 5.60 | 5.37 | 5.12 | 4.85 |
| | 7 | 8.07 | 6.54 | 5.89 | 5.52 | 5.29 | 5.12 | 4.90 | 4.67 | 4.42 | 4.14 |
| | 8 | 7.57 | 6.06 | 5.42 | 5.05 | 4.82 | 4.65 | 4.43 | 4.20 | 3.95 | 3.67 |
| | 9 | 7.21 | 5.71 | 5.08 | 4.72 | 4.48 | 4.32 | 4.10 | 3.87 | 3.61 | 3.33 |
| | 10 | 6.94 | 5.46 | 4.83 | 4.47 | 4.24 | 4.07 | 3.85 | 3.62 | 3.37 | 3.08 |
| | 11 | 6.72 | 5.26 | 4.63 | 4.28 | 4.04 | 3.88 | 3.66 | 3.43 | 3.17 | 2.88 |
| | 12 | 6.55 | 5.10 | 4.47 | 4.12 | 3.89 | 3.73 | 3.51 | 3.28 | 3.02 | 2.72 |
| | 14 | 6.30 | 4.86 | 4.24 | 3.89 | 3.66 | 3.50 | 3.29 | 3.05 | 2.79 | 2.49 |
| | 16 | 6.12 | 4.69 | 4.08 | 3.73 | 3.50 | 3.34 | 3.12 | 2.89 | 2.63 | 2.32 |
| | 18 | 5.98 | 4.56 | 3.95 | 3.61 | 3.38 | 3.22 | 3.01 | 2.77 | 2.50 | 2.19 |
| | 20 | 5.87 | 4.46 | 3.86 | 3.51 | 3.29 | 3.13 | 2.91 | 2.68 | 2.41 | 2.09 |
| | 25 | 5.69 | 4.29 | 3.69 | 3.35 | 3.13 | 2.97 | 2.75 | 2.51 | 2.24 | 1.91 |
| | 30 | 5.57 | 4.18 | 3.59 | 3.25 | 3.03 | 2.87 | 2.65 | 2.41 | 2.14 | 1.79 |
| | 40 | 5.42 | 4.05 | 3.46 | 3.13 | 2.90 | 2.74 | 2.53 | 2.29 | 2.01 | 1.64 |
| | 60 | 5.29 | 3.93 | 3.34 | 3.01 | 2.79 | 2.63 | 2.41 | 2.17 | 1.88 | 1.48 |
| | ∞ | 5.02 | 3.69 | 3.12 | 2.79 | 2.57 | 2.41 | 2.19 | 1.94 | 1.64 | 1.00 |

(c) 99% points

| | | 1 | 2 | 3 | 4 | 5 | 6 | 8 | 10 | 12 | ∞ |
|---|---|---|---|---|---|---|---|---|---|---|---|
| | 1 | 4052 | 4999 | 5403 | 5625 | 5764 | 5859 | 5981 | 6106 | 6235 | 6366 |
| | 2 | 98.50 | 99.00 | 99.17 | 99.25 | 99.30 | 99.33 | 99.37 | 99.42 | 99.46 | 99.50 |
| | 3 | 34.12 | 30.82 | 29.46 | 28.71 | 28.24 | 27.91 | 27.49 | 27.05 | 26.60 | 26.13 |
| | 4 | 21.20 | 18.00 | 16.69 | 15.98 | 15.52 | 15.21 | 14.80 | 14.37 | 13.93 | 13.46 |
| | 5 | 16.26 | 13.27 | 12.06 | 11.39 | 10.97 | 10.67 | 10.29 | 9.89 | 9.47 | 9.02 |
| | 6 | 13.74 | 10.92 | 9.78 | 9.15 | 8.75 | 8.47 | 8.10 | 7.72 | 7.31 | 6.88 |
| | 7 | 12.25 | 9.55 | 8.45 | 7.85 | 7.46 | 7.19 | 6.84 | 6.47 | 6.07 | 5.65 |
| | 8 | 11.26 | 8.65 | 7.59 | 7.01 | 6.63 | 6.37 | 6.03 | 5.67 | 5.28 | 4.86 |
| | 9 | 10.56 | 8.02 | 6.99 | 6.42 | 6.06 | 5.80 | 5.47 | 5.11 | 4.73 | 4.31 |
| | 10 | 10.04 | 7.56 | 6.55 | 5.99 | 5.64 | 5.39 | 5.06 | 4.71 | 4.33 | 3.91 |
| | 11 | 9.65 | 7.21 | 6.22 | 5.67 | 5.32 | 5.07 | 4.74 | 4.40 | 4.02 | 3.60 |
| | 12 | 9.33 | 6.93 | 5.95 | 5.41 | 5.06 | 4.82 | 4.50 | 4.16 | 3.78 | 3.36 |
| | 14 | 8.86 | 6.51 | 5.56 | 5.04 | 4.69 | 4.46 | 4.14 | 3.80 | 3.43 | 3.00 |
| | 16 | 8.53 | 6.23 | 5.29 | 4.77 | 4.44 | 4.20 | 3.89 | 3.55 | 3.18 | 2.75 |
| | 18 | 8.29 | 6.01 | 5.09 | 4.58 | 4.25 | 4.01 | 3.71 | 3.37 | 3.00 | 2.57 |
| | 20 | 8.10 | 5.85 | 4.94 | 4.43 | 4.10 | 3.87 | 3.56 | 3.23 | 2.86 | 2.42 |
| | 25 | 7.77 | 5.57 | 4.68 | 4.18 | 3.86 | 3.63 | 3.32 | 2.99 | 2.62 | 2.17 |
| | 30 | 7.56 | 5.39 | 4.51 | 4.02 | 3.70 | 3.47 | 3.17 | 2.84 | 2.47 | 2.01 |
| | 40 | 7.31 | 5.18 | 4.31 | 3.83 | 3.51 | 3.29 | 2.99 | 2.66 | 2.29 | 1.80 |
| | 60 | 7.08 | 4.98 | 4.13 | 3.65 | 3.34 | 3.12 | 2.82 | 2.50 | 2.12 | 1.60 |
| | ∞ | 6.63 | 4.60 | 3.78 | 3.32 | 3.02 | 2.80 | 2.51 | 2.18 | 1.79 | 1.00 |

Degree of freedom of numerator ($v_1$). Degrees of freedom of denominator ($v_2$).

The table gives for various degrees of freedom, $v_1$, $v_2$, the values of $F$ such that

$$P = 100 \int_0^F h(z|v_1, v_2)\, dz$$

where $h(z|v_1, v_2)$ is the probability density function of the $F$-ratio (6.6).

Table C.6 *Percentage points of the studentized range* $q(\alpha\%, n, v)$
(*a*) 5% points

| $v$ | $n$ 2 | 3 | 4 | 5 | 6 | 7 | 8 | 10 | 12 | 20 |
|---|---|---|---|---|---|---|---|---|---|---|
| 5 | 3.64 | 4.60 | 5.22 | 5.67 | 6.03 | 6.33 | 6.58 | 6.99 | 7.32 | 8.21 |
| 6 | 3.46 | 4.34 | 4.90 | 5.30 | 5.63 | 5.90 | 6.12 | 6.49 | 6.79 | 7.59 |
| 7 | 3.34 | 4.16 | 4.68 | 5.06 | 5.36 | 5.61 | 5.82 | 6.16 | 6.43 | 7.17 |
| 8 | 3.26 | 4.04 | 4.53 | 4.89 | 5.17 | 5.40 | 5.60 | 5.92 | 6.18 | 6.87 |
| 9 | 3.20 | 3.95 | 4.41 | 4.76 | 5.02 | 5.24 | 5.43 | 5.74 | 5.98 | 6.64 |
| 10 | 3.15 | 3.88 | 4.33 | 4.65 | 4.91 | 5.12 | 5.30 | 5.60 | 5.83 | 6.47 |
| 11 | 3.11 | 3.82 | 4.26 | 4.57 | 4.82 | 5.03 | 5.20 | 5.49 | 5.71 | 6.33 |
| 12 | 3.08 | 3.77 | 4.20 | 4.51 | 4.75 | 4.95 | 5.12 | 5.39 | 5.61 | 6.21 |
| 13 | 3.06 | 3.73 | 4.15 | 4.45 | 4.69 | 4.88 | 5.05 | 5.32 | 5.53 | 6.11 |
| 14 | 3.03 | 3.70 | 4.11 | 4.41 | 4.64 | 4.83 | 4.99 | 5.25 | 5.46 | 6.03 |
| 15 | 3.01 | 3.67 | 4.08 | 4.37 | 4.59 | 4.78 | 4.94 | 5.20 | 5.40 | 5.96 |
| 20 | 2.95 | 3.58 | 3.96 | 4.23 | 4.45 | 4.62 | 4.77 | 5.01 | 5.20 | 5.71 |
| 30 | 2.89 | 3.49 | 3.85 | 4.10 | 4.30 | 4.46 | 4.60 | 4.82 | 5.00 | 5.47 |
| 60 | 2.83 | 3.40 | 3.74 | 3.98 | 4.16 | 4.31 | 4.44 | 4.65 | 4.81 | 5.24 |
| 120 | 2.80 | 3.36 | 3.68 | 3.92 | 4.10 | 4.24 | 4.36 | 4.56 | 4.71 | 5.13 |
| $\infty$ | 2.77 | 3.31 | 3.63 | 3.86 | 4.03 | 4.17 | 4.29 | 4.47 | 4.62 | 5.01 |

(*b*) 1% points

| $v$ | $n$ 2 | 3 | 4 | 5 | 6 | 7 | 8 | 10 | 12 | 20 |
|---|---|---|---|---|---|---|---|---|---|---|
| 5 | 5.70 | 6.98 | 7.80 | 8.42 | 8.91 | 9.32 | 9.67 | 10.24 | 10.70 | 11.93 |
| 6 | 5.24 | 6.33 | 7.03 | 7.56 | 7.97 | 8.32 | 8.61 | 9.10 | 9.48 | 10.54 |
| 7 | 4.95 | 5.92 | 6.54 | 7.01 | 7.37 | 7.68 | 7.94 | 8.37 | 8.71 | 9.65 |
| 8 | 4.75 | 5.64 | 6.20 | 6.62 | 6.96 | 7.24 | 7.47 | 7.86 | 8.18 | 9.03 |
| 9 | 4.60 | 5.43 | 5.96 | 6.35 | 6.66 | 6.91 | 7.13 | 7.49 | 7.78 | 8.57 |
| 10 | 4.48 | 5.27 | 5.77 | 6.14 | 6.43 | 6.67 | 6.87 | 7.21 | 7.49 | 8.23 |
| 11 | 4.39 | 5.15 | 5.62 | 5.97 | 6.25 | 6.48 | 6.67 | 6.99 | 7.25 | 7.95 |
| 12 | 4.32 | 5.05 | 5.50 | 5.84 | 6.10 | 6.32 | 6.51 | 6.81 | 7.06 | 7.73 |
| 13 | 4.26 | 4.96 | 5.40 | 5.73 | 5.98 | 6.19 | 6.37 | 6.67 | 6.90 | 7.55 |
| 14 | 4.21 | 4.89 | 5.32 | 5.63 | 5.88 | 6.08 | 6.26 | 6.54 | 6.77 | 7.39 |
| 15 | 4.17 | 4.84 | 5.25 | 5.56 | 5.80 | 5.99 | 6.16 | 6.44 | 6.66 | 7.26 |
| 20 | 4.02 | 4.64 | 5.02 | 5.29 | 5.51 | 5.69 | 5.84 | 6.09 | 6.28 | 6.82 |
| 30 | 3.89 | 4.45 | 4.80 | 5.05 | 5.24 | 5.40 | 5.54 | 5.76 | 5.93 | 6.41 |
| 60 | 3.76 | 4.28 | 4.59 | 4.82 | 4.99 | 5.13 | 5.25 | 5.45 | 5.60 | 6.01 |
| 120 | 3.70 | 4.20 | 4.50 | 4.71 | 4.87 | 5.01 | 5.12 | 5.30 | 5.44 | 5.83 |
| $\infty$ | 3.64 | 4.12 | 4.40 | 4.60 | 4.76 | 4.88 | 4.99 | 5.16 | 5.29 | 5.65 |

Table C.7  *Coefficients* $\{a_{n-i+1}\}$ *for the Shapiro-Wilk W test for normality, for* $n = 2(1)50$.

| *i* | *n* 2 | 3 | 4 | 5 | 6 | 7 | 8 | 9 | 10 |
|---|---|---|---|---|---|---|---|---|---|
| 1 | 0.7071 | 0.7071 | 0.6872 | 0.6646 | 0.6431 | 0.6233 | 0.6052 | 0.5888 | 0.5739 |
| 2 | — | .0000 | .1677 | .2413 | .2806 | .3031 | .3164 | .3244 | .3291 |
| 3 | — | — | — | .0000 | .0875 | .1401 | .1743 | .1976 | .2141 |
| 4 | — | — | — | — | — | .0000 | .0561 | .0947 | .1224 |
| 5 | — | — | — | — | — | — | — | .0000 | .0399 |

| *i* | *n* 11 | 12 | 13 | 14 | 15 | 16 | 17 | 18 | 19 | 20 |
|---|---|---|---|---|---|---|---|---|---|---|
| 1 | 0.5601 | 0.5475 | 0.5359 | 0.5251 | 0.5150 | 0.5056 | 0.4968 | 0.4886 | 0.4808 | 0.4734 |
| 2 | .3315 | .3325 | .3325 | .3318 | .3306 | .3290 | .3273 | .3253 | .3232 | .3211 |
| 3 | .2260 | .2347 | .2412 | .2460 | .2495 | .2521 | .2540 | .2553 | .2561 | .2565 |
| 4 | .1429 | .1586 | .1707 | .1802 | .1878 | .1939 | .1988 | .2027 | .2059 | .2085 |
| 5 | .0695 | .0922 | .1099 | .1240 | .1353 | .1447 | .1524 | .1587 | .1641 | .1686 |
| 6 | 0.0000 | 0.0303 | 0.0539 | 0.0727 | 0.0880 | 0.1005 | 0.1109 | 0.1197 | 0.1271 | 0.1334 |
| 7 | — | — | .0000 | .0240 | .0433 | .0593 | .0725 | .0837 | .0932 | .1013 |
| 8 | — | — | — | — | .0000 | .0196 | .0359 | .0496 | .0612 | .0711 |
| 9 | — | — | — | — | — | — | .0000 | .0163 | .0303 | .0422 |
| 10 | — | — | — | — | — | — | — | — | .0000 | .0140 |

| *i* | *n* 21 | 22 | 23 | 24 | 25 | 26 | 27 | 28 | 29 | 30 |
|---|---|---|---|---|---|---|---|---|---|---|
| 1 | 0.4643 | 0.4590 | 0.4542 | 0.4493 | 0.4450 | 0.4407 | 0.4366 | 0.4328 | 0.4291 | 0.4254 |
| 2 | .3185 | .3156 | .3126 | .3098 | .3069 | .3043 | .3018 | .2992 | .2968 | .2944 |
| 3 | .2578 | .2571 | .2563 | .2554 | .2543 | .2533 | .2522 | .2510 | .2499 | .2487 |
| 4 | .2119 | .2131 | .2139 | .2145 | .2148 | .2151 | .2152 | .2151 | .2150 | .2148 |
| 5 | .1736 | .1764 | .1787 | .1807 | .1822 | .1836 | .1848 | .1857 | .1864 | .1870 |
| 6 | 0.1399 | 0.1443 | 0.1480 | 0.1512 | 0.1539 | 0.1563 | 0.1584 | 0.1601 | 0.1616 | 0.1630 |
| 7 | .1092 | .1150 | .1201 | .1245 | .1283 | .1316 | .1346 | .1372 | .1395 | .1415 |
| 8 | .0804 | .0878 | .0941 | .0997 | .1046 | .1089 | .1128 | .1162 | .1192 | .1219 |
| 9 | .0530 | .0618 | .0696 | .0764 | .0823 | .0876 | .0923 | .0965 | .1002 | .1036 |
| 10 | .0263 | .0368 | .0459 | .0539 | .0610 | .0672 | .0728 | .0778 | .0822 | .0862 |
| 11 | 0.0000 | 0.0122 | 0.0228 | 0.0321 | 0.0403 | 0.0476 | 0.0540 | 0.0598 | 0.0650 | 0.0697 |
| 12 | — | — | .0000 | .0107 | .0200 | .0284 | .0358 | .0424 | .0483 | .0537 |
| 13 | — | — | — | — | .0000 | .0094 | .0178 | .0253 | .0320 | .0381 |
| 14 | — | — | — | — | — | — | .0000 | .0084 | .0159 | .0227 |
| 15 | — | — | — | — | — | — | — | — | .0000 | .0076 |

Table C.7 (*Contd.*)

| | n | | | | | | | | | |
|---|---|---|---|---|---|---|---|---|---|---|
| i | 31 | 32 | 33 | 34 | 35 | 36 | 37 | 38 | 39 | 40 |
| 1 | 0.4220 | 0.4188 | 0.4156 | 0.4127 | 0.4096 | 0.4068 | 0.4040 | 0.4015 | 0.3989 | 0.3964 |
| 2 | .2921 | .2898 | .2876 | .2854 | .2834 | .2813 | .2794 | .2774 | .2755 | .2737 |
| 3 | .2475 | .2463 | .2451 | .2439 | .2427 | .2415 | .2403 | .2391 | .2380 | .2368 |
| 4 | .2145 | .2141 | .2137 | .2132 | .2127 | .2121 | .2116 | .2110 | .2104 | .2098 |
| 5 | .1874 | .1878 | .1880 | .1882 | .1883 | .1883 | .1883 | .1881 | .1880 | .1878 |
| 6 | 0.1641 | 0.1651 | 0.1660 | 0.1667 | 0.1673 | 0.1678 | 0.1683 | 0.1686 | 0.1689 | 0.1691 |
| 7 | .1433 | .1449 | .1463 | .1475 | .1487 | .1496 | .1505 | .1513 | .1520 | .1526 |
| 8 | .1243 | .1265 | .1284 | .1301 | .1317 | .1331 | .1344 | .1356 | .1366 | .1376 |
| 9 | .1066 | .1093 | .1118 | .1140 | .1160 | .1179 | .1196 | .1211 | .1225 | .1237 |
| 10 | .0899 | .0931 | .0961 | .0988 | .1013 | .1036 | .1056 | .1075 | .1092 | .1108 |
| 11 | 0.0739 | 0.0777 | 0.0812 | 0.0844 | 0.0873 | 0.0900 | 0.0924 | 0.0947 | 0.0967 | 0.0986 |
| 12 | .0585 | .0629 | .0669 | .0706 | .0739 | .0770 | .0798 | .0824 | .0848 | .0870 |
| 13 | .0435 | .0485 | .0530 | .0572 | .0610 | .0645 | .0677 | .0706 | .0733 | .0759 |
| 14 | .0289 | .0344 | .0395 | .0441 | .0484 | .0523 | .0559 | .0592 | .0622 | .0651 |
| 15 | .0144 | .0206 | .0262 | .0314 | .0361 | .0404 | .0444 | .0481 | .0515 | .0546 |
| 16 | 0.0000 | 0.0068 | 0.0131 | 0.0187 | 0.0239 | 0.0287 | 0.0331 | 0.0372 | 0.0409 | 0.0444 |
| 17 | — | — | .0000 | .0062 | .0119 | .0172 | .0220 | .0264 | .0305 | .0343 |
| 18 | — | — | — | — | .0000 | .0057 | .0110 | .0158 | .0203 | .0244 |
| 19 | — | — | — | — | — | — | .0000 | .0053 | .0101 | .0146 |
| 20 | — | — | — | — | — | — | — | — | .0000 | .0049 |

| | n | | | | | | | | | |
|---|---|---|---|---|---|---|---|---|---|---|
| i | 41 | 42 | 43 | 44 | 45 | 46 | 47 | 48 | 49 | 50 |
| 1 | 0.3940 | 0.3917 | 0.3894 | 0.3872 | 0.3850 | 0.3830 | 0.3808 | 0.3789 | 0.3770 | 0.3751 |
| 2 | .2719 | .2701 | .2684 | .2667 | .2651 | .2635 | .2620 | .2604 | .2589 | .2574 |
| 3 | .2357 | .2345 | .2334 | .2323 | .2313 | .2302 | .2291 | .2281 | .2271 | .2260 |
| 4 | .2091 | .2085 | .2078 | .2072 | .2065 | .2058 | .2052 | .2045 | .2038 | .2032 |
| 5 | .1876 | .1874 | .1871 | .1868 | .1865 | .1862 | .1859 | .1855 | .1851 | .1847 |
| 6 | 0.1693 | 0.1694 | 0.1695 | 0.1695 | 0.1695 | 0.1695 | 0.1695 | 0.1693 | 0.1692 | 0.1691 |
| 7 | .1531 | .1535 | .1539 | .1542 | .1545 | .1548 | .1550 | .1551 | .1553 | .1554 |
| 8 | .1384 | .1392 | .1398 | .1405 | .1410 | .1415 | .1420 | .1423 | .1427 | .1430 |
| 9 | .1249 | .1259 | .1269 | .1278 | .1286 | .1293 | .1300 | .1306 | .1312 | .1317 |
| 10 | .1123 | .1136 | .1149 | .1160 | .1170 | .1180 | .1189 | .1197 | .1205 | .1212 |
| 11 | 0.1004 | 0.1020 | 0.1035 | 0.1049 | 0.1062 | 0.1073 | 0.1085 | 0.1095 | 0.1105 | 0.1113 |
| 12 | .0891 | .0909 | .0927 | .0943 | .0959 | .0972 | .0986 | .0998 | .1010 | .1020 |
| 13 | .0782 | .0804 | .0824 | .0842 | .0860 | .0876 | .0892 | .0906 | .0919 | .0932 |
| 14 | .0677 | .0701 | .0724 | .0745 | .0765 | .0783 | .0801 | .0817 | .0832 | .0846 |
| 15 | .0575 | .0602 | .0628 | .0651 | .0673 | .0694 | .0713 | .0731 | .0748 | .0764 |
| 16 | 0.0476 | 0.0506 | 0.0534 | 0.0560 | 0.0584 | 0.0607 | 0.0628 | 0.0648 | 0.0667 | 0.0685 |
| 17 | .0379 | .0411 | .0442 | .0471 | .0497 | .0522 | .0546 | .0568 | .0588 | .0608 |
| 18 | .0283 | .0318 | .0352 | .0383 | .0412 | .0439 | .0465 | .0489 | .0511 | .0532 |
| 19 | .0188 | .0227 | .0263 | .0296 | .0328 | .0357 | .0385 | .0411 | .0436 | .0459 |
| 20 | .0094 | .0136 | .0175 | .0211 | .0245 | .0277 | .0307 | .0335 | .0361 | .0386 |
| 21 | 0.0000 | 0.0045 | 0.0087 | 0.0126 | 0.0163 | 0.0197 | 0.0229 | 0.0259 | 0.0288 | 0.0314 |
| 22 | — | — | .0000 | .0042 | .0081 | .0118 | .0153 | .0185 | .0215 | .0244 |
| 23 | — | — | — | — | .0000 | .0039 | .0076 | .0111 | .0143 | .0174 |
| 24 | — | — | — | — | — | — | .0000 | .0037 | .0071 | .0104 |
| 25 | — | — | — | — | — | — | — | — | .0000 | .0035 |

Table C.8 *Percentage points of the W test\* for n = 3(1)50*

| | Level | | | | | | | | |
|---|---|---|---|---|---|---|---|---|---|
| $n$ | **0.01** | **0.02** | **0.05** | **0.10** | **0.50** | **0.90** | **0.95** | **0.98** | **0.99** |
| 3 | 0.753 | 0.756 | 0.767 | 0.789 | 0.959 | 0.998 | 0.999 | 1.000 | 1.000 |
| 4 | .687 | .707 | .748 | .792 | .935 | .987 | .992 | .996 | .997 |
| 5 | .686 | .715 | .762 | .806 | .927 | .979 | .986 | .991 | .993 |
| 6 | 0.713 | 0.743 | 0.788 | 0.826 | 0.927 | 0.974 | 0.981 | 0.986 | 0.989 |
| 7 | .730 | .760 | .803 | .838 | .928 | .972 | .979 | .985 | .988 |
| 8 | .749 | .778 | .818 | .851 | .932 | .972 | .978 | .984 | .987 |
| 9 | .764 | .791 | .829 | .859 | .935 | .972 | .978 | .984 | .986 |
| 10 | .781 | .806 | .842 | .869 | .938 | .972 | .978 | .983 | .986 |
| 11 | 0.792 | 0.817 | 0.850 | 0.876 | 0.940 | 0.973 | 0.979 | 0.984 | 0.986 |
| 12 | .805 | .828 | .859 | .883 | .943 | .973 | .979 | .984 | .986 |
| 13 | .814 | .837 | .866 | .889 | .945 | .974 | .979 | .984 | .986 |
| 14 | .825 | .846 | .874 | .895 | .947 | .975 | .980 | .984 | .986 |
| 15 | .835 | .855 | .881 | .901 | .950 | .975 | .980 | .984 | .987 |
| 16 | 0.844 | 0.863 | 0.887 | 0.906 | 0.952 | 0.976 | 0.981 | 0.985 | 0.987 |
| 17 | .851 | .869 | .892 | .910 | .954 | .977 | .981 | .985 | .987 |
| 18 | .858 | .874 | .897 | .914 | .956 | .978 | .982 | .986 | .988 |
| 19 | .863 | .879 | .901 | .917 | .957 | .978 | .982 | .986 | .988 |
| 20 | .868 | .884 | .905 | .920 | .959 | .979 | .983 | .986 | .988 |
| 21 | 0.873 | 0.888 | 0.908 | 0.923 | 0.960 | 0.980 | 0.983 | 0.987 | 0.989 |
| 22 | .878 | .892 | .911 | .926 | .961 | .980 | .984 | .987 | .989 |
| 23 | .881 | .895 | .914 | .928 | .962 | .981 | .984 | .987 | .989 |
| 24 | .884 | .898 | .916 | .930 | .963 | .981 | .984 | .987 | .989 |
| 25 | .888 | .901 | .918 | .931 | .964 | .981 | .985 | .988 | .989 |
| 26 | 0.891 | 0.904 | 0.920 | 0.933 | 0.965 | 0.982 | 0.985 | 0.988 | 0.989 |
| 27 | .894 | .906 | .923 | .935 | .965 | .982 | .985 | .988 | .990 |
| 28 | .896 | .908 | .924 | .936 | .966 | .982 | .985 | .988 | .990 |
| 29 | .898 | .910 | .926 | .937 | .966 | .982 | .985 | .988 | .990 |
| 30 | .900 | .912 | .927 | .939 | .967 | .983 | .985 | .988 | .900 |
| 31 | 0.902 | 0.914 | 0.929 | 0.940 | 0.967 | 0.983 | 0.986 | 0.988 | 0.990 |
| 32 | .904 | .915 | .930 | .941 | .968 | .983 | .986 | .988 | .990 |
| 33 | .906 | .917 | .931 | .942 | .968 | .983 | .986 | .989 | .990 |
| 34 | .908 | .919 | .933 | .943 | .969 | .983 | .986 | .989 | .990 |
| 35 | .910 | .920 | .934 | .944 | .969 | .984 | .986 | .989 | .990 |
| 36 | 0.912 | 0.922 | 0.935 | 0.945 | 0.970 | 0.984 | 0.986 | 0.989 | 0.990 |
| 37 | .914 | .924 | .936 | .946 | .970 | .984 | .987 | .989 | .990 |
| 38 | .916 | .925 | .938 | .947 | .971 | .984 | .987 | .989 | .990 |
| 39 | .917 | .927 | .939 | .948 | .971 | .984 | .987 | .989 | .991 |
| 40 | .919 | .928 | .940 | .949 | .972 | .985 | .987 | .989 | .991 |
| 41 | 0.920 | 0.929 | 0.941 | 0.950 | 0.972 | 0.985 | 0.987 | 0.989 | 0.991 |
| 42 | .922 | .930 | .942 | .951 | .972 | .985 | .987 | .989 | .991 |
| 43 | .923 | .932 | .943 | .951 | .973 | .985 | .987 | .990 | .991 |
| 44 | .924 | .933 | .944 | .952 | .973 | .985 | .987 | .990 | .991 |
| 45 | .926 | .934 | .945 | .953 | .973 | .985 | .988 | .990 | .991 |
| 46 | 0.927 | 0.935 | 0.945 | 0.953 | 0.974 | 0.985 | 0.988 | 0.990 | 0.991 |
| 47 | .928 | .936 | .946 | .954 | .974 | .985 | .988 | .990 | .991 |
| 48 | .929 | .937 | .947 | .954 | .974 | .985 | .988 | .990 | .991 |
| 49 | .929 | .937 | .947 | .955 | .974 | 985 | .988 | .990 | .991 |
| 50 | .930 | .938 | .947 | .955 | .974 | .985 | .988 | .990 | .991 |

\*Based on fitted Johnson (1949) $S_B$ approximation, see Shapiro & Wilk (1965) for details.

Table C.9 *Expected normal order statistics $u_{i,n}$*

| i | n=2 | 3 | 4 | 5 | 6 | 7 | 8 | 9 | 10 | 11 | 12 |
|---|---|---|---|---|---|---|---|---|---|---|---|
| 1 | 0.564 | 0.846 | 1.029 | 1.163 | 1.267 | 1.352 | 1.424 | 1.485 | 1.539 | 1.586 | 1.629 |
| 2 | | .000 | 0.297 | 0.495 | 0.642 | 0.757 | 0.852 | 0.932 | 1.001 | 1.062 | 1.116 |
| 3 | | | | .000 | .202 | .353 | .473 | .572 | 0.656 | 0.729 | 0.793 |
| 4 | | | | | | .000 | .153 | .275 | .376 | .462 | .537 |
| 5 | | | | | | | | 0.000 | 0.123 | 0.225 | 0.312 |
| 6 | | | | | | | | | | .000 | .103 |

| i | n=13 | 14 | 15 | 16 | 17 | 18 | 19 | 20 | 21 | 22 | 23 | 24 | 25 |
|---|---|---|---|---|---|---|---|---|---|---|---|---|---|
| 1 | 1.668 | 1.703 | 1.736 | 1.766 | 1.794 | 1.820 | 1.844 | 1.867 | 1.89 | 1.91 | 1.93 | 1.95 | 1.97 |
| 2 | 1.164 | 1.208 | 1.248 | 1.285 | 1.319 | 1.350 | 1.380 | 1.408 | 1.43 | 1.46 | 1.48 | 1.50 | 1.52 |
| 3 | 0.850 | 0.901 | 0.948 | 0.990 | 1.029 | 1.066 | 1.099 | 1.131 | 1.16 | 1.19 | 1.21 | 1.24 | 1.26 |
| 4 | .603 | .662 | .715 | .763 | 0.807 | 0.848 | 0.886 | 0.921 | 0.95 | 0.98 | 1.01 | 1.04 | 1.07 |
| 5 | 0.388 | 0.456 | 0.516 | 0.570 | 0.619 | 0.665 | 0.707 | 0.745 | 0.78 | 0.82 | 0.85 | 0.88 | 0.91 |
| 6 | .190 | .267 | .335 | .396 | .451 | .502 | .548 | .590 | .63 | .67 | .70 | .73 | .76 |
| 7 | .000 | .088 | .165 | .234 | .295 | .351 | .402 | .448 | .49 | .53 | .57 | .60 | .64 |
| 8 | | | .000 | .077 | .146 | .208 | .264 | .315 | .36 | .41 | .45 | .48 | .52 |
| 9 | | | | | .000 | .069 | .131 | .187 | .24 | .29 | .33 | .37 | .41 |
| 10 | | | | | | | 0.000 | 0.062 | 0.12 | 0.17 | 0.22 | 0.26 | 0.30 |
| 11 | | | | | | | | | .00 | .06 | .11 | .16 | .20 |
| 12 | | | | | | | | | | | .00 | .05 | .10 |
| 13 | | | | | | | | | | | | | .00 |

Table C. 9. (*Contd.*)

| i | 26 | 28 | 30 | 32 | 34 | 36 | 38 | 40 | 42 | 44 | 46 | 48 | 50 |
|---|----|----|----|----|----|----|----|----|----|----|----|----|----|
| 1 | 1.98 | 2.01 | 2.04 | 2.07 | 2.09 | 2.12 | 2.14 | 2.16 | 2.18 | 2.20 | 2.22 | 2.23 | 2.25 |
| 2 | 1.54 | 1.58 | 1.62 | 1.65 | 1.68 | 1.70 | 1.73 | 1.75 | 1.78 | 1.80 | 1.82 | 1.84 | 1.85 |
| 3 | 1.29 | 1.33 | 1.36 | 1.40 | 1.43 | 1.46 | 1.49 | 1.52 | 1.54 | 1.57 | 1.59 | 1.61 | 1.63 |
| 4 | 1.09 | 1.14 | 1.18 | 1.22 | 1.25 | 1.28 | 1.32 | 1.34 | 1.37 | 1.40 | 1.42 | 1.44 | 1.46 |
| 5 | 0.93 | 0.98 | 1.03 | 1.07 | 1.11 | 1.14 | 1.17 | 1.20 | 1.23 | 1.26 | 1.28 | 1.31 | 1.33 |
| 6 | .79 | .85 | 0.89 | 0.94 | 0.98 | 1.02 | 1.05 | 1.08 | 1.11 | 1.14 | 1.17 | 1.19 | 1.22 |
| 7 | .67 | .73 | .78 | .82 | .87 | 0.91 | 0.94 | 0.98 | 1.01 | 1.04 | 1.07 | 1.09 | 1.12 |
| 8 | .55 | .61 | .67 | .72 | .76 | .81 | .85 | .88 | 0.91 | 0.95 | 0.98 | 1.00 | 1.03 |
| 9 | .44 | .51 | .57 | .62 | .67 | .71 | .75 | .79 | .83 | .86 | .89 | 0.92 | 0.95 |
| 10 | 0.34 | 0.41 | 0.47 | 0.53 | 0.58 | 0.63 | 0.67 | 0.71 | 0.75 | 0.78 | 0.81 | 0.84 | 0.87 |
| 11 | .24 | .32 | .38 | .44 | .50 | .54 | .59 | .63 | .67 | .71 | .74 | .77 | .80 |
| 12 | .14 | .22 | .29 | .36 | .41 | .47 | .51 | .56 | .60 | .64 | .67 | .70 | .74 |
| 13 | .05 | .13 | .21 | .28 | .34 | .39 | .44 | .49 | .53 | .57 | .60 | .64 | .67 |
| 14 | | .04 | .12 | .20 | .26 | .32 | .37 | .42 | .46 | .50 | .54 | .58 | .61 |
| 15 | | | 0.04 | 0.12 | 0.18 | 0.24 | 0.30 | 0.35 | 0.40 | 0.44 | 0.48 | 0.52 | 0.55 |
| 16 | | | | .04 | .11 | .17 | .23 | .28 | .33 | .38 | .42 | .46 | .49 |
| 17 | | | | | .04 | .10 | .16 | .22 | .27 | .32 | .36 | .40 | .44 |
| 18 | | | | | | .03 | .10 | .16 | .21 | .26 | .30 | .34 | .38 |
| 19 | | | | | | | .03 | .09 | .15 | .20 | .25 | .29 | .33 |
| 20 | | | | | | | | 0.03 | 0.09 | 0.14 | 0.19 | 0.24 | 0.28 |
| 21 | | | | | | | | | .03 | .09 | .14 | .18 | .23 |
| 22 | | | | | | | | | | .03 | .08 | .13 | .18 |
| 23 | | | | | | | | | | | .03 | .08 | .13 |
| 24 | | | | | | | | | | | | .03 | .07 |
| 25 | | | | | | | | | | | | | 0.03 |

Table C.10 *Critical values of Stefansky's MNR test*
$\alpha = 0.05$

| C | R | | | | | | |
|---|---|---|---|---|---|---|---|
| | 3 | 4 | 5 | 6 | 7 | 8 | 9 |
| 3 | 0.648 | 0.645 | 0.624 | 0.600 | 0.577 | 0.555 | 0.535 |
| 4 | | 0.621 | 0.590 | 0.561 | 0.535 | 0.513 | 0.493 |
| 5 | | | 0.555 | 0.525 | 0.499 | 0.477 | 0.457 |
| 6 | | | | 0.495 | 0.469 | 0.447 | 0.428 |
| 7 | | | | | 0.444 | 0.423 | 0.405 |
| 8 | | | | | | 0.402 | 0.385 |
| 9 | | | | | | | 0.368 |

$\alpha = 0.01$

| C | R | | | | | | |
|---|---|---|---|---|---|---|---|
| | 3 | 4 | 5 | 6 | 7 | 8 | 9 |
| 3 | 0.660 | 0.675 | 0.664 | 0.646 | 0.626 | 0.606 | 0.587 |
| 4 | | 0.665 | 0.640 | 0.613 | 0.588 | 0.565 | 0.544 |
| 5 | | | 0.608 | 0.578 | 0.551 | 0.527 | 0.506 |
| 6 | | | | 0.546 | 0.519 | 0.495 | 0.475 |
| 7 | | | | | 0.492 | 0.469 | 0.449 |
| 8 | | | | | | 0.446 | 0.426 |
| 9 | | | | | | | 0.407 |

Reprinted by permission from Stefansky, *Technometrics*, **14** (1972).

Table C.11. *Orthogonal polynomials*

(for equidistant values of the explanatory variable)

| | n = 3 | | n = 4 | | | n = 5 | | | | n = 6 | | | | |
|---|---|---|---|---|---|---|---|---|---|---|---|---|---|---|
| | $f_1$ | $f_2$ | $f_1$ | $f_2$ | $f_3$ | $f_1$ | $f_2$ | $f_3$ | $f_4$ | $f_1$ | $f_2$ | $f_3$ | $f_4$ | $f_5$ |
| | | | | | | | | | | −5 | +5 | −5 | +1 | −1 |
| | | | −3 | +1 | −1 | −2 | +2 | −1 | +1 | −3 | −1 | +7 | −3 | +5 |
| | −1 | +1 | −1 | −1 | +3 | −1 | −1 | +2 | −4 | −1 | −4 | +4 | +2 | −10 |
| | 0 | −2 | +1 | −1 | −3 | 0 | −2 | 0 | +6 | +1 | −4 | −4 | +2 | +10 |
| | +1 | +1 | +3 | +1 | +1 | +1 | −1 | −2 | −4 | +3 | −1 | −7 | −3 | −5 |
| | | | | | | +2 | +2 | +1 | +1 | +5 | +5 | +5 | +1 | +1 |
| $\sum f_i^2$ | 2 | 6 | 20 | 4 | 20 | 10 | 14 | 10 | 70 | 70 | 84 | 180 | 28 | 252 |
| $\lambda_i$ | 1 | 3 | 2 | 1 | $\frac{10}{3}$ | 1 | 1 | $\frac{5}{6}$ | $\frac{35}{12}$ | 2 | $\frac{3}{2}$ | $\frac{5}{3}$ | $\frac{7}{12}$ | $\frac{21}{10}$ |

| n = 7 | | | | | n = 8 | | | | | n = 9 | | | | |
|---|---|---|---|---|---|---|---|---|---|---|---|---|---|---|
| $f_1$ | $f_2$ | $f_3$ | $f_4$ | $f_5$ | $f_1$ | $f_2$ | $f_3$ | $f_4$ | $f_5$ | $f_1$ | $f_2$ | $f_3$ | $f_4$ | $f_5$ |
| 0 | −4 | 0 | +6 | 0 | | | | | | 0 | −20 | 0 | +18 | 0 |
| +1 | −3 | −1 | +1 | +5 | +1 | −5 | −3 | +9 | +15 | +1 | −17 | −9 | +9 | +9 |
| +2 | 0 | −1 | −7 | −4 | +3 | −3 | −7 | −3 | +17 | +2 | −8 | −13 | −11 | +4 |
| +3 | +5 | +1 | +3 | +1 | +5 | +1 | −5 | −13 | −23 | +3 | +7 | −7 | −21 | −11 |
| | | | | | +7 | +7 | +7 | +7 | +7 | +4 | +28 | +14 | +14 | +4 |
| 28 | 84 | 6 | 154 | 84 | 168 | 168 | 264 | 616 | 2184 | 60 | 2772 | 990 | 2002 | 468 |
| 1 | 1 | $\frac{1}{6}$ | $\frac{7}{12}$ | $\frac{7}{20}$ | 2 | 1 | $\frac{2}{3}$ | $\frac{7}{12}$ | $\frac{7}{10}$ | 1 | 3 | $\frac{5}{6}$ | $\frac{7}{12}$ | $\frac{3}{20}$ |

Since the tables are symmetrical, the tables for n = 7, 8 and 9 have been shortened.

| n = 10 | | | | | n = 11 | | | | | n = 12 | | | | |
|---|---|---|---|---|---|---|---|---|---|---|---|---|---|---|
| $f_1$ | $f_2$ | $f_3$ | $f_4$ | $f_5$ | $f_1$ | $f_2$ | $f_3$ | $f_4$ | $f_5$ | $f_1$ | $f_2$ | $f_3$ | $f_4$ | $f_5$ |
| 1 | −4 | −12 | 18 | 6 | 0 | −10 | 0 | 6 | 0 | 1 | −35 | −7 | 28 | 20 |
| 3 | −3 | −31 | 3 | 11 | 1 | −9 | −14 | 4 | 4 | 3 | −29 | −19 | 12 | 44 |
| 5 | −1 | −35 | −17 | 1 | 2 | −6 | −23 | −1 | 4 | 5 | −17 | −25 | −13 | 29 |
| 7 | 2 | −14 | −22 | −14 | 3 | −1 | −22 | −6 | −1 | 7 | 1 | −21 | −33 | −21 |
| 9 | 6 | 42 | 18 | 6 | 4 | 6 | −6 | −6 | −6 | 9 | 25 | −3 | −27 | −57 |
| | | | | | 5 | 15 | 30 | 6 | 3 | 11 | 55 | 33 | 33 | 33 |
| 330 | 132 | 8580 | 2860 | 780 | 110 | 858 | 4290 | 286 | 156 | 572 | 12 012 | 5148 | 8008 | 15 912 |
| 2 | $\frac{1}{2}$ | $\frac{5}{2}$ | $\frac{5}{12}$ | $\frac{1}{16}$ | 1 | 1 | $\frac{5}{6}$ | $\frac{1}{12}$ | $\frac{1}{40}$ | 2 | 3 | $\frac{2}{3}$ | $\frac{7}{24}$ | $\frac{3}{20}$ |

Table C.11  (*Contd.*)

| $f_1$ | $f_2$ | $f_3$ | $f_4$ | $f_5$ | $f_6$ |
|---|---|---|---|---|---|

**n = 13**

| $f_1$ | $f_2$ | $f_3$ | $f_4$ | $f_5$ | $f_6$ |
|---|---|---|---|---|---|
| −6 | 22 | −11 | 99 | −22 | 22 |
| −5 | 11 | 0 | −66 | 33 | −55 |
| −4 | 2 | 6 | −96 | 18 | 8 |
| −3 | −5 | 8 | −54 | −11 | 43 |
| −2 | −10 | 7 | 11 | −26 | 22 |
| −1 | −13 | 4 | 64 | −20 | −20 |
| 0 | −14 | 0 | 84 | 0 | −40 |

| 182 | | 572 | | 6188 | |
|---|---|---|---|---|---|
| | 2002 | | 68 068 | | 14 212 |
| 1 | 1 | $\frac{1}{6}$ | $\frac{7}{12}$ | $\frac{7}{120}$ | $\frac{11}{360}$ |

**n = 14**

| $f_1$ | $f_2$ | $f_3$ | $f_4$ | $f_5$ | $f_6$ |
|---|---|---|---|---|---|
| −13 | 13 | −143 | 143 | −143 | 143 |
| −11 | 7 | −11 | −77 | 187 | −319 |
| −9 | 2 | 66 | −132 | 132 | −11 |
| −7 | −2 | 98 | −92 | −28 | 227 |
| −5 | −5 | 95 | −13 | −139 | 185 |
| −3 | −7 | 67 | 63 | −145 | −25 |
| −1 | −8 | 24 | 108 | −60 | −200 |

| 910 | | 97 240 | | 235 144 | |
|---|---|---|---|---|---|
| | 728 | | 136 136 | | 497 420 |
| 2 | $\frac{1}{2}$ | $\frac{5}{2}$ | $\frac{7}{12}$ | $\frac{7}{20}$ | $\frac{77}{720}$ |

**n = 15**

| $f_1$ | $f_2$ | $f_3$ | $f_4$ | $f_5$ | $f_6$ |
|---|---|---|---|---|---|
| −7 | 91 | −91 | 1001 | −1001 | 143 |
| −6 | 51 | −13 | −429 | 1144 | −286 |
| −5 | 19 | 35 | −869 | 979 | −55 |
| −4 | −8 | 58 | −704 | 44 | 176 |
| −3 | −29 | 61 | −249 | −751 | 197 |
| −2 | −44 | 49 | 251 | −1000 | 50 |
| −1 | −53 | 27 | 621 | −675 | −125 |
| 0 | −56 | 0 | 756 | 0 | −200 |

| 280 | | 39 780 | | 10 581 480 | |
|---|---|---|---|---|---|
| | 37 128 | | 6 466 460 | | 426 360 |
| 1 | 3 | $\frac{5}{6}$ | $\frac{35}{12}$ | $\frac{21}{20}$ | $\frac{11}{120}$ |

**n = 16**

| $f_1$ | $f_2$ | $f_3$ | $f_4$ | $f_5$ | $f_6$ |
|---|---|---|---|---|---|
| −15 | 35 | −455 | 273 | −143 | 65 |
| −13 | 21 | −91 | −91 | 143 | −117 |
| −11 | 9 | 143 | −221 | 143 | −39 |
| −9 | −1 | 267 | −201 | 33 | 59 |
| −7 | −9 | 301 | −101 | −77 | 87 |
| −5 | −15 | 265 | 23 | −131 | 45 |
| −3 | −19 | 179 | 129 | −115 | −25 |
| −1 | −21 | 63 | 189 | −45 | −75 |

| 1360 | | 1 007 760 | | 201 552 | |
|---|---|---|---|---|---|
| | 5712 | | 470 288 | | 77 520 |
| 2 | 1 | $\frac{10}{3}$ | $\frac{7}{12}$ | $\frac{1}{10}$ | $\frac{1}{60}$ |

Table C.11  (*Contd.*)

$n = 17$

| $f_1$ | $f_2$ | $f_3$ | $f_4$ | $f_5$ | $f_6$ |
|---|---|---|---|---|---|
| $-8$ | 40 | $-28$ | 52 | $-104$ | 104 |
| $-7$ | 25 | $-7$ | $-13$ | 91 | $-169$ |
| $-6$ | 12 | 7 | $-39$ | 104 | $-78$ |
| $-5$ | 1 | 15 | $-39$ | 39 | 65 |
| $-4$ | $-8$ | 18 | $-24$ | $-36$ | 128 |
| $-3$ | $-15$ | 17 | $-3$ | $-83$ | 93 |
| $-2$ | $-20$ | 13 | 17 | $-88$ | 2 |
| $-1$ | $-23$ | 7 | 31 | $-55$ | $-85$ |
| 0 | $-24$ | 0 | 36 | 0 | $-120$ |
| 408 | | 3876 | | 100 776 | |
| | 7752 | | 16 796 | | 178 296 |
| 1 | 1 | $\frac{1}{6}$ | $\frac{1}{12}$ | $\frac{1}{20}$ | $\frac{1}{60}$ |

$n = 18$

| $f_1$ | $f_2$ | $f_3$ | $f_4$ | $f_5$ | $f_6$ |
|---|---|---|---|---|---|
| $-17$ | 68 | $-68$ | 68 | $-884$ | 442 |
| $-15$ | 44 | $-20$ | $-12$ | 676 | $-650$ |
| $-13$ | 23 | 13 | $-47$ | 871 | $-377$ |
| $-11$ | 5 | 33 | $-51$ | 429 | 169 |
| $-9$ | $-10$ | 42 | $-36$ | $-156$ | 481 |
| $-7$ | $-22$ | 42 | $-12$ | $-588$ | 439 |
| $-5$ | $-31$ | 35 | 13 | $-733$ | 145 |
| $-3$ | $-37$ | 23 | 33 | $-583$ | $-209$ |
| $-1$ | $-40$ | 8 | 44 | $-220$ | $-440$ |
| 1938 | | 23 256 | | 6 953 544 | |
| | 23 256 | | 28 424 | | 2 941 884 |
| 2 | $\frac{3}{2}$ | $\frac{1}{3}$ | $\frac{1}{12}$ | $\frac{3}{10}$ | $\frac{11}{140}$ |

Table C.11  (*Contd.*)

$n = 19$

| $f_1$ | $f_2$ | $f_3$ | $f_4$ | $f_5$ | $f_6$ |
|---|---|---|---|---|---|
| −9 | 51 | −204 | 612 | −102 | 1326 |
| −8 | 34 | − 68 | − 68 | 68 | − 1768 |
| −7 | 19 | 28 | −388 | 98 | −1222 |
| −6 | 6 | 89 | −453 | 58 | 234 |
| −5 | − 5 | 120 | −354 | − 3 | 1235 |
| −4 | −14 | 126 | −168 | − 54 | 1352 |
| −3 | −21 | 112 | 42 | − 79 | 729 |
| −2 | −26 | 83 | 227 | − 74 | − 214 |
| −1 | −29 | 44 | 352 | − 44 | − 1012 |
| 0 | −30 | 0 | 396 | 0 | − 1320 |
| 570 | | 213 180 | | 89 148 | |
| | 13 566 | | 2 288 132 | | 24 515 700 |
| 1 | 1 | $\frac{5}{6}$ | $\frac{7}{12}$ | $\frac{1}{40}$ | $\frac{11}{120}$ |

$n = 20$

| $f_1$ | $f_2$ | $f_3$ | $f_4$ | $f_5$ | $f_6$ |
|---|---|---|---|---|---|
| −19 | 57 | −969 | 1938 | − 1938 | 1938 |
| −17 | 39 | −357 | − 102 | 1122 | − 2346 |
| −15 | 23 | 85 | − 1122 | 1802 | − 1870 |
| −13 | 9 | 377 | − 1402 | 1222 | 6 |
| −11 | − 3 | 539 | − 1187 | 187 | 1497 |
| − 9 | −13 | 591 | − 687 | − 771 | 1931 |
| − 7 | −21 | 553 | − 77 | − 1351 | 1353 |
| − 5 | −27 | 445 | 503 | − 1441 | 195 |
| − 3 | −31 | 287 | 948 | − 1076 | − 988 |
| − 1 | −33 | 99 | 1188 | − 396 | − 1716 |
| 2660 | | 4 903 140 | | 31 201 800 | |
| | 17 556 | | 22 881 320 | | 49 031 400 |
| 2 | 1 | $\frac{10}{3}$ | $\frac{35}{24}$ | $\frac{7}{20}$ | $\frac{11}{120}$ |

Table C.12  *Orthogonal polynomial functions*

$$f_1(x) = \lambda_1 x$$
$$f_2(x) = \lambda_2 \{x^2 - \tfrac{1}{12}(n^2 - 1)\}$$
$$f_3(x) = \lambda_3 \{x^3 - \tfrac{1}{20}(3n^2 - 7)x\}$$
$$f_4(x) = \lambda_4 \{x^4 - \tfrac{1}{14}(3n^2 - 13)x^2 + \tfrac{3}{560}(n^2 - 1)(n^2 - 9)\}$$
$$f_5(x) = \lambda_5 \{x^5 - \tfrac{5}{18}(n^2 - 7)x^3 + \tfrac{1}{1008}(15n^4 - 230n^2 + 407)x\}$$

Table C.13  *Coefficients of orthogonal polynomial functions*

| $n$ | $f_1$ $x$ | $f_2$ $x^2$ | *const.* | $f_3$ $x^3$ | $x$ |
|---|---|---|---|---|---|
| 3 | 1 | 3 | $-2$ | — | — |
| 4 | 2 | 1 | $-\frac{5}{4}$ | $\frac{10}{3}$ | $-\frac{41}{6}$ |
| 5 | 1 | 1 | $-2$ | $\frac{5}{6}$ | $-\frac{17}{6}$ |
| 6 | 2 | $\frac{3}{2}$ | $-\frac{35}{8}$ | $\frac{5}{3}$ | $-\frac{101}{12}$ |
| 7 | 1 | 1 | $-4$ | $\frac{1}{6}$ | $-\frac{7}{6}$ |
| 8 | 2 | 1 | $-\frac{21}{4}$ | $\frac{2}{3}$ | $-\frac{37}{6}$ |
| 9 | 1 | 3 | $-20$ | $\frac{5}{6}$ | $-\frac{59}{6}$ |
| 10 | 2 | $\frac{1}{2}$ | $-\frac{33}{8}$ | $\frac{5}{3}$ | $-\frac{293}{12}$ |
| 11 | 1 | 1 | $-10$ | $\frac{5}{6}$ | $-\frac{89}{6}$ |
| 12 | 2 | 3 | $-\frac{143}{4}$ | $\frac{2}{3}$ | $-\frac{85}{6}$ |
| 13 | 1 | 1 | $-14$ | $\frac{1}{6}$ | $-\frac{25}{6}$ |
| 14 | 2 | $\frac{1}{2}$ | $-\frac{65}{8}$ | $\frac{5}{3}$ | $-\frac{581}{12}$ |
| 15 | 1 | 3 | $-56$ | $\frac{5}{6}$ | $-\frac{167}{6}$ |
| 16 | 2 | 1 | $-\frac{85}{4}$ | $\frac{10}{3}$ | $-\frac{761}{6}$ |
| 17 | 1 | 1 | $-24$ | $\frac{1}{6}$ | $-\frac{43}{6}$ |
| 18 | 2 | $\frac{3}{2}$ | $-\frac{323}{8}$ | $\frac{1}{3}$ | $-\frac{193}{12}$ |
| 19 | 1 | 1 | $-30$ | $\frac{5}{6}$ | $-\frac{269}{6}$ |
| 20 | 2 | 1 | $-\frac{133}{4}$ | $\frac{10}{3}$ | $-\frac{1193}{6}$ |

Arguments are $x = t - (n + 1)/2$ for $t = 1, 2, \ldots, n$.

*Example* For $n = 5$  $f_1 = x$,  $f_2 = x^2 - 2$,  $f_3 = \frac{5}{6}x^3 - \frac{17}{6}x$.
For further entries to Tables C.12 and C.13, see *Biometrika Tables*.

# Index